Chaos and Fractals

Chaos and Fractals

An Elementary Introduction

David P. Feldman

College of the Atlantic, Bar Harbor, Maine, USA

OXFORD

UNIVERSITY PRESS

OXFORD
UNIVERSITY PRESS

Great Clarendon Street, Oxford, OX2 6DP,
United Kingdom

Oxford University Press is a department of the University of Oxford.
If furthers the University's objective of excellence in research, scholarship,
and education by publishing worldwide. Oxford is a registered trade mark of
Oxford University Press in the UK and in certain other countries

British Library Cataloguing in Publication Data

Data available

Library of Congress Cataloging in Publication Data
Library of Congress Control Number: 2012941969

ISBN 978–0–19–956643–3 (Hbk.)
ISBN 978–0–19–956644–0 (Pbk.)

Printed and bound by
CPI Group (UK) Ltd, Croydon, CR0 4YY

For my parents

Preface

> I would...urge that people be introduced to [chaos] early in their mathematical education. [Chaos] can be studied phenomenologically by iterating it on a calculator, or even by hand. Its study does not involve as much conceptual sophistication as does elementary calculus. Such study would greatly enrich the student's intuition. Not only in research, but also in the everyday world of politics and economics, we would all be better off if more people realised that simple nonlinear systems do not necessarily possess simple dynamical properties.
>
> Robert May, 1976 (May, 1976)

Such was the conclusion to Robert May's influential 1976 paper, "Simple mathematical models with very complicated dynamics," (May, 1976) published in *Nature*, perhaps the most prestigious scientific journal in the world. May believes, as do I, that chaos is well suited for introductory classes and that its inclusion in mathematics curricula has far-reaching implications. Nevertheless, more than thirty years after May's paper, there are still only a handful of introductory textbooks on chaos, fractals, and non-linear dynamics. And none of these textbooks are suited for students who are not focusing their studies on mathematics or science. This book aims to fill this void.

Understanding chaos requires much less advanced mathematics than other current areas of physics research such as general relativity or particle physics. Observing chaos and fractals requires no specialized equipment; chaos is seen in scores of everyday phenomena—a boiling pot of water, a dripping faucet, shifting weather patterns. And fractals are almost ubiquitous in the natural world. Thus, it is possible to teach the central ideas and insights of chaos in a rigorous, genuine, and relevant way to students with relatively little mathematics background.

I have found that chaos and fractals are wonderful topics for capturing the imagination of a wide range of students while (re)introducing them to some of the basic notions of algebra and functions. I am convinced that it is possible to present the excitement and novelty of chaos and fractals in an intellectually honest and rigorous way to those possessing only a background in algebra. My experience has been that students find it easy to make connections among chaos and fractals, everyday experience, and other academic or creative interests.

There are three main audiences for this book. First, I believe that it is well suited for use in a college math or physics course for students who are not math or science majors and who have not necessarily taken calculus. Indeed, I have developed this book out of such a course that I have offered at the College of the Atlantic. It could also serve as a text for an interdisciplinary first-year seminar. Second, I believe that this text is well suited for a mathematics course for high school students. For students who are not interested in or are not quite ready for a calculus or statistics class, a course in chaos and fractals could be a good alternative. Third, I have aimed to make this book suitable for self-study. Anyone with some modest experience with algebra should be able to use this book to learn some of the technical details and foundations of chaos and fractals.

 In writing this book I have two central goals. First, I want the reader to gain a solid understanding of the basic mathematical ideas behind chaos and fractals. As part of this, I hope readers get to share some of the amazement and wonder experienced by the scientists and mathematicians who first viewed chaos or realized the intricate depth and beauty of fractals. Chaos and fractals provide important tools and constitute a different way of looking at physical and mathematical phenomena. By any measure, chaos and fractals are an important part of the current scientific and mathematical landscape.

Chaos and fractals are also increasingly part of the popular landscape. They have sparked popular imagination, and deservedly so. Ideas and phenomena from chaos and fractals have appeared in works of fiction and non-fiction. Many of these applications of and references to chaos and fractals are careful and well justified. Others are based on fundamental misunderstandings. And of course, not everybody agrees on which is which, and there are many applications and invocations of chaos and fractals which do not fit neatly into either extreme. In this book I try largely to steer clear of such controversies. Rather, my aim is to help readers obtain the technical background to confidently enter into these discussions and decide for themselves.

 The second main goal of this text is to provide a fun, engaging, and relevant context in which readers may improve their foundational mathematics, problem-solving skills, and confidence doing quantitative and analytical work. Reading and working through this book will be a good review of some basic algebra, including linear functions, exponents, logarithms, functional notation, and complex numbers. Equally important, throughout the book I emphasize a variety of different types of graphs to visualize and gain insight into chaos and fractals. This book will thus give readers considerable practice using many different sorts of graphs.

Advice and Notes for Students

This book uses no mathematics beyond basic algebra. A number of topics, including exponents, linear functions, and logarithms, are reviewed in appendices. If you had an algebra class at some point in your life and did fairly well, you almost surely have the math background needed for this book. You need not have enjoyed algebra. You need not have felt perfectly happy about it. And it is okay if you have forgotten almost all of it. It is helpful if you have seen functional notation—i.e., $f(x)$—before, but this is not a requirement. As long as you have had algebra at some point before and, at least at the time, felt like you generally understood it, my experience teaching this material has shown that you should be fine.

I have found that the most important prerequisite is a willingness to engage the material, to work hard, and to try to put aside any anxieties and preconceptions that you might have about mathematics. The study of chaos is very different than most other math you have been exposed to, and this book is probably unlike other textbooks you have used. As much as possible, approach this book with a sense of anticipation and adventure. With a little intellectual initiative I think you will find many opportunities to make connections between the material in this book and other interests of yours, even if you have had a difficult relationship with math in the past and are only taking a course that uses this book to meet a graduation requirement of some sort.

Here are some additional thoughts on the text:

(1) Students have reported that the book starts a little slow. It takes around seven or eight chapters to get to the topic of chaos. I do not think there is any way around this. In the past, students have found that it is worth the wait.

(2) Much of what we will do with chaos later on is graphical in nature. Given this, the early chapters emphasize graphical interpretations and constructions. This may seem tedious. But experience has shown that investing some time early in the course to learn various graphical constructions pays off down the road.

(3) This book is not a systematic review of algebra or trigonometry; it is not explicitly designed to prepare you for further math classes. Nevertheless, it will review many algebra topics, and working through the problems should strengthen your algebra skills. It also provides students with practice analyzing graphs and visually representing numerical relationships.

(4) To get as much as possible out of this text you will need to work with some web-based computer programs. No computer experience is necessary, but you will need to have access to a computer that has internet access and a web browser. You will also need a calculator. You do not need a scientific calculator, but one with logarithms would be helpful.

Exercises

There are exercises at the end of most chapters. Some of these are exploratory in nature, and as such might be rather different from the math or physics homework encountered in other classes. Rather than doing a quick calculation and getting a simple answer, these exercises ask students to explore, make observations, and look for patterns. Such exercises are best approached as if they were short laboratory exercises in a science class.

There are a few exercises in the text that require a good bit of algebra or are more conceptually advanced. These are marked with a ♮ symbol. Most of the exercises are designed to be straightforward, but a few are open-ended and possibly even deliberately ambiguous, with the intention of leading you to think critically about a particular situation or phenomenon. Particularly important exercises are marked with a ⋆. The text refers to these exercises in later chapters, so I strongly suggest that you try them. In many cases they present an opportunity for you to discover something interesting for yourself before I explain it in subsequent chapters, giving away the punch-line and possibly a fun surprise.

Advice and Notes for Instructors

Part I of this text introduces discrete dynamical systems, including graphical iteration techniques and an initial exploration of the logistic equation. In Part II I use one-dimensional maps to introduce and explore the main notions of chaos—aperiodic behavior, sensitive dependence on initial conditions, bifurcations, invariant densities and ergodicity, and the universality of the period-doubling route to chaos. In many ways differential equations are more fundamental mathematical objects; in most situations iterated maps are used as approximations to the continuous flow generated by a differential equation. However, one-dimensional maps have the benefit that they are much more accessible to students without a calculus background. In Part III I introduce fractals, including the self-similarity and box-counting dimensions as well as discussions of random fractals and power laws. This part of the book also gives an example of a simple process for which there is not a well-defined average and contains a discussion of countable vs. uncountable infinities. Part IV covers Julia sets and the Mandelbrot set, while Part V covers a number of dynamical systems beyond the one-dimensional maps covered previously. These include two-dimensional, discrete dynamical systems; cellular automata; and differential equations in one, two, and three dimensions.

There are many possible paths through the book. I consider Chapters 0–11 and 15–18 to be core material, providing a solid introduction to the central ideas of dynamical systems and fractals. I cover these chapters almost every time I teach the class, and then choose from other chapters depending on time and student interest. A course that focuses on fractals could easily begin with Parts III and IV. A course that focuses on dynamical systems can easily skip most of III and all of IV and cover Part V more fully.

Depending on the preparation of your students, you may be able to go through the first several chapters fairly quickly. I would encourage you, though, to not go too quickly. If students get lost in a course like this, it can be difficult to get them back on track. While the first several chapters might seem simple, they can be quite difficult for students who have been away from algebra for several years. How to pace the first few weeks of the course is tricky. Just about every time I have taught this material my students and I have wished we had more time at the end of the course to explore some additional interesting and fun material, but at the same time many students report feeling that it would have been difficult to have gone much faster during the initial stages of the course.

You will notice that the chapters are not of uniform length and that a few are quite short. I have found that introducing some concepts on their own in a concise fashion is quite helpful for students. For example, time series plots are covered in their own four-page chapter rather than being a section in a longer chapter on graphical iteration. I am convinced that this makes these topics easier for students to learn, and that these educational benefits justify the non-uniformity in chapter length.

Students are introduced to many different sorts of graphs in the course of this text: plots of functions, time series plots, "cobweb" diagrams, bifurcation diagrams, phase lines, histograms, Julia sets, and the Mandelbrot set. Students very often have surprising difficulty interpreting these different graphs and keeping them straight. Students will need to be reminded repeatedly how to make and interpret these graphs. Be sure to assign a good amount of graphical homework. Doing so will be essential for students if they are to really understand the material.

I have tried to develop end-of-chapter exercises in which students are led to discover some key ideas or interesting phenomena, which are often discussed in more detail in the subsequent chapter. These exercises are indicated by the symbol $\star$. I strongly encourage you to assign them and to not discuss them in class until students have explored them on their own. They can also make good lab or group-work exercises. Sprinkled throughout the text there are a handful of exercises that are algebra-intensive or require a little bit more mathematical sophistication. These exercises, indicated by the symbol $\sharp$, might be good for more advanced students or if you are teaching a course that emphasizes algebra skills. Finally, in many of the later chapters there are only only a handful of exercises. I found it difficult to come up with meaningful exercises on some of the more advanced material. When I teach the the course this usually is not a problem, as students are working on term projects at this point, and so I need to assign them a little less homework than in the first part of the course.

Although some students occasionally find it initially frustrating, I would be sure to assign problems in which students use some of the simple web-based programs to make time series plots, bifurcation diagrams, and the like. Most of the homework problems that use these programs are quite simple, but I think it is very important for students to use the programs themselves and explore. If at all possible, do not just use the programs as passive demonstrations in class. Students learn a topic better if they have played a role in discovering an idea or phenomenon for themselves.

I have prepared a longer guide for instructors, including suggestions for classroom activities. I have also prepared a solutions manual for the end-of-chapter exercises. Both are available upon request. The website for the book, `http://chaos.coa.edu` contains further information and resources for teaching using this text.

Books that Complement this Text

I was first introduced to the excitement and fun of the topics in this book when I read James Gleick's, *Chaos: Making a New Science* (Gleick, 1987) while a senior in college. More than twenty years later, this book still reads well and is engaging and interesting. Gleick brings to life both the science and the scientists who helped shape the modern fields of chaos and fractals. When I teach the class from which this text was developed, I have always had students read Gleick's book. As a result, this book can function as an accompaniment or complement to Gleick's. If you are teaching a class, or are setting off on your own to learn about chaos and fractals, I would strongly suggest pairing my book with Gleick's.

There are two other books that I think are particularly well suited for pairing with this text. Ian Stewart's *Does God Play Dice?* (2002) is at a similar level to Gleick's book but is much less journalistic and sensational. Stewart's explanations are excellent, and he does an impressive job of putting developments in chaos and fractals into context. Also highly recommended is Stephen Kellert's *In the Wake of Chaos* (1993). This short book is a very clear and well argued analysis of the epistemological and philosophical implications of chaos. A number of suggestions for additional reading, including more advanced texts, are given in Appendix C. Also, at the end of some chapters I have included references to topics particular to that chapter.

How to Contact Me

I have set up a website for the book at `http://chaos.coa.edu/` where you will find my contact information and an updated list of errata. I would be grateful to hear about your experiences using this book and to receive suggestions for improvement.

David P. Feldman
Mount Desert, Maine, USA, August 2011

Acknowledgments

My knowledge of chaos and fractals has been shaped by many collaborators and colleagues over the years. First among these is Jim Crutchfield—mentor, collaborator, and friend. I have benefited from conversations about chaos and fractals with many friends and collaborators, including Bai-lin Hao, Cosma Shalizi, and Karl Young. Several anonymous reviewers at Oxford University Press gave extremely thoughtful and constructive comments on my initial book proposal.

I am grateful for the support and encouragement offered by friends and colleagues at the College of the Atlantic. I am also grateful for the hospitality extended to me by the Santa Fe Institute, and the friends and colleagues I have met there over the years. I especially thank Cris Moore, who provided advice and encouragement and introduced me to my editor at Oxford University Press. I feel fortunate to be able to spend time at two institutions, COA and SFI, that are so committed to interdisciplinary work. The final edits of this book were completed while I was a U.S. Fulbright Lecturer in Rwanda. I thank the Fulbright Scholarship Board and U.S. Department of State for their support and the Applied Physics Department at the Kigali Institute of Science and Technology for their hospitality.

This book was produced exclusively with open-source, freely available software: the book was typeset using LaTeX; figures were produced with xfig, inkscape, GIMP, and gnuplot; gnumeric was used for some data analysis; and all programming was done in either python or C++ using the g++ compiler. All applications were run on various personal computers running Fedora Core or Ubuntu versions of the Linux operating system. I thank the developers and maintainers of this code and all those who contribute to open source projects. I am also grateful for resources such as Google Scholar, Google Books, the citeulike.org bibliography manager, and Wikipedia. They all have been incredibly valuable and have saved me a great deal of time.

Early portions of this project were supported by the Maine Space Grant Consortium. The College of the Atlantic faculty development fund provided generous support at several stages in the book's writing. Portions of this book were written during visits to the Santa Fe Institute and the Complexity Sciences Center in the Department of Physics at the University of California Davis. I thank SFI and the CSC for their hospitality.

I have drawn great inspiration from my students at College of the Atlantic. Their energy and enthusiasm for chaos and fractals has been remarkable. I am grateful for their encouragement and thoughtful comments on drafts of this book. Teaching assistants at COA have provided input, feedback, encouragement, and advice. I especially thank Mikus Abolins-Abols, Adrianna Beaudette, Iris Lowery, Dale Quinby, and Amy Wesolowski. Mark Feldman, Todd Little-Siebold, and John Visvader each read several chapters and provided valuable feedback. Adrianna Beaudette read the full manuscript; her thoughtful critiques improved the book considerably.

I thank Sönke Adlung at Oxford University Press for his cheerful patience and encouragement throughout all stages of the book's preparations. Jessica White, April Warman, and Clare Charles at Oxford also provided helpful guidance. Catherine Cragg at OUP and Vijayasankar Natesan at SPi patiently and expertly answered a ridiculous number of questions about LaTeX and font sizes. I thank them both for their assistance in preparing the final manuscript.

 I am tremendously grateful to Nikki McClure for making available the butterfly image that is on the cover of this book. Chaos and fractals show us that simple processes can lead to surprise and beauty. Nikki demonstrates this same principle in her artwork. Check out her work at `http://www.nikkimcclure.com/`.

Additional inspiration and energy was provided by Armin van Buuren and *A State of Trance*, Morning Glory Bakery, and Ohori's coffee.

Finally, I thank my family, and especially Doreen Stabinsky, about whom I should surely say something touching and sentimental, but I don't know what to say or where to begin. Any words I come up with seem insufficient. So a simple and heartfelt "thank you" will have to suffice.

Contents

Part I

Introducing Discrete Dynamical Systems

Opening Remarks

0

In 1975 the word *chaos* was first used in a technical sense to describe a type of irregular behavior seen in mathematical systems (Li and Yorke, 1975). That same year, Benoît Mandelbrot coined the term *fractal* (Mandelbrot, 1975) to describe mathematical and natural objects that are self-similar: made up of the same motif repeated across scales, large and small. Since 1975 these two ideas—chaos and fractals—have become an important part of the physical, natural, and social sciences. The concepts, ideas, tools, and mindset associated with chaos and fractals are unarguably an important part of modern science, and they have been borrowed and appropriated by other academic disciplines. Chaos and fractals are also fixtures within popular culture, receiving mention in plays and film and in popular scientific writing.

Why are chaos and fractals now a standard part of scientific and non-scientific vocabulary? And what are chaos and fractals, anyway? My aim is to develop answers to these questions gradually throughout this book. To really dig into chaos and fractals—to understand what they are, what they are not, why they matter, and how fascinating and fun they can be—requires building up some mathematical tools. Nevertheless, a few introductory remarks will help set the stage.[1] Chaos and fractals are two distinct ideas, although they are often taught together and there are relationships between them. I will begin with a few comments on chaos, and then move to fractals.

[1] You can skip these remarks if you want. If you are eager to get started, go ahead and jump to Chapter 1.

0.1 Chaos

Chaos is a phenomenon encountered in science and mathematics wherein a deterministic (rule-based) system behaves unpredictably. That is, a system which is governed by fixed, precise rules, nevertheless behaves in a way which is, for all practical purposes, unpredictable in the long run. The mathematical use of the word "chaos" does not align well with its more common usage to indicate lawlessness or the complete absence of order. On the contrary, mathematically chaotic systems are, in a sense, perfectly ordered, despite their apparent randomness. This seems like nonsense, but it is not. The first two parts of this book are largely concerned with explaining this apparent paradox.

The phenomenon of chaos is usually considered to be part of the field of study known as **dynamical systems**, an interdisciplinary area that lies mainly at the intersection of physics and mathematics, but also includes researchers from biology, economics, and elsewhere. Dynamical

Dynamical Systems

[2]The field of dynamical systems is sometimes referred to simply as *dynamics*. It is also called *nonlinear dynamics*, to emphasize that chaos and related phenomena arise only in nonlinear systems. I will use all terms interchangeably, although I will mainly use dynamical systems.

[3]Actually, most researchers would say they study nonlinear dynamics or dynamical systems, which are generally viewed as a more respectable and scientific term than chaos. The term "chaos" is somewhat informal and perhaps is tainted by the fact that it has been misused.

Nonlinear Dynamics

systems is the study of systems or processes that change over time. This includes examining particular systems or areas of application, as well as looking at systems more broadly and abstractly to develop generally applicable tools for studying dynamical systems or to classify different sorts of behavior.[2]

Although the technical use of the term chaos originated in 1975, research in dynamical systems dates back at least to Henri Poincaré and the early 1900s. However, it was not until the 1960s and '70s that researchers and ideas from a variety of different fields coalesced. Since at least the 1990s, chaos has been recognized as an area of study; there are multiple books on the subject, conferences that bring together researchers who study chaos, and scientists and mathematicians who would say that chaos is their primary area of study.[3]

Chaos is just one phenomenon out of many that are encountered in the study of dynamical systems. In addition to behaving chaotically, systems may show fixed equilibria, simple periodic cycles, and more complicated behaviors that defy easy categorization. The study of dynamical systems holds many surprises and shows that the relationships between order and disorder, simplicity and complexity, can be subtle, and counterintuitive. My aim in this book is to introduce readers to these relationships.

The term "chaos" has somewhat of a dual life. It refers to a particular type of dynamical behavior, but it is also often used as a general shorthand for the study of dynamical systems. This book is about chaos in its more general sense. We will encounter not just chaotic behavior, but a host of other dynamical behaviors, too.

0.2   Fractals

Shifting gears for a moment, a few words about fractals. Fractals are objects which are self-similar. Small parts of a fractal look like larger parts. For example, a tree is a fractal, since if you break off a branch of the tree, it resembles the entire tree in miniature. In contrast, a person is not a fractal; an arm does not look like a small copy of the person. A person is not self-similar, but a tree is. Fractal objects are characterized by their fractal dimension which, very roughly speaking, is related to their degree of branching and the extent to which the features at successive scales are related.

The study of fractals gives us a quantitative language to describe the myriad of self-similar shapes found in the natural world, including: mountain ranges; river basins; clouds and lightning; trees, ferns, and other plants; and vascular systems in plants and animals. Fractals need not be natural objects; they can be human-made and can also unfold in time in addition to space. For example, the sizes of earthquakes, the populations of cities, the frequency of words within a text, and the distribution of the number of links into web pages all can be usefully viewed as fractals. A few simple tools and ideas for analyzing fractals prove to be surprisingly powerful and flexible.

Fractals are different from chaos. Fractals are self-similar geometric objects, while chaos is a type of deterministic yet unpredictable dynamical behavior. Nevertheless, the two ideas or areas of study have several interesting and important links. Fractal objects at first blush seem intricate and complex. However, they are often the product of very simple dynamical systems. So the two areas of study—chaos and fractals—are naturally paired, even though they are distinct concepts. The first two parts of this book are exclusively concerned with dynamical systems and chaos. In the first several chapters of Part III I will introduce fractals, and then throughout the rest of the book we will see a number of different ways that chaos and fractals are interwoven.

0.3   The Character of Chaos and Fractals

Are chaos and fractals a big deal? Do they deserve all the hype? Are they a revolution or a paradigm shift? My answers to these questions are, in order: yes, to some extent, and not really. I think that most, but not all, scientists and philosophers of science would answer similarly. Part of the purpose of this book is to provide a clear introduction so that you can form your own answers to these, and related questions. Nevertheless, some introductory remarks are in order to help set the stage for what is to follow.

Chaos is not a theory. The phrase *chaos theory* is used in popular science books and media accounts. However, the phrase is rarely used by scientists and mathematicians, either in papers or books or, in my experience, when talking with each other. What do I mean when I say that chaos theory is not a theory?

It is difficult to precisely define what is meant by a scientific theory, but usually a theory is a concise and consistent body of knowledge that allows one to understand a broad class of natural phenomena. For example, in electrodynamics—the study of electricity and magnetism—just five equations describe the behavior of any arrangement of stationary and moving charges. One can use these equations to calculate the value of electric and magnetic fields and determine the forces these fields exert on other charges.[4]

A scientific theory can provide a broad explanatory framework without being associated with the equations and calculational methods that are typical of physics. For example, the germ theory of disease, the theory of evolution by natural selection, or the theory of plate tectonics, each explain a large body of facts and provide an organizing structure for thinking about epidemiology, biodiversity, and earthquakes and mountain ranges. Note that these theories explain without necessarily predicting. The theory of plate tectonics has led to only little success in predicting the timing of earthquakes, but it certainly helps us explain what earthquakes are and why they occur where they do.

Chaos is not a theory like electrodynamics, quantum mechanics, the germ theory, evolution, or plate tectonics. There is not a chaos equation,

[4] These five equations are the Lorentz force law and the set of four equations known collectively as Maxwell's equations.

nor does chaos quite provide a recipe for calculating physical quantities of interest in the way that quantum mechanics does. There are no axioms or postulates of chaos. Chaos does provide a framework or a mindset or point of view, but it is not as directly explanatory as germ theory or plate tectonics. Chaos is a behavior—a phenomenon—not a causal mechanism.

The situation with fractals is similar. The study of fractals draws one's eye toward patterns and structures that repeat across different length or time scales. There is also a set of analytical tools—mainly calculating various fractal dimensions—that can be used to quantify structural properties of fractals. Fractal dimensions and related quantities have become standard tools used across the sciences. As with chaos, there is not a fractal theory. However, the study of fractals has helped to explain why certain types of shapes and patterns occur so frequently.

I hope these remarks are not deflating or discouraging. I certainly do not intend them to be. But I do want to lay my cards on the table so you are not expecting this book to lay out a neat, comprehensive theory. So what *is* chaos if not a theory? And how do fractals fit into all of this? I will revisit these questions throughout the book, but for now, here is an imperfect analogy to get things started.

In some ways the study of chaos and fractals is like the study of trees. Trees are awesome. They are beautiful, found all over the world, have a wide range of sizes and shapes, but also have some definite similarities. There is no such thing as *tree theory*. But there are theories that apply to trees: evolution, chemistry, and physics all have a lot to say about trees. Likewise, there are certain techniques and methods that are useful for studying trees: measuring their height, taking cores, pressing leaves, using microscopes, and so on. So there is lots to learn about particular trees, trees in general, and also lots to learn about the methods scientists use to study trees. So it is with chaos and fractals.

Are chaos and fractals a revolution or a passing fad? A paradigm shift or nothing but a lot of hot air and hype? While it is unarguable that there has been quite a bit of undeserved hype and pseudoscientific speculation, I certainly think chaos and fractals are very important. The study of chaos shows that simple systems can exhibit complex and unpredictable behavior. This realization both suggests limits on our ability to predict certain phenomena and that complex behavior may have a simple explanation. Fractals give scientists a simple and concise way to qualitatively and quantitatively understand self-similar objects or phenomena. More generally, the study of chaos and fractals hold many fun surprises; it challenges one's intuition about simplicity and complexity, order and disorder.

So let us put aside philosophical concerns for the time being and begin. In the next chapter I discuss a number of different ways of viewing and thinking about functions. In the subsequent chapter we will look at repeatedly applying a function. These iterated functions are a dynamical system. By studying them, we will begin our journey into chaos.

Further Reading

This short chapter attempts to give a very brief conceptual overview of chaos and fractals and to put forth some initial claims about the character of these new scientific and mathematical areas. Other authors have written similar introductions that you might want to consult as well. I highly recommend the first chapter of Stephen Kellert's *Borrowed Knowledge: Chaos Theory and the Challenge of Learning Across Disciplines* (2008). I also recommend: the prologue and first chapter of Kellert's earlier book, *In the Wake of Chaos* (1993); the very short preface to Leonard Smith's *Chaos: A Very Short Introduction* (2007); the preface to *Does God Play Dice?* by Ian Stewart (2002); and also the introductory chapter of Peter Smith's *Explaining Chaos* (1998). The latter book is more mathematically advanced than the others, but still quite clear.

Functions

Before we can get into chaos and fractals, I will need to lay a good bit of groundwork, introducing some key terminology and ideas. This is the first of several chapters that lay this foundation. The starting point is to consider the mathematical idea of a *function*. Functions are the most basic way of mathematically representing a relationship. They will be a key to the first two parts of this book.

In everyday speech, we might say something like "how tired you are is a function of how much you slept last night." Or, "how hungry you are depends on how many cookies you have eaten." These statements suggest that one quantity—tiredness or appetite—depends on another quantity—hours sleeping or numbers of cookies consumed. In this sense, the common usage of the word function aligns with the mathematical use; mathematically a function represents a dependence. However, as is often the case in science and mathematics, the technical meaning of a word is narrower than the common meaning, as we will see below.

To further explore the idea of a function, we now consider several different, complementary ways of looking at functions.

1.1 Functions as Actions

In mathematics, it is useful to think of a function as an action; a function takes a number as input, does something to it, and outputs a new number. This is illustrated in Fig. 1.1. Here, the function is called f. The function takes a number x as input. The function, indicated schematically by a box, then acts on the number x and produces another number as output. This new number is called $f(x)$.

Fig. 1.1 A schematic view of a function f that takes a number x as input, does something to the number, and outputs a new number called $f(x)$.

For concreteness, let us say the function is the action *triple*. That is, the function takes a number and multiplies it by 3. So, for example, if the input is 4, the output would be 12. If the input is 20, the output would be 60. And if the input is 2.7, the output would be 8.1.

There are a number of ways that we can denote this symbolically. One way is as follows:

$$4 \xrightarrow{f} 12 \,, \tag{1.1}$$

$$20 \xrightarrow{f} 60 \,, \tag{1.2}$$

and

$$2.7 \xrightarrow{f} 8.1 \,. \tag{1.3}$$

This notation helps make it clear that the function f takes a 4 and turns it into a 12, takes 20 and turns it into 60, and so on. This is represented pictorially or schematically by Fig. 1.2.

This can also be indicated symbolically:

Fig. 1.2 A schematic view of the triple function. This function takes a number as an input, and outputs that number multiplied by 3.

$$f(4) = 12 \,, \tag{1.4}$$

$$f(20) = 60 \,, \tag{1.5}$$

and

$$f(2.7) = 8.1 \,. \tag{1.6}$$

Read aloud, Eq. (1.4) would read "f of 4 equals 12." What this means is that if the function f gets 4 as input, the output is 12.

1.2   Functions as a Formula

In the preceding example I specified the function f by saying in words what it does. Namely, it triples the input number. We can also specify the function using algebra:

$$f(x) = 3x \,. \tag{1.7}$$

In this equation, x is a placeholder for the number that we input and the right hand side, $3x$, represents the value of the number that the function outputs. In this case, the output is three times the input: take the input x and multiply it by 3. So,

$$f(3) = 3 \times 3 = 9 \,, \tag{1.8}$$

$$f(4.5) = 3 \times 4.5 = 13.5 \,, \tag{1.9}$$

and so on.

By the way, there is nothing special about the letter "x" in Eq. (1.7). I could just as well have written:

$$f(z) = 3z \,, \tag{1.10}$$

or

$$f(q) = 3q \,. \tag{1.11}$$

The letter "x" (or "z" or "q") is just a placeholder. That is, "x" is just a form of shorthand or a nickname for whatever number we use as input for the function.

One final note about formulas: an equation such as Eq. (1.7) is not something that one would solve. In fact, there is nothing to solve for.

The purpose of Eq. (1.7) is to define the function f. It says, for all values of x, what the value of the function is, $f(x)$.

In subsequent chapters I will have more to say about algebraic approaches to functions. For now, the main point is that functions can be described by a formula, as we did in Eq. (1.7).

1.3   Functions are Deterministic

We are now in a position to refine our definition of a function. A function is a rule that assigns an output value $f(x)$ to every input x. This is consistent with the everyday use of the word function: the output $f(x)$ is a function of the input x. The output depends on the input, just as your grade on a test might depend on how many hours you study.

However, there is a crucial difference between the common definition of the word function and its mathematical definition. For a mathematical function, the output is determined *entirely* by the input. This means that if you give a function the same input, you will always get the same output. In contrast, your grade on a test depends on more than just how many hours you study: your grade also might depend on the questions on the test, how tired you are when you take the test, how much beer you had the night before the test, and so on.

Consider again the function $f(x) = 3x$, of Eq. (1.7). The output depends on the input, and the same number input into the function always yields the same output. For example, $f(5) = 15$. And, a few moments later, $f(5)$ is still 15. Doing the same thing gives the same result. Such a function is said to be **deterministic**, because the output is completely determined by the input.

If a function is not deterministic, we would say that it is **stochastic**.[1] This means that the function involves some element of chance. For example, perhaps $h(x) = 2x$ with probability 1/2, and $h(x) = x$ with probability 1/2. So, if one used $x = 5$ as input, approximately half of the time one would get 10 as output, and approximately half of the time one would get 5. Clearly, the input does not completely determine the output.

FUNCTION:
DETERMINISTIC
STOCHASTIC

[1] The root of the word stochastic is *sto-chos*, which means "guess" or "target" in ancient Greek.

1.4   Functions as Graphs

Thus far we have described functions verbally as an action—e.g., triple the input—and symbolically using algebra—$f(x) = 3x$. Another way to describe or specify a function is via a graph. Figure 1.3 shows a graph of the function $f(x) = 3x$. In this case, the function appears as a straight line. (This is not the always the case, as we will see shortly.) The idea here is that the input is plotted on the horizontal axis and the output $f(x)$ is plotted on the vertical axis. For example, if $x = 4$, $f(x) = 12$. Accordingly, the point $(4, 12)$ is on the function's graph. And if $x = 5$, $f(x) = 15$, so $(5, 15)$ is on the graph.

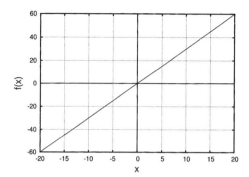

Fig. 1.3 A graph of the function $f(x) = 3x$.

As another example, a different function is shown in Fig. 1.4. In this case we are given the graph of function, but we do not know the formula for the function. Nevertheless, the function shown in Fig. 1.4 is a perfectly legitimate function. We can specify a function graphically just as legitimately as we can specify it using algebra.

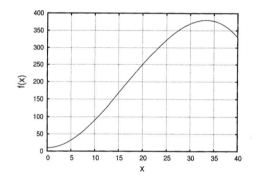

Fig. 1.4 A graphical representation of a function. The algebraic formula for this function is not given.

We can use the graph to read off approximate values for the function for given inputs. For example, we can see that $f(5) \approx 30$ and $f(30) \approx 370$. The symbol "$\approx$" is read "is approximately equal to." You should take a moment to look at the function in Fig 1.4 and verify for yourself the approximate values of $f(5)$ and $f(30)$.

There are many different possible shapes that graphs of functions can take. However, not every curve that you draw corresponds to a function. An example of a plot that is not a function is shown in Fig. 1.5. The reason that this is not a function is that there is not only one output for every input. For example, consider $f(-2)$. Looking at the graph, it appears that there are three possible values for $f(-2)$: 1.2, 1.6, and 2.3. Thus, this is not deterministic, and hence this is not a function. For f to be a function there has to be one and only one output for every input.

There is a geometric way to see that the curve in Fig. 1.5 is not a function: if at any point on a curve a vertical line intersects the curve more than once, then the curve does not describe a function. In the example at hand, if we draw a vertical line at $x = -2$, then that line will intersect the curve three times. In general, any time a vertical line intersects the curve more than once, this indicates that there is more than one output for a single input. Hence, the curve cannot be a function.

You might have encountered this idea before; it commonly goes by the name of **vertical line test.** Knowing the name of the test is not important; the key point is that, by definition, a function can only have one output for every input.

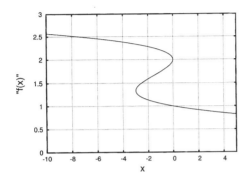

Fig. **1.5** A plot that does not give a function. The reason that this is not a function is that there is not a unique output value for every input value. E.g., the graph indicates three possible outputs for the input -2.

1.5 Functions as Maps

A function is also sometimes referred to as a map or a mapping. This terminology is common in mathematics, but less so in physics or other scientific fields. The idea of a mapping is useful if one wants to think of a function as acting on an entire set of input values. This idea is illustrated in Fig. 1.6. The function f maps input values in the set A to output values in the set B. For example, 2 maps to 8. This is the same as saying $f(2) = 8$.

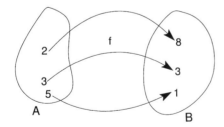

Fig. **1.6** A diagram illustrating a mapping. The function f maps elements of the set A on the left, to the set B.

The statement that the function f maps A to B is denoted symbolically by

$$f : A \mapsto B .$$ (1.12)

This equation would be read "f is a function that maps A to B." I will usually just refer to functions as functions, and not maps. Nevertheless, this is good terminology to know, as it is quite common and you may encounter it elsewhere.

Exercises

(1.1) Let g be the doubling function.

   (a) Calculate:

      (i) $g(3)$

      (ii) $g(0)$

      (iii) $g(17)$

      (iv) $g(0.4)$

      (v) $g(-3)$

   (b) Sketch the graph of g.

   (c) Determine the formula for g.

(1.2) Let h be a function that takes a number, quadruples it, and then subtracts 3.

   (a) Calculate:

      (i) $h(5)$

      (ii) $h(0)$

      (iii) $h(0.5)$

      (iv) $h(-1)$

   (b) Determine the formula for h.

(1.3) Let f be a function that takes a number, subtracts three, and then quadruples it.

   (a) Calculate:

      (i) $f(5)$

      (ii) $f(0)$

      (iii) $f(0.5)$

      (iv) $f(-1)$

   (b) Determine a formula for f.

   (c) Compare your answers to those for Exercise 1.2. Are your answers different? Why or why not?

(1.4) Let $g(x) = 3 + x^2$.

   (a) Evaluate the following

      (i) $g(0)$

      (ii) $g(1)$

      (iii) $g(-1)$

      (iv) $g(2)$

      (v) $g(2+1)$

      (vi) $g(g(1))$

   (b) Does $g(2+1) = g(2) + g(1)$? Should it?

   (c) If $g(x) = 7$, what is x?

   (d) If $g(x) = 0$, what is x?

(1.5) Let $f(x) = 2x$.

   (a) Evaluate the following

      (i) $f(0)$

      (ii) $f(1)$

      (iii) $f(2)$

      (iv) $f(2+1)$

      (v) $f(f(0))$

      (vi) $f(f(1))$

   (b) Does $f(2+1) = f(2) + f(1)$? Should it? Compare with Exercise 1.4b. What is the difference between the two situations?

(1.6) Consider the function shown in Fig 1.7. Calculate

   (a) $f(-5)$

   (b) $f(0)$

   (c) $f(5)$

   (d) $f(10)$

(1.7) Consider the function shown in Fig. 1.7.

   (a) If $f(x) = 7$, what is x?

   (b) If $f(x) = 2$, what is x?

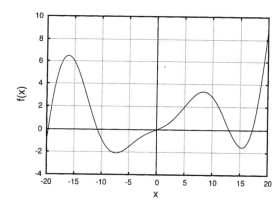

Fig. 1.7 The function for Exercises 1.6 and 1.7.

(1.8) Describe an everyday, "real life" example of a function. Explain how your example fits the criteria for being a function.

(1.9) Consider the function shown in Fig. 1.8. Calculate

  (a) $g(-30)$

  (b) $g(-20)$

  (c) $g(0)$

  (d) $g(10)$

  (e) $g(20)$

(1.10) Consider the function shown in Fig. 1.8.

  (a) If $g(x) = -10$, what is x?

  (b) If $g(x) = 15$, what is x?

  (c) If $g(x) = 50$, what is x?

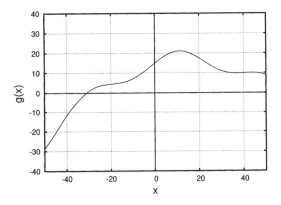

Fig. 1.8 The function for Exercises 1.9 and 1.10.

(1.11) ⋆ Figure 1.9 shows a possible relationship between this year's and next year's population of rabbits on a small coastal island. The reason that the rabbits may be considered to behave this way is as follows. Let us imagine that the rabbits do not have any predators on this island, but that there is a limited amount of food, since the island is small. Suppose there are a lot of rabbits on the island one year, say 100. Then there will not be enough food on the island for all the rabbits, and some will starve. So there will be fewer rabbits in the following year. This is indicated on the graph in Fig. 1.9; if one year there are 100 rabbits, the next year there will be approximately 63 rabbits. On the other hand, suppose there are few rabbits on the island, say 10. Then there will be plenty of food to go around, the well-fed rabbits will reproduce, and there will be more rabbits next year—around 50.

  (a) In 1999 there are 70 rabbits on the island. How many rabbits are there in 2000?

  (b) In 2003 there are 35 rabbits on the island. How many rabbits are there in 2004?

  (c) In 1985 three are 20 rabbits on the island. How many rabbits are there in 1987. Explain your reasoning.

  (d) In 1992 there are 80 rabbits on the island. How many rabbits were there in 1991?

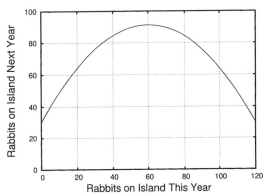

Fig. 1.9 The rabbit population on an island next year as a function of the number of rabbits on the island this year. See Exercise 1.11.

Iterating Functions

2

When I was in high school there were neither cell phones nor pagers nor even graphing calculators. This meant that in class if I was bored I did not have any high-tech amusements. But I did, however, have a basic calculator. One source of entertainment was to enter a number on the calculator, and then hit a function key, like x^2, over and over and over again. Sometimes the number gets too big for the calculator. Other times it eventually goes to zero. Sometimes it gets stuck at 1. Admittedly, this is not the most scintillating of games. But it did help pass the time in some dreary pre-calculus classes.

Despite its apparent simplicity, this process—applying a function over and over—is at the core of the rest of this book. In this chapter I introduce this process, known as iteration, along with some important terminology..

2.1 The Idea of Iteration

Iteration entails doing the same thing again and again using the previous step's output as the next step's input. In other words, we start with a number and apply a function to it to get a new number. Then we take that new number and apply the function to it to get yet another number. Then we apply the function to this new number, and so on.

This process is illustrated schematically in Fig. 2.1. The output of the function is used as input for the next step. This can also be thought of as a feedback process, in which output is used as input. This is what happens, for example, when a microphone and amplifier produces feedback where the microphone picks up some sound, inputs it to the amplifier which amplifies it. The microphone then picks up the amplified sound and inputs it to the amplifier to produce a new sound. This new sound is then picked up by the amplifier which amplifies it, and so on. I suspect that many of you are familiar with the high-pitched squeal that can result from this process.[1]

If you did Exercise 1.11 in Chapter 1, you have already iterated a function.[2] To do part (c) of this exercise, you started with 30 and then applied the function *twice* to this number. The number 30 is the number of rabbits in 1985. The function tells you that in 1986 there were approximately 65 rabbits. Then, you applied the function again, this time using 65 as input. The result is approximately 90 rabbits. The key feature here is that you start with a number and then apply a function repeatedly to it.

[1] You may wonder why audio feedback is usually a single high-pitched tone and not just a loud reproduction of the tones in the original sound. What happens is that there is a particular frequency that is amplified the most by the amplifier. What this frequency is depends on the details of the microphone and amplifier, as well as the acoustics of the room. When the positive feedback loop starts, this one frequency dominates over all others, since at each step of the process it is amplified the most. The result is a loud squeal at that frequency.

[2] If you have not done this exercise, I suggest going back and doing it now before continuing.

Fig. 2.1 A schematic view of an iterated function. The function f takes a number x as input, does something to the number, and outputs a new number called $f(x)$. This process is then repeated, or iterated: the output $f(x)$ is used as input. Compare this to Fig 1.1 in Chapter 1.

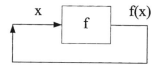

In this case, the function was specified by a graph and not an equation. But we can easily do the same iteration process symbolically using an equation. Suppose our function is the tripling function, $f(x) = 3x$. Let us use 2 as our input. We then apply the function and get $f(2) = 3 \times 2 = 6$. We then take 6, our output, and use it as input for the function to get $f(6) = 3 \times 6 = 18$. We then repeat the process: $f(18) = 3 \times 18 = 54$. Here is another, perhaps clearer, way to see this process:

$$2 \xrightarrow{f} 6 \xrightarrow{f} 18 \xrightarrow{f} 54 \xrightarrow{f} 162 \cdots \tag{2.1}$$

If we started with a different number, we would get a different series of outputs:

$$0.5 \xrightarrow{f} 1.5 \xrightarrow{f} 4.5 \xrightarrow{f} 13.5 \xrightarrow{f} 40.5 \cdots \tag{2.2}$$

2.2 Some Vocabulary and Notation

The next two sections are concerned with developing some notation and vocabulary that will be useful for describing the results of iteration in the subsequent chapters. We begin with another example: the squaring function, $g(x) = x^2$. If we start, say, with 3, we will get 9 and 81 and 6561 and so on.

$$3 \xrightarrow{g} 9 \xrightarrow{g} 81 \xrightarrow{g} 6561 \cdots . \tag{2.3}$$

Table 2.1 The orbit of $g(x) = x^2$ for the seed $x_0 = 3$.

x_0	3
x_1	9
x_2	81
x_3	6561

ORBIT:
RESULTS

The number we start with, 3 in this particular case, is known as the **initial condition** or the **seed**. Very often the initial condition is denoted x_0, which is read "x zero" or "x naught." The next value is denoted x_1, and then x_2, and so on. It is often convenient to show this in a table, as is done in Table 2.1.

Iterating a function produces a sequence of numbers. A sequence is simply a list with an order to it. This sequence is often called an **itinerary**. This is consistent with the everyday usage of the term; an itinerary is a list, in order, of all the places visited along a journey. The mathematical usage of itinerary is similar; the itinerary of 3 is a list of the results, in order, that one gets from applying the function again and again. Another word for itinerary is **orbit**. Orbit and itinerary, in their mathematical usages, are synonymous. Throughout this book I will use the terms orbit and itinerary interchangeably; you will encounter both if you read other books or papers on chaos.

Note that for a given seed, you will get different itineraries depending on the function you iterate. To make this clear, one often refers to "the orbit of x_0 *under* f" to remind us that the orbit (or itinerary) depends on f in addition to x_0. For example, we would say that Eq. (2.3) gives the orbit of 3 under x^2.

2.3 Iterated Function Notation

We can also indicate the process of iteration by making use of functional notation. The first step when iterating is to apply the function, let us call it f, to the seed x_0 to obtain x_1:

$$x_1 = f(x_0) \, . \tag{2.4}$$

The next step is to apply f to x_1, yielding x_2:

$$x_2 = f(x_1) \, . \tag{2.5}$$

We can also combine these two equations, plugging in Eq. (2.4) to Eq. (2.5), as follows:

$$x_2 = f(x_1) = f(f(x_0)) \, . \tag{2.6}$$

At first blush, $f(f(x_0))$ might look funny. But this expression actually summarizes the idea of iteration in a nice, compact form. What $f(f(x))$ means is: start with x, apply f to get $f(x)$, then apply f again to get $f(f(x))$. The end result, $f(f(x))$ is a number that equals x after f has been applied to it twice. Note that, as is always the case with nested parentheses, one starts with the innermost expression and evaluates outward. For example, if f is the squaring function, then $f(f(5)) = f(5^2) = f(25) = 25^2 = 625$.

We can do a similar thing for x_3, the third iterate:

$$x_3 = f(f(f(x))) \, , \tag{2.7}$$

and the fourth iterate,

$$x_4 = f(f(f(f(x)))) \, , \tag{2.8}$$

and so on. I like this notation, because it makes it quite clear that x_4 is obtained by starting with x_0 and then applying f four times. However, writing all those f's can be time-consuming, and if we get too many of them it will start to look ridiculous. For example,

$$x^{11} = f(f(f(f(f(f(f(f(f(f(f(x))))))))))) \, . \tag{2.9}$$

This equation is true, but it is not that useful.

So we need some better notation. The standard thing to do is to indicate multiple applications of f as follows:

$$f(f(x)) = f^{(2)}(x) \, , \tag{2.10}$$

$$f(f(f(x))) = f^{(3)}(x) \, , \tag{2.11}$$

and, in general, for n applications of f:

$$f(\underbrace{f(f(\cdots f(x))))}_{n \text{ times}} = f^{(n)}(x) \,. \tag{2.12}$$

This notation can be potentially misleading, because it looks like the n in $f^{(n)}$ is an exponent. But this is not the case. The expression $f^{(n)}(x)$ means f applied to x a total of n times. It does *not* mean $f(x)$ times itself n times.[3] For example, for $n = 2$,

$$f^{(2)} = f(f(x)) \,, \tag{2.13}$$

but

$$f^{(2)} \neq f(x) \times f(x) \,. \tag{2.14}$$

The quantity on the right-hand side of the above equation would be denoted

$$f(x) \times f(x) = (f(x))^2 \,. \tag{2.15}$$

If you are unclear about this, be sure to try Exercise 2.5 at the end of the chapter.

2.4 Algebraic Expressions for Iterated Functions

Let us go back and think about the second iterate, x_2:

$$x_2 = f(f(x)) = f^{(2)}(x) \,. \tag{2.16}$$

This notation helps make it clear that we can view $f^{(2)}$ as a function in its own right. After all, $f^{(2)}$ takes a number as input and returns some number as output. And if f is deterministic—one always gets the same output for the same input—then it follows that $f^{(2)}$ is also deterministic. The same holds true for $f^{(3)}$, and $f^{(4)}$, and so on.

At this point it is perhaps reasonable to wonder: if I have a formula for $f(x)$, can I figure out a formula for $f^{(2)}(x)$? The answer to this question is yes. It is a little bit messy, but it can be done.[4] As an example, let us consider $g(x) = 2x^2 - 3$. We start with

$$g^{(2)} = g(g(x)) \,. \tag{2.17}$$

Let us plug in for the inner $g(x)$—remember that we always evaluate compound expressions by starting on the inside and working our way out:

$$g^{(2)} = g(2x^2 - 3) \tag{2.18}$$

Now we apply g to $2x^2 - 3$:

$$g^{(2)} = 2(2x^2 - 3)^2 - 3 \,. \tag{2.19}$$

[3]Another notation for this that you might have seen is $f(f(x)) = (f \circ f)(x)$. The idea here is that $f \circ f$ is the function that consists of "two f's". This is often said "f composed with f." This is common notation, and you may have seen this before in a pre-calculus class. However, I will not use this notation in this book.

[4]This example is mostly to illustrate the general point that this sort of thing is possible. If you do not follow all the steps, do not worry. The details of the algebra are not essential to what follows. On the other hand, this sort of manipulation is excellent algebra practice for those who are so inclined.

Squaring the polynomial, we get

$$g^{(2)} = 2(2x^2 - 3)(2x^2 - 3) - 3 = 2(4x^4 - 12x^2 + 9) - 3 . \quad (2.20)$$

Simplifying, this becomes

$$g^{(2)} = 8x^4 - 24x^2 + 18 - 3 . \quad (2.21)$$

Simplifying further, we get

$$g^{(2)} = 8x^4 - 24x^2 + 15 . \quad (2.22)$$

If you find yourself confused by the operations on exponents that just occurred, you might want to look at Appendix A.1, which is a brief review of the properties of exponents.

So, we have figured out an expression for $g^{(2)}(x)$ given $g(x)$. What about higher iterates, such as $g^{(3)}(x)$ or $g^{(4)}(x)$? We could follow the same procedure. It turns out, as the previous example suggests, that doing so is difficult algebraically and often is not that useful. For example, the eleventh iterate of $g(x) = 2x^2 - 3$ is given by

$$
\begin{aligned}
g^{(11)} = \ & +2147483648\, x^{32} - 51539607552\, x^{30} \\
& + 573378134016\, x^{28} - 3923452624896\, x^{26} \\
& + 18475540021248\, x^{24} - 63464986902528\, x^{22} \\
& + 164457502212096\, x^{20} - 327851849023488\, x^{18} \\
& + 508055167303680\, x^{16} - 613957375623168\, x^{14} \\
& + 576610919055360\, x^{12} - 416437490221056\, x^{10} \\
& + 226741238065152\, x^{8} - 89984826961920\, x^{6} \\
& + 24551831324160\, x^{4} - 4115586931200\, x^{2} \\
& + 319384296447 . \quad (2.23)
\end{aligned}
$$

Quite a mess.[5]

[5] I used a computer to figure this out; I did not do it by hand.

Clearly, Eq. (2.23) is, at best, unwieldy. It might be useful in a computer's brain, but there is not much that humans can do with this—looking at it does not convey much specific meaning. In contrast, though, the idea behind $g^{(11)}$ is very simple: give me a number x and apply g to it eleven times.

What I am trying to suggest is that the equation for iterates, especially higher-order ones such as Eq. (2.23), are not that useful or informative. We will encounter this general phenomenon frequently throughout the book: there exist many ideas that are precisely mathematically defined, but which are best thought of in non-algebraic terms.

2.5 Why Iteration?

You may be wondering why we are worrying about iteration. Who cares, anyway? This is certainly a fair question, and I have several answers. The first answer might seem like a bit of a cop-out: why not? Iterating functions is perhaps fun, or is an interesting game or a brain-teaser. Mathematics does not necessarily need to justify itself by appealing to usefulness or the real world. Math can be about puzzles or games or looking for amusing or surprising patterns. In this line of thinking, asking "why iteration?" is akin to asking "why chess?" or "why crossword puzzles?"

A second answer is that many systems in the real world can be thought of as being governed by an equation that is applied over and over. One example was the population of rabbits on an island, mentioned in Exercise 1.11 in Chapter 1, where the population of the rabbits next year is a simple function of the rabbit population this year. Of course no one really believes the world is this simple—even the world of rabbits on an island. But perhaps this rabbit function is a reasonable enough approximation that we can learn something from this approach.

But it is not just about rabbits. The laws of physics themselves can be viewed as a sort of function. Objects change their state of motion due to forces acting on them, and these forces are determined by (usually) well understood rules. So an object is acted upon by a force, and it moves. It is acted upon by the force again. (The force might be different now, because the object has moved.) So the object moves. And the force acts. And the object moves. And all the while it is the same rules, the same functions, that are at play.[6]

Thus, much of physics can be seen as a iterative process: an object or a bunch of objects have some initial condition or seed. The laws of Newtonian physics are applied over and over, and the objects end up somewhere else. This might sound almost disheartening. Is this all there is to the physical universe in which we make our home? Perhaps. But we will see that even this arguably over-simplified view of the universe holds some intriguing surprises. In Chapter 8 I will discuss the Newtonian worldview in more detail and how the study of chaos—among several other discoveries—altered our idea of predictability and the implications of a mechanical universe.

Iterated functions are an example of what mathematicians call **dynamical systems.** A dynamical system is just a generic name for some variable or set of variables that change over time. There are many different types of dynamical systems—the iterated functions introduced above are just one type among many. Dynamical systems is now generally recognized as a branch of applied mathematics that studies properties of how systems change over time. Scientists and mathematicians in this field are interested both in general questions about what sorts of change are typical for different types of systems, as well as applications to particular systems of physical or biological interest. In this book I will largely take the former approach and focus on general properties of dynamical systems. We will begin in the next chapter, where we will consider different long-term behaviors for iterated functions.

[6]It is important to underscore that iteration is not mere *repetition*, doing the same thing again and again. Rather, iteration entails a closed loop—doing the same thing again and again, but using the output of the previous step as the input for next step.

Exercises

(2.1) Let g be the doubling function. Determine the first five numbers in the orbit for the following seeds:

(a) $x_0 = -2$

(b) $x_0 = -0.5$

(c) $x_0 = 0$

(d) $x_0 = 0.5$

(e) $x_0 = 2$

(2.2) Let $f(x) = \sqrt{x}$. Determine the first five numbers in the orbit for the following seeds:

(a) $x_0 = 0$

(b) $x_0 = \frac{1}{2}$

(c) $x_0 = 1$

(d) $x_0 = 2$

(e) $x_0 = 4$

(2.3) Consider the function $f(x) = \sqrt{x}$.

(a) Complete the following table for f.

x_0	9
x_1	
x_2	
x_3	
x_4	

(b) Determine a formula for $f^{(2)}(x)$. (Hint: it may help to write $\sqrt{x}$ as $x^{\frac{1}{2}}$. A review of the properties of exponents can be found in Appendix A.1.)

(c) Determine a formula for $f^{(3)}(x)$.

(d) Determine a formula for $f^{(4)}(x)$.

(e) Determine a formula for $f^{(n)}(x)$, the nth iterate of x.

(2.4) Consider the function $f(x) = \frac{1}{2}x + 4$.

(a) Complete the following table for f.

x_0	2
x_1	
x_2	
x_3	
x_4	

(2.5) Let $f(x) = 3x - 1$

(a) Calculate $f^{(2)}(1)$ and $(f(1))^2$.

(b) Are the two quantities equal? Should they be?

(2.6) Consider the function $f(x) = (x + 3)^2$.

(a) Complete the following table for f.

x_0	2
x_1	
x_2	
x_3	
x_4	

(2.7) Let $f(x) = x^2$. Determine an algebraic expression for:

(a) $f^{(2)}(x)$

(b) $f^{(3)}(x)$

(c) $f^{(n)}(x)$

(2.8) Let $h(x) = 3x - 1$. Determine the numerical value of:

(a) $h^{(2)}(1)$

(b) $h^{(2)}(3)$

(c) $h^{(4)}\left(\frac{2}{3}\right)$

(d) $h^{(3)}(2)$

(2.9) Let $g(x) = x^2 + 1$. Determine the numerical value of:

(a) $g^{(2)}(1)$

(b) $g^{(2)}(3)$

(c) $g^{(4)}(0)$

(d) $g^{(3)}(2)$

(2.10) Let $f(x) = x^2 - 1$. Determine an algebraic expression for $f^{(2)}(x)$.

(2.11) Let $f(x) = 3x - 1$. Determine an algebraic expression for $f^{(2)}(x)$.

(2.12) Let us consider again the rabbit population function from the previous chapter. The function is shown again in Fig. 2.2. Call the population function P. Determine the following:

(a) $P(50)$

(b) $P^{(2)}(75)$

(c) $P^{(3)}(10)$

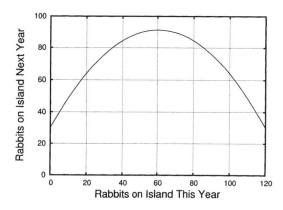

Fig. 2.2 The rabbit population on an island next year as a function of the number of rabbits on the island this year. See Exercise 2.12.

(2.13) Consider the function f shown in Fig. 2.3. Determine:

    (a) $f(-5)$

    (b) $f^{(2)}(-5)$

    (c) The first four iterates of 0.

    (d) The first four iterates of -15.

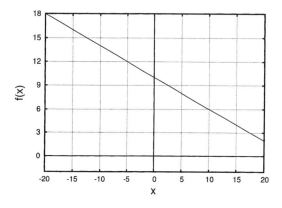

Fig. 2.3 The function for Exercise 2.13.

(2.14) Consider the function $f(x)$ shown in Fig. 2.4. Determine:

    (a) The first three iterates of 0.5.

    (b) The first three iterates of 2.0.

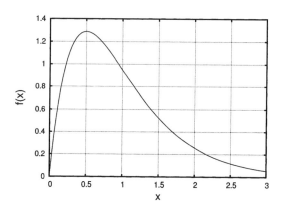

Fig. 2.4 The function for Exercise 2.14.

(2.15) ⋆ Let g be the squaring function: $g(x) = x^2$.

    (a) Calculate the first five iterates of 2.

    (b) What do you think happens to $g^{(n)}(2)$ as n gets large?

    (c) Calculate the first five iterates of 1.

    (d) What do you think happens to $g^{(n)}(1)$ as n gets large?

    (e) Calculate the first five iterates of $\frac{3}{4}$.

    (f) What do you think happens to $g^{(n)}(\frac{3}{4})$ as n gets large?

(2.16) ⋆ Let h be the square root function: $h(x) = \sqrt{x}$.

    (a) Calculate the first five iterates of 2.

    (b) What do you think happens to $h^{(n)}(2)$ as n gets large?

    (c) Calculate the first five iterates of 1.

    (d) What do you think happens to $h^{(n)}(1)$ as n gets large?

    (e) Calculate the first five iterates of $\frac{3}{4}$.

    (f) What do you think happens to $h^{(n)}(\frac{3}{4})$ as n gets large?

Qualitative Dynamics: The Fate of the Orbit

3

3.1 Dynamical Systems

The iterated functions of the previous chapter are an example of a dynamical system, a mathematical system that changes over time. There are many other sorts of dynamical systems. Some vary continuously in time, like the altitude of an airplane or the temperature of a cup of coffee. The height and temperature are continually changing. Iterated functions are discrete, in the sense that they change suddenly at fixed intervals. There are also dynamical systems where the variable is something more complicated than a single number that could represent population, temperature, or altitude. The thing that changes could be a set of numbers or an entire two-dimensional image.

For the first two parts of this book I will focus exclusively on iterated functions. These dynamical systems are the simplest to analyze and conceptualize, yet nevertheless show almost all of the fun and interesting features that are found in more complicated dynamical systems. We will encounter some of these other dynamical systems in Part V of the book. For now, though, let us return to our exploration of iterated functions.

In the previous chapter our main concern was determining the orbit, or itinerary, for a particular initial condition. For example, for the function $g(x) = 3x^2 - 1$ and the seed $x_0 = 1$, we have

$$1 \longrightarrow 2 \longrightarrow 11 \longrightarrow 362 \longrightarrow 393131 \longrightarrow \cdots . \qquad (3.1)$$

This process is straightforward enough, at least if you have a calculator at hand.

In this chapter we will take a more global, qualitative view. Rather than asking about the particulars of an orbit, we will ask about its long-term behavior: do the numbers get bigger and bigger, or smaller and smaller, or something else? Put another way, rather than just paying attention to one particular initial condition, we will look to make statements about a whole bunch—perhaps even all—initial conditions at once.

3.2   Dynamics of the Squaring Function

We begin by considering an extended example: the squaring function $f(x) = x^2$. Let us choose a seed, say 2, and see what happens under iteration:

$$2 \longrightarrow 4 \longrightarrow 16 \longrightarrow 256 \longrightarrow 65536 \longrightarrow \cdots . \qquad (3.2)$$

We can see that the numbers are getting bigger. And as we keep squaring, they will keep getting bigger.

The same thing will happen if we use 3 as our seed; square 3, and keep on squaring, and pretty soon you will have a really big number. In fact, this will be the case for any seed greater than 1. Any time we multiply together two numbers, assuming that those numbers are positive and larger than 1, the result is a larger number. So any seed larger than 1 will get bigger and bigger.

There are a number of equivalent ways of expressing the idea that a seed gets bigger and bigger. Typically we say that 2 (or whatever the seed is) **tends toward infinity**. This indicates that the iterates grow without bound; there is no limit to how large the orbits become. We might also say, somewhat more colloquially, that the iterates **go to infinity**. One could also say in some contexts that the iterates **diverge**. All of these are equivalent; I will usually use the phrase "tends to infinity" in this book.[1]

Thus far, we have established that for any seed larger than 1, the orbit tends to infinity. What about other seeds? If we square a number between 0 and 1, the number gets smaller. E.g.,

$$\left(\frac{1}{2}\right)^2 = \left(\frac{1}{2}\right)\left(\frac{1}{2}\right) = \frac{1}{4} . \qquad (3.3)$$

If we were to iterate $x_0 = \frac{1}{2}$ we would get:

$$\frac{1}{2} \longrightarrow \frac{1}{4} \longrightarrow \frac{1}{16} \longrightarrow \frac{1}{256} \longrightarrow \frac{1}{65536} \longrightarrow \cdots . \qquad (3.4)$$

Or, using decimals instead of fractions:

$$0.5 \longrightarrow 0.25 \longrightarrow 0.0625 \longrightarrow 0.00391 \longrightarrow 0.0000153 \longrightarrow \cdots . \quad (3.5)$$

So, numbers that start between 0 and 1 get smaller and smaller, closer and closer to zero. We thus say that the orbit of $\frac{1}{2}$ **tends toward zero** or **approaches zero**.

What about the number zero itself? Zero squared is zero. So if we iterate zero we do not go anywhere:

$$0 \longrightarrow 0 \longrightarrow 0 \longrightarrow \cdots . \qquad (3.6)$$

We would thus say that 0 is a **fixed point** because it is unchanged by the function. Symbolically, $f(0) = 0$. And what about initial condition $x_0 = 1$? This is also a fixed point, because 1 squared is 1: $f(1) = 1$.

[1] Note that infinity is not a number; it is a more abstract notion that captures the idea of unlimited growth. As such, it does not really make sense to write $x = \infty$. Instead, one might write "$x \to \infty$ as n gets large" to indicate that the values of x_n keep on growing as n gets bigger and bigger.

3.3   The Phase Line

We have now figured out what happens to all non-negative initial conditions: if x_0 is between 0 and 1, the orbit approaches zero; if x_0 is larger than 1, then the orbit tends toward infinity; and if the initial condition is 0 or 1, then the orbit is fixed. We can summarize this state of affairs quite succinctly using a graphical device known as the **phase line**.

The phase line for the squaring function is shown in Fig. 3.1; the phase line contains information about what happens to all initial conditions. We can see that seeds larger than 1 are pushed to the right, toward infinity. And seeds between 0 and 1 are pushed toward zero. The two fixed points, 0 and 1, are shown as small circles. For discrete-time systems such as this, the phase line can be potentially misleading. The value of the variable x does not slide or flow continuously along the line. Rather it jumps. For example, when we square 2 using the squaring rule it jumps immediately to 4. It does not slide along the line, passing through intermediate values on the way to 4.

Fig. 3.1 The phase line for the function $f(x) = x^2$, for non-negative x. Zero and 1 are fixed points. Seeds less than 1 approach 0. Seeds greater than 1 tend toward infinity.

The phase line summarizes in graphical form the qualitative behavior of the orbits for all initial conditions. By qualitative behavior, I mean that we can tell if an orbit flies off to infinity, stays put, or gets pulled toward zero. We cannot get detailed quantitative information from a phase plot. For example, we cannot use Fig. 3.1 to figure out $f^{(3)}(2.2)$. But this is not a serious drawback. Often this long-term behavior—the orbit's fate—is all that we are interested in. For example, we might want to know if a population of rabbits on an island dies off, grows without bound, or stays at some equilibrium value. This information is usually more important than the particular sequence of population values that forms the itinerary. This is especially the case if our model is only approximate or if we cannot measure the value of x accurately.

3.4   Fixed Points via Algebra

We now focus on some methods for finding fixed points of a function. As discussed above, a fixed point of a function f is an input x that yields the same output. This sentence can be written as an equation. Let us denote a fixed point by x^*. Then, the equation for a fixed point is:

$$f(x^*) = x^* . \tag{3.7}$$

We use the symbol x^* to remind us that Eq. (3.7) is not true in general. It is only true for some special values of x—namely, fixed points—and we denote this special value by x^*.

We can use Eq. (3.7) to solve for the fixed point(s) of a given function. For example, suppose that $f(x) = 7x + 4$. The fixed point equation for this function is

$$7x^* + 4 = x^* . \tag{3.8}$$

To find the fixed point, we need to solve for x^*. In other words, we need to isolate x^* on one side of the equation. The rules of the game are that we can manipulate the equation in any way we want, as long as we do the same thing to each side.[2]

Let us start by subtracting x^* from each side:

$$7x^* + 4 - x^* = x^* - x^* . \tag{3.9}$$

Simplifying, this becomes:

$$6x^* + 4 = 0 . \tag{3.10}$$

Now, subtract 4 from each side:

$$6x^* + 4 - 4 = 0 - 4 , \tag{3.11}$$

to obtain

$$6x^* = -4 . \tag{3.12}$$

Dividing both sides by 6, we get

$$x^* = -\frac{4}{6} . \tag{3.13}$$

We have thus found our fixed point: $x^* = -\frac{4}{6}$, or, $x^* = -\frac{2}{3}$.

We can easily check to see if we have done the algebra right by plugging back into Eq. (3.8) and seeing if the equation is true.

$$7\frac{-2}{3} + 4 = -\frac{2}{3} \ ? \tag{3.14}$$

Let us multiply each term in this equation by 3.

$$-14 + 12 = -2 \ ? \tag{3.15}$$

Since $12 - 14$ is -2, we see that the equation is indeed true. This confirms that $x^* = -\frac{2}{3}$ really is a fixed point of f.

In the above examples there were two equations involving the function f that look similar but have different meanings:

$$f(x) = 7x + 4 , \tag{3.16}$$

and

$$f(x^*) = x^* . \tag{3.17}$$

Equation (3.16) defines the function. It says that "f is the function that takes x, multiplies it by 7, and then adds 4." This is true for any x. Equation (3.17) says that x^*, when used as an input for f, yields x^*

[2]This is not quite true. We cannot divide both sides by zero. But other than that, any algebraic operation is fair game.

again. This equation is not true for all x; there are only special values of x—the fixed point(s) x^*—for which Eq. (3.17) is true.

To summarize, the main point of the above example is that we can use the fixed-point equation, Eq. (3.7), together with some algebra, to determine the fixed point. This method will pretty much always work, as long as the function is not too complex. The function could be such that using algebra to find the fixed point is forbiddingly difficult, e.g., if $f(x) = 4x^4 - 17x^3 + x^2 - x + 32$. There are some functions which have a fixed point, but it is impossible to solve for the fixed point using algebra; graphical or numerical techniques are needed instead.[3] Also, there is nothing that says that a function has to have a fixed point. Similarly, there could be functions that have multiple fixed points.

[3] I will not discuss this case in the text, but you can explore this phenomenon in Exercise 3.17 at the end of this chapter.

3.5 Fixed Points Graphically

There is also a graphical way to find fixed points. Consider the function $f(x)$ shown in Fig. 3.2. Take a moment to find the fixed point of $f(x)$ using this plot. Remember that a fixed point is just a number x such that $f(x)$ is the same as x. I suggest that you do this now before reading further.

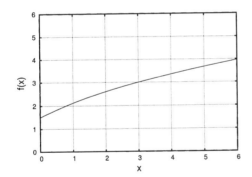

Fig. 3.2 A graphical representation of a function, used to illustrate the process of finding a fixed point.

You should have found that there is a fixed point around $x = 3$, since it appears that $f(3) \approx 3$. This is the only fixed point for this function. For example, 2 is not a fixed point, because $f(2) \approx 2.6$.

You probably found the fixed point on the graph by scanning up and down the horizontal and vertical axes, looking for a point or points on the curve that have the same x and y coordinates. This search process can be made much easier by the following trick. In addition to the function $f(x)$, let us plot the $y = x$ line on the same axes. This is shown in Fig. 3.3. As you can see, the fixed point, $x = 3$, that we found by squinting and searching, occurs exactly when the $y = x$ line crosses the $f(x)$ curve.

To see why this is, let us think about what the line $y = x$ means. First, note that there are an infinite number of points that make up this line. What do all these points have in common? They have the same x and y value. If we think of y as the output of a function, then the $y = x$

line is the graph of a function which has the property that its output is always the same as the input. In other words, *every* point on the $y = x$ line is fixed. Thus, the $y = x$ line draws out for us all possible fixed points.

So, when the $f(x)$ curve intercepts the $y = x$ curve, we know right away that that point is fixed—i.e., that the input and the output are the same. We now have a graphical way to find fixed points of a function if

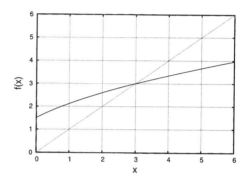

Fig. 3.3 The same function as in Fig. 3.2, along with the $y = x$ line. The line $y = x$ is the thin, straight line. The fixed point of $f(x)$ occurs when the $f(x)$ curve crosses the $y = x$ line.

we have access to the function's graph. Namely, we draw the $y = x$ line, and then the fixed points occur where the function crosses $y = x$.

3.6 Types of Fixed Points

Let us return to the example that we used to start this chapter: the squaring function $f(x) = x^2$. Recall that there are two fixed points: 0 and 1. These two fixed points are different in character: numbers close to 1 get pushed away from it, while numbers close to zero get pulled closer to 0. This is illustrated in Fig. 3.1, which shows the phase line for $f(x) = x^2$.

This difference is a crucial one. The orbit of 1 is fixed, and so it will remain at 1 as we iterate the function. However, suppose there is a slight fluctuation—some small external influence that makes the value of x deviate slightly from 1. Then, when the function is iterated, the orbit will either go to zero or tend toward infinity, depending on whether the fluctuation moves x below or above 1.

Thus, if we are at $x = 1$, then the situation is like a marble delicately balanced on the top of an upturned bowl, as sketched in Fig. 3.4. The marble is at rest—it is at a fixed point. But a small disturbance, such as a little gust of wind or a slight vibration of the bowl, will lead to the marble leaving its perch on the top of the bowl (its fixed point), and it will move to the left or to the right, never to return.

So, fixed points like $x = 1$ in our example of the squaring function are classified as **unstable.** What this means is that if x is somehow moved away from the fixed point, even a very little bit, the orbit of this new x will move *away* from the fixed point. Such a fixed point is also called a **repellor,** because nearby orbits are repelled by it.

Fig. 3.4 A schematic illustration of an unstable fixed point. The marble is at an equilibrium, or fixed, position. However, a small perturbation will cause the marble to move away from the fixed point. This type of a fixed point is also known as a repellor.

On the other hand, the fixed point 0 is very different. Here, if we get "bumped" off the fixed point, the subsequent orbit moves us back to the original fixed point. Fixed points with this property are called **attractors** or **stable**. Returning to the marble for a moment, an attractive or stable fixed point is illustrated in Fig. 3.5. Here, the marble is at the bottom of a bowl. A small change in the marble's position will result in the marble getting pulled back to its original location.

The distinction between stable and unstable fixed points is an important one. In real systems (or computer simulations), one does not expect to observe or encounter unstable fixed point, for the simple reason that unstable fixed points do not stick around for long. A tiny bump or a nudge, and the orbit will move away from the fixed point.

In everyday experience it is very rare that we encounter a marble perched on an upturned bowl as in Fig. 3.4. It is much more common to see a marble at the bottom of a bowl, as in Fig 3.5. Actually, one probably rarely sees marbles in bowls at all, but one can think of the marbles metaphorically.

Finally, there is another type of fixed point that I should mention. One could have a fixed point for which it is the case that if one moves away from the fixed point the resulting orbits neither move away from the fixed point (as is the case for a repellor) nor back toward the fixed point (as is the case for an attractor). Such a fixed point is called **neutral**. An illustration of a neutral fixed point is shown in Fig. 3.6. We will not encounter neutral fixed points very often, but it is useful terminology to know. Note that a neutral fixed point is, in a sense, a type of fixed point that is in between being stable and unstable.

To summarize, in this chapter I have introduced a number of key terms and ideas: stable and unstable fixed points, and the notion that some orbits tend toward infinity. I also introduced the phase line as a convenient graphic that summarizes the global behavior of an iterated function. For the most part, the ideas and terminology of this chapter are pretty straightforward—the technical meaning of the terms is not very different from their everyday meaning. Nevertheless, this vocabulary is quite standard, and we will need it to describe the dynamical systems to come.

Fig. 3.5 A schematic illustration of a stable fixed point. The marble is at an equilibrium, or fixed, position. A small perturbation will cause the marble to move back toward the fixed point. This type of a fixed point is also known as an attractor.

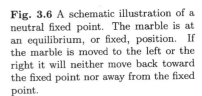

Fig. 3.6 A schematic illustration of a neutral fixed point. The marble is at an equilibrium, or fixed, position. If the marble is moved to the left or the right it will neither move back toward the fixed point nor away from the fixed point.

Exercises

(3.1) Consider the square root function, $f(x) = \sqrt{x}$.

    (a) Determine the phase line for $f(x)$ for non-negative x. Explain your reasoning carefully.

    (b) Determine all fixed points, if any, of $f(x)$.

    (c) What is the stability of these fixed points?

(3.2) Consider the cubing function, $h(x) = x^3$.

    (a) Determine the phase line for $h(x)$. Consider both positive and negative x. Explain your reasoning carefully.

    (b) Determine all fixed points, if any, of $h(x)$.

    (c) What is the stability of these fixed points?

(3.3) Find the fixed point(s), if any, of $f(x) = 2x - 5$.

(3.4) Find the fixed point(s), if any, of $g(x) = \frac{1}{2}x + 4$.

(3.5) Find the fixed point(s), if any, of $h(x) = x^2 - 1$.

(3.6) Find the fixed point(s), if any, of $f(x) = x^2 + 1$.

(3.7) Find the fixed point(s), if any, of $g(x) = x - 3$.

(3.8) Find the fixed point(s), if any, of $h(x) = x^3$.

(3.9) Consider the function $g(x) = x + 2$.

    (a) Determine the phase line for $g(x)$. Explain your reasoning carefully.

    (b) Determine all fixed points, if any, of $g(x)$.

    (c) What is the stability of these fixed points?

(3.10) Consider the function $f(x) = \frac{1}{2}x - 2$.

    (a) Show that $x = -4$ is a fixed point.

    (b) Determine the first several iterates of $x_0 = -3.9$.

    (c) Determine the first several iterates of $x_0 = -4.1$.

    (d) What do your answers to the above two questions let you conclude about the stability of the fixed point $x = 4$?

(3.11) Determine all fixed points for the function shown in Fig. 3.7.

(3.12) Determine all fixed points for the function shown in Fig. 3.8.

(3.13) (a) Describe a real-life example of a stable equilibrium or fixed point.

    (b) Describe a real-life example of an unstable equilibrium or fixed point.

(3.14) Draw the graph of a function that has:

    (a) no fixed points

    (b) two fixed points

    (c) thirteen fixed points

(3.15) Draw the graph of a function that has:

    (a) one stable fixed point

    (b) one unstable fixed point

(3.16) ⋆ Consider the following function: $f(x) = x^2 - 3$.

    (a) Determine the first several iterates of $x_0 = 1$.

    (b) How would you describe the long-term behavior of $x_0 = 1$?

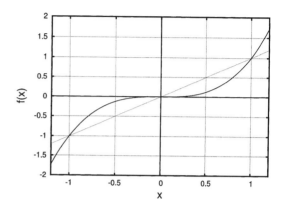

Fig. 3.7 The function for Exercise 3.11. The line $y = x$ is the thin, straight line.

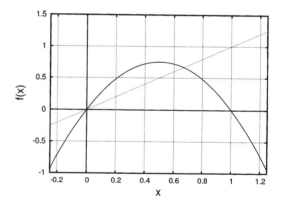

Fig. 3.8 The function for Exercise 3.12. The line $y = x$ is the thin, straight line.

(3.17) ♯ Consider the function $h(x) = 2^x$. Find the fixed points, if any, for $h(x)$. Graphing the function may help.

Time Series Plots

In this short chapter I will introduce a way to visualize and think about the orbits of a function. As usual, we begin with an example.

4.1 Examples of Time Series Plots

Consider the square root function, $f(x) = \sqrt{x}$. The orbit of 4 is:

$$4 \longrightarrow 2 \longrightarrow 1.4142 \longrightarrow 1.1892 \longrightarrow 1.0905 \longrightarrow \cdots \ . \tag{4.1}$$

The orbit is approaching 1, as you probably saw in Exercise 3.1 of Chapter 3. Displaying a list of numbers as in Eq. (4.1) is fine, but often a clearer way to see the behavior of the orbit is via a **time series plot**.

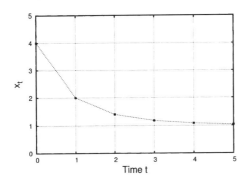

Fig. 4.1 A time series plot of the itinerary of 4 for the function $f(x) = \sqrt{x}$. The itinerary approaches the fixed point at $x = 1$. Numerical values for the orbit are given in Eq. (4.1).

Such a plot for the orbit of Eq. (4.1) is shown in Fig. 4.1. On the vertical axis, we plot the successive values in the orbit: 4.0, 2.0, 1.4142, and so on. On the horizontal axis we plot time, where the time is taken to be the number of the "stop" on the itinerary. For example, 1.4142 is the second stop in the itinerary, so we plot 2 on the horizontal axis and 1.4142 on the vertical axis. Note that the first number in the itinerary is x_0, the starting point, so time is starting at zero and not at 1.

We can also plot several different orbits on the same axes. This is done in Fig. 4.2, in which I have plotted the orbits for four different seeds: 4.0, 2.0, 0.5, and 0.0625. As we have seen, the square root function has an attracting fixed point at 1. This can be visualized on a plot such as that of Fig. 4.2. We can clearly see that the orbits of all these initial conditions are pulled toward the fixed point. Note that the difference between orbits shrinks as time moves forward; x values that were once far apart get closer and closer together.

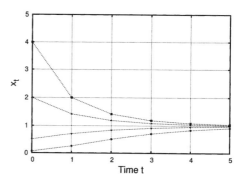

Fig. 4.2 A time series plot of the itineraries for four different seeds for the square root function. The four seeds are $4, 2, 0.5$, and 0.0625. Note that orbits are pulled toward the fixed point, $x = 1$, and that orbits that are initially far apart get closer together as time progresses.

Let us now consider another example, the function $g(x) = -\frac{1}{2}x + 2$. The time series for the seed $x_0 = 4$ is shown in Fig. 4.3. Looking at the figure, we suspect that there is a fixed point near $x = 1.3$. One can verify using algebra that there is indeed a fixed point at $x = \frac{4}{3}$; see Exercise 4.1 of this chapter. The orbit is getting closer to $x \approx 1.3$ as it oscillates around it, so the fixed point is an attractor.

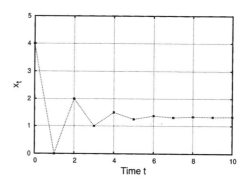

Fig. 4.3 The time series plot for the seed $x_0 = 4$ for the function $g(x) = -\frac{1}{2}x + 2$. The itinerary oscillates as it approaches the attractor at $x = \frac{4}{3}$.

Finally, I should mention that one can go directly from a graph of a function to a time series plot. The time series plot might be approximate, but this is all we need to describe the qualitative dynamics of the system. For example, consider the function whose graph is shown in Fig. 4.4. Let us choose an initial condition of $x_0 = 1.4$, near the fixed point that occurs near $x = 1.5$. Reading successive values off the graph, we obtain

$$1.4 \longrightarrow 1.3 \longrightarrow 1.1 \longrightarrow 0.8 \longrightarrow 0.5 \cdots . \tag{4.2}$$

Plotting this itinerary as a time series plot, we obtain Fig. 4.5

Time series plots can reveal behavior that is richer than that seen in the two examples considered above, as we will see frequently later in the book. In the next chapter we shall see how to go more directly from the graph of a function to a time series plot. In particular, we will learn a nice graphical technique that allows us to skip the step of writing down the numbers for the orbit, as we did in Eq. (4.2).

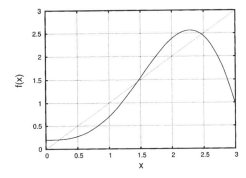

Fig. 4.4 The graph of a function $f(x)$. The time series for this function for $x_0 = 1.4$ is shown in Fig.4.5.

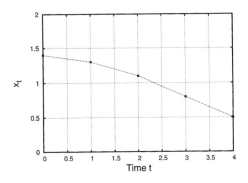

Fig. 4.5 The time series for $x_0 = 1.4$ for the function $f(x)$ given graphically in Fig. 4.4.

Exercises

(4.1) Verify algebraically that $g(x) = -\frac{1}{2}x + 2$ has a fixed point at $x = \frac{4}{3}$.

(4.2) Consider the squaring function, $f(x) = x^2$.

    (a) On the same axes, plot the first several elements of the time series plot for the seeds 0.8, 0.9, 1.1, and 1.2.

    (b) What does this plot tell you about the stability of $x = 1$?

(4.3) Consider the cubing function, $f(x) = x^3$.

    (a) On the same axes, plot the first several elements of the time series plot for the seeds 0.8, 0.9, 1.1, and 1.2.

    (b) What does this plot tell you about the stability of $x = 1$?

(4.4) Consider the linear function, $g(x) = 2x - 1$.

    (a) On the same axes, plot the first several elements of the time series plot for the seeds 0.8, 0.9, 1.1, and 1.2.

    (b) What does this plot tell you about the stability of $x = 1$?

(4.5) Consider the function, $g(x) = -x + 2$.

    (a) Sketch the first several elements of the time series plot for $x_0 = 0$.

    (b) Sketch the first several elements of the time series plot for $x_0 = 1$.

    (c) Sketch the first several elements of the time series plot for $x_0 = -1$.

(4.6) ⋆ Consider the function $f(x) = x^2 - 3$.

    (a) Sketch the first several elements of the time series plot for $x_0 = 1$.

    (b) How would you describe this behavior?

(4.7) Consider the function $f(x) = \frac{1}{2}x - 2$.

    (a) Show that $x = -4$ is a fixed point.

    (b) Determine the first several iterates of $x_0 = -3.9$.

(c) Determine the first several iterates of $x_0 = -4.1$.

(d) What do your answers to the above two questions let you conclude about the stability of the fixed point $x = -4$?

(b) Sketch the first several elements in the time series plot for $x_0 = 2.8$.

(c) Sketch the first several elements in the time series plot for $x_0 = 3.2$.

(d) What is the stability of the fixed point(s) you found?

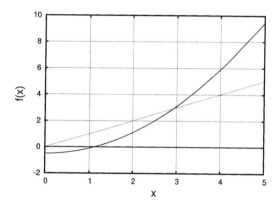

Fig. 4.6 The function for Exercise 4.8. The line $y = x$ is the thin, straight line.

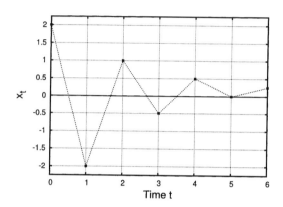

Fig. 4.7 The time series for Exercise 4.9.

(4.8) Consider the function shown in Fig. 4.6.

(a) Determine all fixed points for the function.

(4.9) A time series plot is shown in Fig. 4.7. What is the orbit for the seed $x_0 = 2$?

Graphical Iteration

In this chapter we continue the introduction to discrete dynamical systems by introducing another tool for visualizing and analyzing iterated functions. The benefit of this new tool is that it will let us quickly determine the stability of all of a function's fixed points.

5.1 An Initial Example

We will start by considering the function shown in Fig. 5.1. Also plotted in Fig. 5.1 is the line $y = x$. Recall that fixed points occur where the $y = x$ line intersects $f(x)$. We thus see that the function in Fig. 5.1 has fixed points at $x = 0$ and $x = 3$.

Suppose we are interested in the orbit of $x_0 = 1$. We can figure this out from the graph of $f(x)$. The first thing we do is to read $f(x_0)$ off of the graph. Doing so, we find that $f(x_0) = x_1 \approx 1.7$. To get the next iterate x_2, we need to use 1.7 as our next input. So, we find 1.7 on the x-axis, then follow our eyes up to the function, and see that $f(1.7) \approx 2.3$.

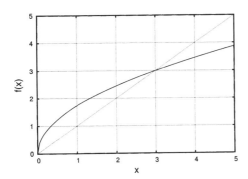

Fig. 5.1 The graph of a function $f(x)$.

Imagine doing this iteration directly on the graph. We start with x_0 as input and move up to get $f(1) \approx 1.7$. Trace this out with your finger or a pen or pencil:[1] start at 1 on the x-axis, and then move straight up to the function. To iterate, we need to take the output, 1.7, and use it as the input for the function. So we need to get to 1.7 on the horizontal axis, as this is where inputs to the function are. Move your finger straight to the right and then down to 1.7 on the x-axis. Note that you make your downward turn exactly at the $y = x$ line.

Now that you are at $x = 1.7$ on the x-axis, move straight up to the function at around 2.3. We now need to use 2.3 as our input. So, as

[1] I suggest actually doing this as you read along.

before, move straight to the right and then head straight down to 2.3 on the x-axis. Note, again, that you made the downward turn exactly at $y = x$. Now move straight up to the function to get $f(2.3)$. We see that we are now at approximately 2.6. Our approximate orbit thus far is

$$1 \longrightarrow 1.7 \longrightarrow 2.3 \longrightarrow 2.6 \cdots . \tag{5.1}$$

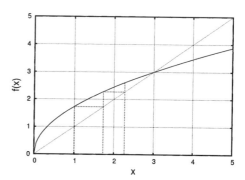

Fig. 5.2 The graph of a function $f(x)$. The dashed lines show the tracings one would make to determine the first few elements in the orbit of $x_0 = 1$, as described in the text.

The tracings that you would make as you use the graph to determine the first few iterates of $x_0 = 1$ are shown in Fig. 5.2. Take a moment and confirm again that these tracings—the dashed lines in Fig. 5.2—are what you follow when iterating using 1 as a seed.

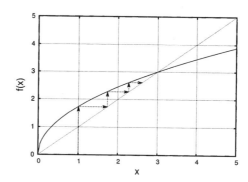

Fig. 5.3 The graph of a function $f(x)$. The arrows show the tracings one would make to determine the first few elements in the orbit of $x_0 = 1$. This figure is identical to Fig. 5.2, except that the redundant portions of the dashed lines have been removed and arrowheads have been added.

Next, note that a portion of the dashed line is redundant. There is really no need to drop back down to the x-axis every time. Instead, we can go right from the $y = x$ line to the function $f(x)$. This is shown in Fig. 5.3. We can clearly see the orbit being drawn toward $x = 3$. Evidently 3 is an attracting fixed point.

5.2 The Method of Graphical Iteration

The above example suggests the following general method for determining the itinerary of a seed x_0 for a function $f(x)$.

(1) Start with the seed x_0 on the x-axis.
(2) Move up to the function $f(x)$.

(3) Move horizontally (left or right) to the $y = x$ line.
(4) Move vertically (up or down) to the function $f(x)$.
(5) Repeat steps 3-4 again and again.

This process is a nice quick way to determine an itinerary. Often, by choosing a few representative initial conditions, we can quickly determine the qualitative behavior of the dynamics.

When using this method, it is easy to make a mistake and do it "backwards." Remember that you first go to the function, and then the $y = x$ line. Doing this the other way around will "iterate" the seed backwards in time instead of forward. After you have drawn the lines for the graphical method, I suggest putting a few arrows on your diagram so that it is easier to see in which direction the orbit is headed. I have done this on Fig. 5.3, and it is easy to see that the orbit is being drawn toward the fixed point at $x = 3$.

The graphical method described above is not new; it is the exact same iteration process we have been doing all along. The graphical approach is just a different representation of the iteration of the last few chapters. I hope that introducing the process of graphical iteration in some detail has helped make this clear. It can take a few moments of pondering to see this connection. If you do not see it right away, try working through the example again.

5.3   Further Examples

This section contains two more examples of graphical iteration. This should give you some additional experience with this method. Along the way I will point out a few of the method's subtleties.

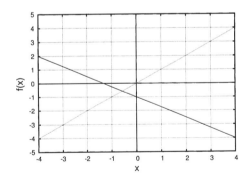

Fig. 5.4 The graph of the linear function $f(x) = -\frac{3}{4}x - 1$. Before proceeding, use this plot to graphically iterate the seed $x_0 = 3$. The line $y = x$ is the thin dashed line.

First, consider the function $f(x) = -\frac{3}{4}x - 1$ shown in Fig. 5.4. Let us try using the graphical technique to iterate the initial condition $x_0 = 3$. Before going on, I suggest using a pencil and trying this out for yourself on Fig. 5.4.

The result of graphically iterating $x_0 = 3$ is shown in Fig. 5.5. There is a fixed point where the lines $f(x)$ and $y = x$ intersect. From the graph, this appears to be at around $x = -0.5$. Using algebra (see Exercise 5.1), one finds that the fixed point occurs at $x^* = -\frac{4}{7} \approx -0.571$. The orbit of 3 gets closer and closer to the fixed point, but moves from one side of the fixed point to the other while doing so. In Fig. 5.5 the orbit appears to spiral in to the fixed point.

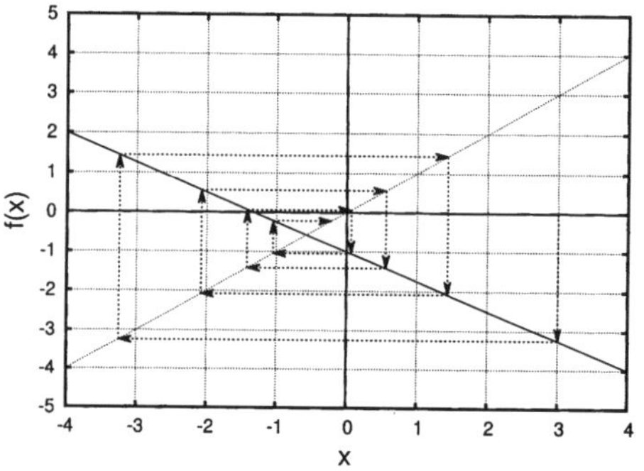

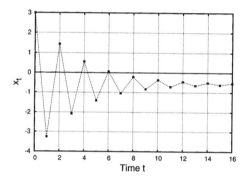

Fig. 5.5 Graphically iterating the seed $x_0 = 3$ for the linear function $f(x) = -\frac{3}{4}x - 1$. The line $y = x$ is the thin dashed line. There is an attracting fixed point at $x^* = -\frac{4}{7} \approx -0.571$. Note that the itinerary oscillates around the attracting fixed point as it moves toward it. See also Fig. 5.6.

Fig. 5.6 A time series plot for the initial condition $x_0 = 3$ for the linear function $f(x) = -\frac{3}{4}x - 1$. Compare to Fig. 5.5. The orbit approaches the fixed point $x^* \approx -0.571$, oscillating around it as it does so.

This oscillatory behavior can also be seen by looking at the time series plot. Such a plot is shown in Fig. 5.6, in which one can clearly see that the orbit of $x_0 = 3$ oscillates around the fixed point $x^* \approx -0.57$ while approaching it. The fixed point $x^* = -\frac{4}{7}$ is attracting, or stable.

The results of graphically iterating a function are sometimes referred to as **cobweb diagrams**. The reason for this is that sometimes they look like cobwebs, as in Fig. 5.5. Although this terminology is moderately widespread, I think it can be somewhat misleading, since not all graphical iteration processes end up looking like cobwebs. Sometimes one gets staircases, as was the case in Fig. 5.3. I will not use the term cobweb diagram, but you might encounter it elsewhere.

As a final example, consider the function shown in Fig. 5.7. This function has fixed points at $x = 1$ and $x = 0$. We will use the graphical methods of this chapter to determine the stability of $x = 1$. To do so, perform graphical iteration for two different seeds, one a little bit larger than 1, and one a little bit smaller. Before going on, I suggest doing this yourself on Fig. 5.7.

The result of doing this is shown in Fig. 5.8. We can see that orbits are pushed away from the fixed point at $x^* = 1$. Hence, this lets us conclude that $x = 1$ is a repelling, or unstable, fixed point. We can also

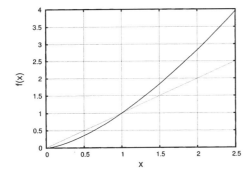

Fig. 5.7 The graph of a function $f(x)$. We will use graphical iteration to investigate the stability of the fixed point at $x = 1$. Before proceeding, try graphically iterating two different seeds, one a little bit larger than 1, and one a little bit smaller. The line $y = x$ is the thin dashed line.

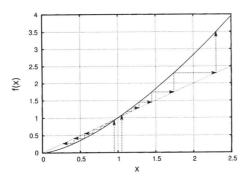

Fig. 5.8 Using graphical iteration to determine the stability of the fixed point at $x = 1$. We can see that orbits that start close to $x = 1$ are pushed away from it. Hence, $x = 1$ is a repelling, or unstable, fixed point. The time series plots for the two orbits iterated graphically are shown in Fig. 5.9.

see that $x = 0$ is an attracting fixed point, and that orbits that start above $x = 1$ will tend toward infinity.

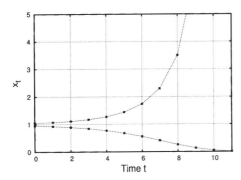

Fig. 5.9 Time series plots for the two orbits determined graphically in Fig. 5.8. The fixed point at $x = 1$ is unstable, or repelling.

A complementary view of this can be seen in Fig. 5.9, in which I have plotted the time series for the two initial conditions used in Fig. 5.8. Again, we can see that $x = 1$ is repelling, $x = 0$ is attracting, and that orbits above $x = 1$ will tend toward infinity.

Exercises

(5.1) Verify that the function $f(x) = -\frac{3}{4}x - 1$ has a fixed point at $x^* = -\frac{4}{7}$.

(5.2) Figure 5.10 shows a plot of the function $f(x) = 1.5x(1 - x)$.

    (a) Use the plot to determine approximate values for all fixed points of $f(x)$.

    (b) Graphically iterate the seed $x_0 = 0.1$.

    (c) Graphically iterate the seed $x_0 = 0.8$.

    (d) What do you conclude about the stability of the fixed point near $x = 0.35$?

    (e) What is the stability of the fixed point at $x = 0$?

    (f) Use algebra to find the fixed points exactly.

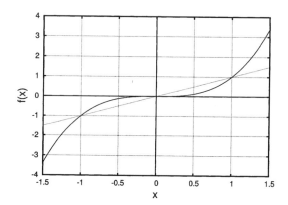

Fig. 5.11 The function for Exercise 5.3.

(5.4) Figure 5.12 shows a plot of the function $f(x) = 3.2x(1 - x)$.

    (a) Use the plot to determine approximate values for all fixed points of $f(x)$.

    (b) Graphically iterate the seed $x_0 = 0.1$.

    (c) Graphically iterate the seed $x_0 = 0.8$.

    (d) What does this let you conclude about the stability of the fixed point near $x = 0.7$?

    (e) Use algebra to find the fixed point(s) exactly.

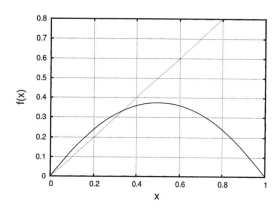

Fig. 5.10 A plot of $f(x) = 1.5x(1 - x)$, the function for Exercise 5.2.

(5.3) Consider the function shown in Fig. 5.11.

    (a) Choose several initial conditions and graphically iterate.

    (b) What do your results let you conclude about the stability of the three fixed points?

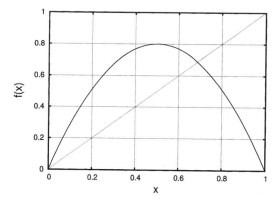

Fig. 5.12 A plot of $f(x) = 3.2x(1 - x)$, the function for Exercise 5.4.

(5.5) Consider the function $f(x) = x^2 - 3$, whose graph is shown in Fig. 5.13.

    (a) Use the graph to determine approximate values for all fixed points of $f(x)$.

    (b) Graphically iterate the seed $x_0 = 1$.

    (c) How would you describe this behavior?

    (d) Is this behavior stable? To check, try graphically iterating an initial condition near to, but not exactly at, $x_0 = 1$.

(5.6) Consider the function $f(x) = -\frac{3}{2}x + 2$, whose graph is shown in Fig. 5.14

    (a) Determine the stability of the fixed point by graphically iterating an initial condition very near to, but not exactly at, the fixed point.

    (b) What does your graphical iteration let you conclude about the fixed point's stability?

    (c) Determine the fixed point using algebra.

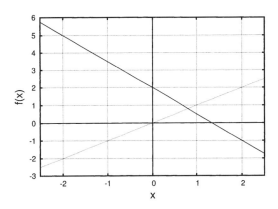

Fig. 5.14 The function $f(x) = -\frac{3}{2}x + 2$, for Exercise 5.6.

(5.7) Consider the function $f(x) = x - 2$, whose graph is shown in Fig. 5.15

    (a) Choose an arbitrary initial condition and iterate it graphically. What long-term behavior do you observe?

    (b) What does your graphical iteration let you conclude about the existence of fixed points and the long-term behavior of the itinerary?

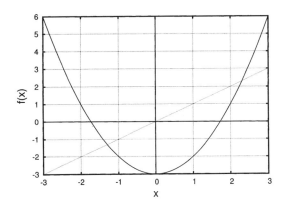

Fig. 5.13 A plot of $f(x) = x^2 - 3$, the function for Exercise 5.5.

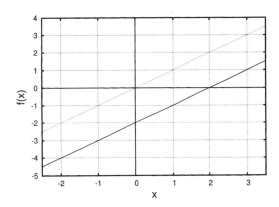

Fig. 5.15 The function $f(x) = x - 2$, for Exercise 5.7.

Iterating Linear Functions

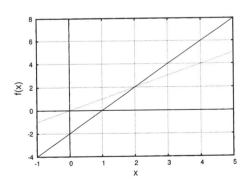

6

In this chapter we will take a systematic look at the different sorts of behavior that arise when iterating linear functions—a particularly simple, and important, family of functions. Our main tool in this investigation will be the graphical iteration techniques covered in the last chapter. In so doing we will gain some valuable insights that apply to more than just linear functions.

Linear functions have the form:

$$f(x) = mx + b \,. \tag{6.1}$$

Their graph is a straight line. The slope is given by m and the y-intercept is b. A brief review of linear functions and their graphs can be found in Appendix A.3.

6.1 A Series of Examples

In this section we will consider a number of different linear functions. Each such function has at most one fixed point, and we will use graphical techniques to examine this fixed point's stability. Our main goal will be to determine what properties of a line lead to its fixed point being stable or unstable.

For our first function, let us look at $f(x) = 2x - 2$. The graph of this function is shown in Fig. 6.1, along with the $y = x$ line. This function has a fixed point at $x = 2$, since the function and the $y = x$ line intersect here. To investigate the stability of this function, we use the graphical methods developed in the previous chapter. Specifically, we iterate two seeds, one on each side of the fixed point. The results of doing this are shown in Fig. 6.2, where we see that the fixed point, $x = 2$, is unstable;

Fig. 6.1 The graph of the function $f(x) = 2x - 2$. The thin dashed line is $y = x$.

orbits that start close to it are pushed away. So $f(x) = 2x - 2$ has one unstable fixed point at $x = 2$.

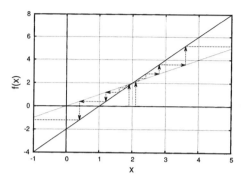

Fig. 6.2 The graph of the function $f(x) = 2x - 2$. The thin dashed line is $y = x$. Graphical iteration shows that the fixed point is unstable.

Let us try another example: $f(x) = 2x + 1$. This function is plotted in Fig. 6.3. Examining the figure, we see that there is a fixed point at $x = -1$. Graphical iteration shows that this fixed point is unstable; initial conditions close to the fixed point are pushed away. So, as with our previous example, the function has one unstable fixed point.

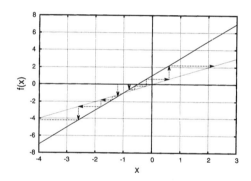

Fig. 6.3 The graph of the linear function $f(x) = 2x + 1$. Graphical iteration shows that the fixed point at $x = -1$ is unstable. Note the similarity to Fig. 6.2.

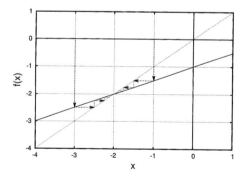

Fig. 6.4 The graph of the linear function $f(x) = \frac{1}{2}x - 1$. Graphical iteration shows that the fixed point at $x = -2$ is stable.

Let us do one more example. Consider the linear function $f(x) = \frac{1}{2}x - 1$, shown in Fig. 6.4. For this function there is a fixed point at $x = -2$. Graphical iteration lets us see that this fixed point is stable. The seeds $x_0 = -3$ and $x_0 = -1$ are both pulled toward the fixed point.

What is different about the three examples shown in Figs. 6.2–6.4? What property of the line determines whether the fixed point is stable or unstable? Note that the functions in Figs. 6.2 and 6.3 both have unstable fixed points, and they have the same slope. The function in Fig. 6.4 has a different slope, and its fixed point is stable.

The key feature turns out to be whether or not the slope is greater or less than 1, because this determines how the linear function intersects the $y = x$ line—from below or from above. If the slope is greater than 1, then the situation will look like that in Fig. 6.2 or 6.3, and the fixed point will be unstable, or repelling. On the other hand, if the slope is less than 1, then the fixed point is stable, or attracting, and the situation looks like that of Fig. 6.4. This is illustrated schematically in parts (1) and (2) of Fig. 6.5.

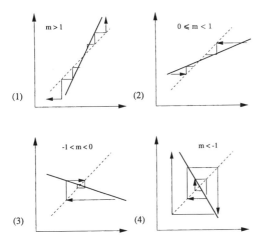

Fig. 6.5 A schematic illustration of the four main types of fixed points. Part (1): repelling, or unstable, $m > 1$; Part (2): attracting, or stable, $0 \leq m < 1$; Part (3): attracting, or stable, with oscillations, $-1 < m < 0$; and Part (4): repelling, or unstable, with oscillations $m < -1$.

But what happens if the slope is negative? To investigate this, we examine $f(x) = -\frac{1}{2}x - 1$, plotted in Fig. 6.6. Note that this is the same function as that of Fig. 6.4, except that the slope is now negative. Graphical iteration shows that orbits are pulled toward the fixed point at $x = -\frac{2}{3}$ and that the orbit oscillates from one side to the other of the fixed point. This can also be seen in Fig. 6.7, in which I show the time series plot for the orbit of $x_0 = -3$ under $f(x) = -\frac{1}{2}x - 1$.

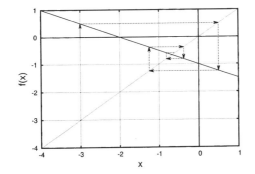

Fig. 6.6 The graph of the linear function $f(x) = -\frac{1}{2}x - 1$. Graphical iteration shows that the fixed point at $x = -\frac{2}{3}$ is stable, since nearby orbits are pulled toward the fixed point. Note, however, that the orbit oscillates from one side of the fixed point to the other as it is getting pulled toward it. The time series for the seed $x_0 = -3$ iterated graphically here is shown in Fig. 6.7.

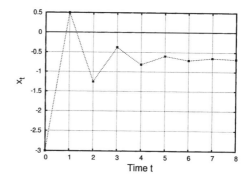

Fig. 6.7 The time series for the seed $x_0 = -3$ for the the linear function $f(x) = -\frac{1}{2}x - 1$, shown in Fig. 6.6. The orbit is being pulled toward the attracting fixed point at $x = -\frac{2}{3}$.

In general, if the slope of our linear function is negative, the orbits will oscillate around the fixed point. If the slope is between 0 and -1, then the fixed point is an attractor, as we saw in Fig. 6.6. However, if the slope is less than -1, then the fixed point is a repellor. An example of this can be found in Exercise 6.1.

Thus far we have found four basic behaviors for the fixed point of a linear function $f(x) = mx + b$. Each behavior corresponds to different ranges of values for the slope m:

(1) $m > 1$: Unstable, or repelling, fixed point.

(2) $0 \leq m < 1$: Stable, or attracting, fixed point.

(3) $-1 < m < 0$: Stable, or attracting, fixed point. Orbit oscillates about fixed point.

(4) $m < -1$: Unstable, or repelling, fixed point. Orbit oscillates about fixed point.

These four different behaviors are illustrated in Fig. 6.5. Note that "$0 \leq m < 1$" is read, "m is less than one and greater than or equal to 0." We have not explicitly considered the $m = 0$ case; you will do so in Exercise 6.5.

6.2 Slopes of $+1$ or -1

We have not yet considered what happens for slopes of $+1$ or -1. We do so in this section. We will start by considering $f(x) = -x + 3$, graphed in Fig. 6.8. This function has a slope of -1. Graphical iteration reveals that every initial condition oscillates forever around the fixed point at $x = \frac{3}{2}$. Orbits are neither drawn toward the fixed point, nor pushed away. Fixed points that have this property are said to be **neutral**. Such fixed points are neither attracting nor repelling.

A time series plot for the seed $x_0 = 3$ is shown in Fig. 6.9. This graph provides another way of seeing that the orbit oscillates and does not approach the fixed point. The orbit shown in the figure is **periodic**, since it repeats every two time steps. Specifically, we would say that the itinerary of $x_0 = 3$ is periodic with period 2.

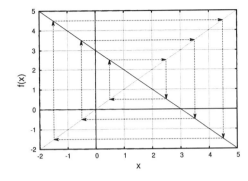

Fig. 6.8 The graph of the linear function $f(x) = -x + 3$. There is a fixed point at $x = \frac{3}{2}$. Graphical iteration shows that all initial conditions oscillate around the fixed point. However, the orbit gets neither pulled toward the fixed point nor pushed away. Such fixed points are called neutral. The time series for the seed $x_0 = 3$ is shown in Fig. 6.9.

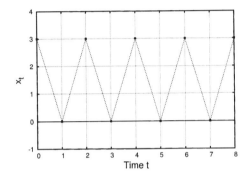

Fig. 6.9 The time series for the seed $x_0 = 3$ for the the linear function $f(x) = -x + 3$, shown in Fig. 6.8.

Our last case to consider is that of lines with a slope of $m = 1$. There are actually two sub-cases to consider. First, if the y-intercept is zero, then the function is simply $f(x) = x$. For such a function *all* points are fixed! The function $f(x) = x$ simply takes whatever you give it as input and returns the same thing as output: $f(17) = 17$, $f(3) = 3$, $f(0.613) = 0.613$, and so on. Hence, any input is a fixed point. The fixed points are all neutral, since orbits are not pushed away or pulled toward any fixed points. In fact, for this function the orbits do not go anywhere—all seeds remain fixed.

On the other hand, consider what happens if the y-intercept is not zero. As an example, let us consider $f(x) = x + 7$. This function has no fixed points; there is no number that has the property that adding seven to it returns the same number. Accordingly, the fixed point equation for this function $f(x) = x$ has no solutions. All initial conditions will grow forever and will approach infinity. Exercises 6.6 and 6.7 give you a chance to investigate this graphically.

To summarize our results for slopes m of 1 or −1:

(1) If $m = -1$ there is one neutral fixed point. All other points are periodic with period 2.

(2) If $m = 1$,

    (a) and the y-intercept $b = 0$, then all points are fixed.

    (b) and the y-intercept $b \neq 0$, then there are no fixed points.

Exercises

(6.1) Consider the function $f(x) = -2x + 3$.

    (a) Sketch the function.

    (b) Find the fixed point from the graph.

    (c) Find the fixed point using algebra.

    (d) Using graphical iteration, determine the stability of the fixed point.

(6.2) Consider the function $f(x) = -\frac{1}{2}x + 4$.

    (a) Sketch the function.

    (b) Find the fixed point from the graph.

    (c) Find the fixed point using algebra.

    (d) Using graphical iteration, determine the stability of the fixed point.

(6.3) Consider the function $f(x) = 3x - 3$.

    (a) Sketch the function.

    (b) Find the fixed point from the graph.

    (c) Find the fixed point using algebra.

    (d) Using graphical iteration, determine the stability of the fixed point.

(6.4) Consider the function $f(x) = 4x$.

    (a) Sketch the function.

    (b) Find the fixed point from the graph.

    (c) Find the fixed point using algebra.

    (d) Using graphical iteration, determine the stability of the fixed point.

(6.5) Consider a function with a slope of zero.

    (a) Does the function have a fixed point?

    (b) If so, what is the fixed point's stability?

Justfy your answers by showing the results of graphically iterating such a function.

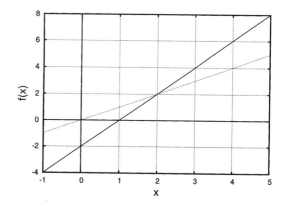

Fig. 6.10 The function for Exercise 6.8.

(6.6) Consider a function with a slope of one and a non-zero intercept such as $f(x) = x + 5$.

    (a) Sketch the function.

    (b) Graphically iterate a convenient seed.

    (c) Sketch the time series plot for this orbit.

    (d) How would you describe the behavior of this orbit?

    (e) Sketch the phase line for this function.

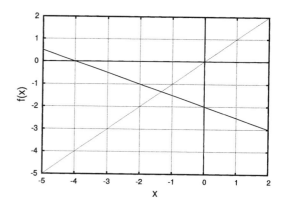

Fig. 6.11 The function for Exercise 6.9.

(6.7) Consider a function with a slope of one and a non-zero intercept such as $f(x) = x - 4$.

    (a) Sketch the function.

    (b) Graphically iterate a convenient seed.

    (c) Sketch the time series plot for this orbit.

    (d) How would you describe the behavior of this orbit?

    (e) Sketch the phase line for this function.

(6.8) Consider the function shown in Fig. 6.10.

    (a) Determine the equation of the function.

    (b) Find the fixed point graphically.

    (c) Find the fixed point using algebra.

    (d) Use graphical iteration to determine the stability of the fixed point.

    (e) Sketch the phase line for this function.

(6.9) Consider the function shown in Fig. 6.11.

    (a) Determine the equation of the function.

    (b) Find the fixed point graphically.

    (c) Find the fixed point using algebra.

    (d) Use graphical iteration to determine the stability of the fixed point.

    (e) Sketch the phase line for this function.

(6.10) ♯ Consider the time series shown in Fig. 6.12. This time series plot was arrived at by iterating a linear function. Determine this function. The exact values for the first few iterates are: $x_0 = 2$, $x_1 = 5$, $x_2 = 6.5$, and $x_3 = 7.25$.

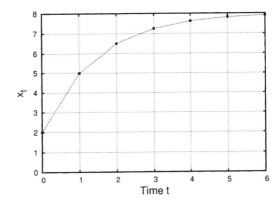

Fig. 6.12 The time series for Exercise 6.10.

(6.11) ♯ Consider the time series shown in Fig. 6.13. This time series plot was arrived at by iterating a linear function. Determine this function. The exact values for the first few iterates are: $x_0 = -1$, $x_1 = 4.25$, $x_2 = 2.9375$, and $x_3 = 3.26562$.

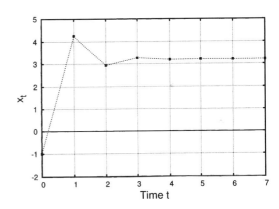

Fig. 6.13 The time series for Exercise 6.11.

Population Models

Thus far we have been concerned with generic functions—functions that are just functions and are not intended to model any particular situation in the real world. In this chapter we will shift momentarily toward a more applied context and will consider two families of functions that are commonly used to model the growth of populations. The latter of two functions, the logistic equation, will be a central item of study in subsequent sections of the book.

7.1 Exponential Growth

Let us imagine that a team of biologists visit an isolated, grassy, warm island in the sea. The biologists have traveled with some of their pet rabbits. One evening, the rabbits and biologists are romping around their campsite, and some of the rabbits wander off and are lost. The next morning, the biologists sail away, leaving some of their rabbit friends behind. In this way a bunch of rabbits come to live on an island that was once rabbit-free.

What will happen to the rabbit population over time? It is reasonable to expect that the rabbit population will grow. Let us try to construct a mathematical model to describe this situation. To simplify things, we will imagine that time is discrete: we will keep track of the rabbit population generation by generation, rather than instant to instant.

To start, let us assume that the population doubles every generation, and that there are initially three rabbits on the island. Then, the subsequent populations are:

$$3 \longrightarrow 6 \longrightarrow 12 \longrightarrow 24 \longrightarrow 48 \longrightarrow 96 \longrightarrow \cdots . \tag{7.1}$$

Clearly, the orbit will tend to infinity. The rabbit population will get larger and larger and larger.

The function that we are iterating in this case is the doubling function: $f(x) = 2x$. The function is plotted in Fig. 7.1. Also shown are the results of graphically iterating the initial population $x_0 = 3$. The itinerary for this seed was given in Eq. (7.1). The corresponding time series is shown in Fig. 7.2. As anticipated, we observe that the rabbit population is growing very quickly.

Because the function $f(x) = 2x$ is so simple, we can determine a convenient expression for $f^{(n)}(x)$, the nth iterate of x. Usually such a formula is not of much use, since we are typically interested in the long-term behavior of an orbit, not the details of particular orbits. However,

Fig. 7.1 The plot of the function $f(x) = 2x$, describing how the population of rabbits one year is related to the population of rabbits the next year. The input x is the population at one generation and the output $f(x)$ is the population at the subsequent generation. Also shown is the effect of graphically iterating the seed $x_0 = 3$. The orbit grows without bound; it tends toward infinity.

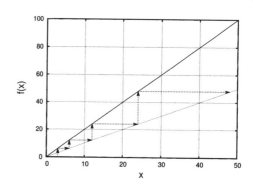

Fig. 7.2 The time series plot for the initial condition $x_0 = 3$ for the doubling function. We interpret this plot as a graph of the rabbit population versus time. The population grows exponentially.

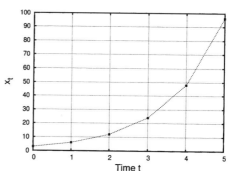

Table 7.1 The orbit of $x_0 = 3$ for the doubling function.

x_0	3
x_1	3×2
x_2	$3 \times 2 \times 2$
x_3	$3 \times 2 \times 2 \times 2$

in this particular instance there is a value to being able to write down an algebraic expression for how the population is growing over time.

To figure out a formula for $f^{(n)}(x)$, let us start by looking at Table 7.1, which contains the numerical values of the orbit. I have written numbers in the right column so as to emphasize that each successive iterate gets multiplied by two. (This is just a restatement of what it means to iterate $f(x) = 2x$; start with a number, and continually multiply it by 2.) This should make it clear that to get the nth iterate we just multiply our seed by 2 a total of n times. This is equivalent to multiplying the seed by 2^n.

Thus,

$$f^{(n)}(x_0) = x_0 2^n . \tag{7.2}$$

In words, this says that if we start with a population of x_0, after n generations the population is $x_0 2^n$. For example, we saw in Eq. (7.1) that the fourth iterate of 3 was 48. Accordingly,

$$f^{(4)}(3) = 3 \times 2^4 = 3 \times 16 = 48 . \tag{7.3}$$

We can easily generalize this result. Suppose the growth rate is not necessarily 2, but is instead given by the quantity r. I.e., we have

$$f(x) = rx , \tag{7.4}$$

where x is the population at one generation and $f(x)$ gives the population at the subsequent generation. The nth iterate of this function is given by

$$f^{(n)}(x) = xr^n . \tag{7.5}$$

This formula can be arrived at by a similar line of reasoning to that which led to Eq. (7.2).

The quantity r in Eq. (7.4) is often referred to as a **parameter**. This is to make clear that it plays a different role than the variable x, which is the population. The population is changing dynamically as the function is iterated, while the parameter r remains constant. One could then start over, change r and ask what effect the change has on the behavior of population. In general, parameters are variables that one can change and experiment with, but that remain constant while a dynamical variable changes.

In any event, Eq. (7.5) says that the population is growing exponentially as it is iterated. The number n of the iterate is interpreted as time. To make this clearer, we rewrite Eq. (7.5) in the following way:

$$P(t) = P_0 r^t , \tag{7.6}$$

where $P(t)$ is the population of rabbits at time t, and P_0 is the initial population. In words, Eq. (7.6) says that the population at time t is equal to the initial population times r^t. The function $P(t)$ is known as an **exponential function** because the variable t is expressed in the exponent.

Note that Eq. (7.4), which gives the population next year as a function of this year's population, is linear. This means that the growth is constant in the sense that at every time step the population is multiplied by r. In contrast, the function $P(t)$, which is the function for the tth iterate—i.e., the population after t years—is exponential. The function $P(t)$ is a formula for the time series plot for the function f.

Varying the Growth Rate

Before moving on to different models of population growth, let us consider the behavior of our model as we change the value of the growth rate r. Our basic model is

$$f(x) = rx , \tag{7.7}$$

where x is the population at the current generation, and $f(x)$ gives the population at the next generation. The quantity r is a parameter: a quantity that we can adjust, depending on the situation we are trying to model. We can think of a parameter as a knob or dial that we can turn to tune the model so it best fits the situation at hand.

As noted above, r is the growth rate of the population; r is the factor by which the population grows every generation. This means that each generation is obtained by multiplying the previous generation by r. This leads to exponential growth, as was seen above:

$$f^{(t)}(x_0) = P(t) = x_0 r^t . \tag{7.8}$$

The picture here is that the function $P(t)$ gives the population at time t, given that the initial population was x_0. This has several distinct behaviors depending on the value of r.

First, as we have seen, if r is greater than one, the population will grow exponentially. On the other hand, if $r < 1$, the population will decrease. For example, suppose that $r = 0.9$. This means that the population every generation is just 90% of the previous generation's population. Equivalently, we can view the population as decreasing by 10% every generation. For example, suppose that the initial population is 100. The itinerary for this initial condition is:

$$100 \longrightarrow 90 \longrightarrow 81 \longrightarrow 72.9 \longrightarrow 65.61 \longrightarrow 59.05 \longrightarrow \cdots , \qquad (7.9)$$

and the population tends toward zero.

Finally, if $r = 1$, then the population does not change from generation to generation; every point is fixed. The function is given by

$$f(x) = x . \qquad (7.10)$$

All x are fixed for this function; the input always equals the output.

To summarize, we have considered a population that grows by a fixed factor every generation. I.e., at each generation the population is multiplied by some number r to get the new population. This model has three different behaviors depending on the value of the parameter r:[1]

(1) $r > 1$: The population grows exponentially.

(2) $r = 1$: The population remains constant.

(3) $0 \leq r < 1$: The population decays exponentially.

The exponential growth model is important, since it is an excellent approximation to many natural phenomena. In addition, exponential growth forms the basis for a slightly more complicated growth model, to be discussed below, that will serve as the key example for the next several chapters.

[1] We will not consider negative r values, as these do not make sense in the context of our model. If we had a negative r, it would lead to a negative population, which is nonsense.

7.2   Modifying the Exponential Growth Model

The model of the previous section leads to unlimited exponential growth. If r is greater than one there will be more and more and more and more rabbits forever. Clearly, this is unrealistic; eventually the population will level off as the rabbits start to run out of food or space. In this section we will construct a new model that has the more realistic property that the population does not grow forever. The result will be an equation that will be our gateway into the study of chaotic systems.

The basic idea that we want to capture in an equation is as follows. If there are few rabbits, there will be lots of food for them, and the next year there will be more rabbits. However, if there are a lot of rabbits, there will not be enough food to go around. The rabbits will be hungry and weak, and as a result, in the following generation there will be fewer rabbits. The population will shrink instead of grow. This decrease in population when there are too many rabbits is what keeps the rabbit population from growing without bound.

How can we express this idea with an equation? Our starting point is the model from the previous section:

$$f(P) = rP \, . \tag{7.11}$$

(I will use P to represent the value of the population here, because later I will use the variable x to represent something else. So I need to save x for later use.) I wish to modify this equation to limit the growth somehow.[2] Let us imagine a certain population A at which there are so many rabbits that they eat all their food, leading to all rabbits starving, and hence zero rabbits at the next generation. I will refer to A as the annihilation parameter. If the population P ever equals A, then the population is doomed; all of the rabbits will die and P will equal zero at the next generation.

The annihilation parameter is incorporated into the basic model of Eq. (7.11) as follows:

$$f(P) = rP \left(1 - \frac{P}{A} \right) \, . \tag{7.12}$$

Let us see what this equation tells us. Suppose that the population P equals the annihilation value A. We would expect this to lead to a zero population. Is this the case? Plugging $P = A$ into Eq. (7.12), we obtain

$$f(A) = rA \left(1 - \frac{A}{A} \right) \, . \tag{7.13}$$

But $A/A = 1$, so this becomes

$$f(A) = rA(1 - 1) \, . \tag{7.14}$$

And, since $1 - 1 = 0$, we have

$$f(A) = 0 \, . \tag{7.15}$$

Thus, if $P = A$, the population is indeed annihilated.

Suppose the population P is small and is not close to the annihilation value A. In this case, the rabbits are not depleting their food source much, and so we would expect them to still grow approximately exponentially. Is this the case in our model? Let us look again at Eq. (7.12). If P is much smaller than A, then P/A is close to zero, and

$$\left(1 - \frac{P}{A} \right) \approx (1 - 0) = 1 \, . \tag{7.16}$$

Thus, Eq. (7.12) becomes

$$f(P) \approx rP \, . \tag{7.17}$$

So, for small P, this model will give exponential growth, just like our previous model did.

This model can also be understood graphically. A plot of Eq. (7.12) is shown in Fig. 7.3. In the graph we can see that if the current population P equals the annihilation parameter A, then the subsequent population

[2] The derivation and discussion that follows is standard. The version presented here closely follows that of Blanchard, Devaney, and Hall (2006, pp. 671–2).

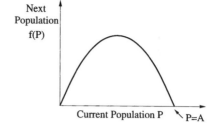

Fig. 7.3 A graph of the function $f(P) = rP(1 - \frac{P}{A})$, given in Eq. (7.12). Note that if $P = A$, then the population at the next generation is zero.

will be zero. We can also see that if the current population is zero then the next population will also be zero. If there are no rabbits, then in the next generation there will still be no rabbits. In between the extremes of $P = 0$ and $P = A$, we can see the scenario discussed above. Namely, if there are many rabbits—i.e., P is close to A—at the next generation there will be fewer rabbits. And if there are few rabbits—P is close to 0—then there will be more rabbits at the next generation.

As a final step in the development of this new model, I will manipulate Eq. (7.12) to put it in a somewhat more convenient and standard form. Repeating Eq. (7.12), we have:

$$f(P) = rP\left(1 - \frac{P}{A}\right).$$ (7.18)

We are interpreting $f(P)$ as the next population and P as the current population. To denote this more explicitly, we can rewrite the above equation as:

$$P_{n+1} = rP_n\left(1 - \frac{P_n}{A}\right),$$ (7.19)

where P_n is the population at generation n, and P_{n+1} is the population at generation $n + 1$.

Equation (7.19) depends on A, the maximum possible number of rabbits. In general, this could be quite a large number. Also, different islands will have different values of A, depending on how big the island is, how nutritious the grass is for the rabbits, and so on. Ultimately, we are interested in the behavior of the rabbit population given that there is some limit to their growth. We are not so interested in the exact value of the doomsday number A, nor are we interested in the particulars of the rabbits. The goal here is to come up with a generic model for populations that have some limit to their growth. We are after the simplest such model so that we can explore its general properties.

With all this in mind, I will divide both sides of Eq. (7.19) by A, for reasons that should hopefully become apparent soon:

$$\frac{P_{n+1}}{A} = \frac{rP_n}{A}\left(1 - \frac{P_n}{A}\right).$$ (7.20)

At first blush, Eq. (7.20) might look like a mess. But let us define a new variable x:

$$x = \frac{P}{A}.$$ (7.21)

In words, x is the population expressed as a fraction of the annihilation parameter. The variable x is thus always between 0 and 1. For example, if $A = 1000$ and there are $P = 600$ rabbits, x would be 0.6. Using this new variable x, Eq. (7.20) takes a much simpler form:

$$x_{n+1} = rx_n(1 - x_n) \,. \tag{7.22}$$

Or, returning to the functional notation that we started with:

$$f(x) = rx(1 - x) \,. \tag{7.23}$$

In this equation x is the current population, expressed as a fraction of the maximum possible population, and $f(x)$ gives the population at the next generation. The variable r is a parameter. As we did in the previous section for the case of exponential growth, we can vary r depending on the situation we are trying to model. We shall see that Eq. (7.23) has rather different properties for different values of r.

7.3   The Logistic Equation

Equation (7.23) is known as the logistic equation. It was first introduced in 1838 by Pierre François Verhulst.[3] Over the next several chapters we will use the logistic equation as an exemplar of a chaotic system. In this section we will begin exploring the surprisingly rich and diverse behaviors exhibited by this simple equation.

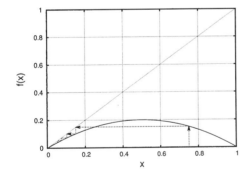

[3] The term "logistic" is something of a mystery. According to Kingsland (1995, p. 66), "Verhulst did not explain his choice of the term 'logistique' for his curve, but in nineteenth-century French the word referred to the art of calculation, as opposed to 'theoretical arithmetic'; it was allied to a type of logarithm used for astronomical calculations. From the context of Verhulst's 1845 memoir, it is likely that he intended to convey the idea of a calculating device, from which one could calculate the saturation level of a population... ." Regardless of its uncertain origin, the use of the term "logistic" to describe equations of the form of Eq. (7.23) and growth curves of the form Fig. 7.7 is ubiquitous.

Fig. 7.4 A graph of the logistic equation, $f(x) = rx(1 - x)$, for $r = 0.8$. The fixed point at $x = 0$ is attracting. All populations will eventually die off. The time series plot for the initial condition iterated graphically is shown in Fig. 7.5.

Let us start by considering what happens when r is less than 1. Since we are interpreting r as a growth rate, we would expect that the population will decay, just as it did for linear growth, $f(x) = rx$, when $r < 1$. Indeed this is the case. In Fig. 7.4 I have shown the logistic equation, Eq. (7.23), for $r = 0.8$. Graphical iteration shows that initial conditions are pulled toward the fixed point at $x = 0$. Thus, $x = 0$ is an attracting fixed point. This means that the population will die off, as expected.

The time series for the orbit iterated graphically in Fig. 7.4 is shown in Fig. 7.5. Again, we can see the orbit approaching zero; the population is dying. Remember that we are measuring the population as a fraction of the maximum possible population. Thus, a population of 0.8 means 80% of the maximum number of rabbits, not 0.8 rabbits.

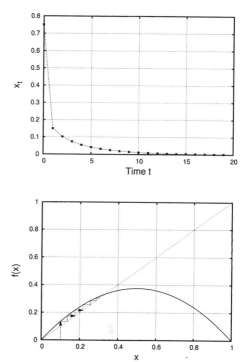

Fig. 7.5 The time series plot for the initial condition $x_0 = 0.75$ for the logistic equation $f(x) = rx(1 - x)$ for $r = 0.8$. This time series plot corresponds to the orbit graphically iterated in Fig. 7.4. We can clearly see that the population dies off.

Fig. 7.6 A graph of the logistic equation, $f(x) = rx(1 - x)$, for $r = 1.5$. Also shown are the graphical iterates of the initial condition $x_0 = 0.1$. The corresponding time series plot is shown in Fig. 7.7. Note that there is a fixed point at $x = 0$, as there was in Fig. 7.4. However, this fixed point is now repelling. There is a stable fixed point at $x \approx 0.33$.

Next, consider a growth parameter r that is greater than 1. For this case presumably the population will not die off. Instead, we might expect it to grow but then level off. In fact, this model was constructed with exactly this sort of behavior in mind. Let us see if this is indeed what occurs. In Fig. 7.6 I have plotted the logistic equation for $r = 1.5$. I have also shown the effect of graphically iterating the initial population $x_0 = 0.1$. The corresponding time series is shown in Fig. 7.7.

As anticipated, we see that the population grows but then reaches a plateau. For this particular r value the population reaches an equilibrium at around $x = 0.3$. (You can verify for yourself using algebra that the fixed point occurs at exactly $x = \frac{1}{3}$; see Exercise 7.6.) Thus, it appears that our modification of the exponential growth model has been

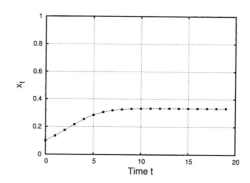

Fig. 7.7 The time series for an orbit of the logistic equation, $f(x) = rx(1 - x)$, for $r = 1.5$. The corresponding graphical iteration is shown in Fig. 7.6. There is a stable fixed point at $x \approx 0.33$.

a success; our new model does not lead to runaway growth, but rather a population that grows and levels off.

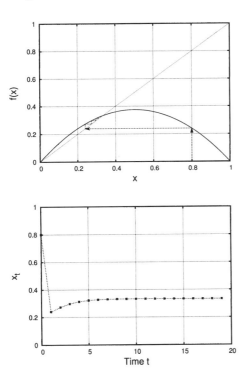

Fig. 7.8 A graph of the logistic equation, $f(x) = rx(1 - x)$, for $r = 1.5$. Also shown is the graphical iteration of the initial condition $x_0 = 0.8$. The corresponding time series plot is shown in the Fig. 7.9. This figure is identical to Fig. 7.6, except that here a different initial condition is used for the graphical iteration.

Fig. 7.9 The time series plot corresponding to the graphical orbit in Fig. 7.8.

What happens, however, if our initial population is above the equilibrium value? The population might decrease to the equilibrium value, or possibly it could grow and then cause the population to die off. Figure 7.8 shows us that it is the former possibility that occurs and not the latter. Figure 7.8 is identical to Fig. 7.6, except that I have used a different initial condition: $x_0 = 0.8$ instead of 0.1. We can see that the population experiences a large decrease in the first generation and then increases to the equilibrium value around $x = 0.3$.

Thus, for this parameter value we would expect to observe a stable population at about a third of the maximum population. Perturbations such as an unusually harsh stretch of weather or an unusually good crop of rabbit food might move the population a little bit away from the equilibrium. So we might observe fluctuations, but we would expect them to be transient. If we observed a sudden change in the population, we would take this as an indication that some external influence was effecting the rabbits. Left to their own devices, the rabbits will quickly reach a stable population size.

As a final example, let us investigate what happens when $r = 3.2$. The growth parameter r is now much larger than in the previous example. Will this lead to a different equilibrium value? Or will the population grow so fast that it reaches the "annihilation value" causing it to go to zero? Or is there some other possibility? Let us find out.

Fig. 7.10 A graph of the logistic equation, $f(x) = rx(1 - x)$, for $r = 3.2$. Also shown are the graphical iterates of the initial condition $x_0 = 0.1$. The corresponding time series plot is shown in Fig. 7.11. The orbit does not reach a single equilibrium value. Rather, it oscillates between two values, $x \approx 0.8$ and $x \approx 0.5$. The behavior is periodic with period 2. This period-2 behavior is stable, or attracting.

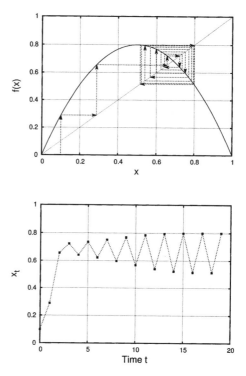

Fig. 7.11 The time series plot for the graphical iteration shown in Fig. 7.10. The function is the logistic equation, $f(x) = rx(1 - x)$, for $r = 3.2$. The orbit does not reach a single equilibrium value. Rather, it oscillates between two values, $x \approx 0.8$ and $x \approx 0.5$. The behavior is periodic with period 2. This period-2 behavior is stable, or attracting.

In Fig. 7.10 I have plotted the logistic equation for $r = 3.2$ and shown the results of graphically iterating the initial condition $x_0 = 0.1$. The corresponding time series is shown in Fig. 7.11. In the time series we see that the population settles into a cycle of period two; the population oscillates between two values, one at $x \approx 0.8$ and one at $x \approx 0.5$. In one generation there are a lot of rabbits—around 80% of the maximum. The rabbits eat a lot of the food on the island, and so in the next generation there is not quite enough food to go around. As a result, there are then fewer rabbits—around 50% of the maximum. Subsequently, the grass recovers, there is plenty of food for the rabbits to eat, and in the next year there are again roughly 80% of the maximum number of rabbits. And so on.

7.4 A Note on the Importance of Stability

The period-two cycle in Figs. 7.10 and 7.11 is stable. Nearby orbits are pulled toward these periodic points. Note that this function also has a fixed point around $x = 0.7$. However, this fixed point is not stable. Rather than pulling nearby orbits toward it, the fixed point pushes them away toward the periodic cycle. We thus would not expect to observe this fixed point if we were studying a real system. Because it is unstable, a tiny fluctuation or perturbation will move the orbit off of the fixed point and it will get pulled in to the cycle of period 2.

There is a bit of subtlety to this that is worth mentioning briefly. The fixed point x around 0.7 is, of course, fixed. There is also an x-value—let us call it x_{-1}—that lands at the fixed point after just one iteration.[4] Moreover, there are two points x_{-2} which go to x_{-1} after one iteration, and thus hit the fixed point after two iterations. And there are four points that go to x_{-2} after one iteration, and thus hit the fixed point after three iterations.

Following this line of reasoning, one can see that there are, in fact, an infinite number of initial conditions which will eventually land exactly on the fixed point x. Given this, does it still make sense to speak of the fixed point as unstable? Do *all* initial conditions eventually end up at x?

Well, just as there are an infinite number of initial conditions that lead to the fixed point, there are also an infinite number of initial conditions that lead to the period-2 cycle. In fact, there are infinitely many more initial conditions that lead to the period-2 cycle than the fixed point. This can be true even though there are infinitely many points that lead to the fixed point.

This seems counterintuitive, but I hope this can be made plausible by the following example. Imagine there is a hail storm and hailstones are falling on a pointed roof. A hailstone could fall on exactly on the pointed edge and remain there. There are an infinite number of locations along the roof point that the hailstones could land. However, it is vastly more likely that the hailstone will fall on one of the sides of the roof and then roll down the side. This is so much more likely that we never even consider the possibility that a hailstone would get stuck on the pointed roof, even though there are an infinite number of locations along the roof point that this could occur.

Similarly, there is such a small probability—vanishingly small—that an initial condition chosen at random just happens to land on the fixed point after iterating for a while, that we do not really need to consider this possibility. The hailstone example is not a perfect analogy for this situation, but it is close enough that I think it gives a good sense of what is going on. The more mathematical way of saying this is that the points that land exactly on the fixed point are a **set of measure zero**. This basically means that such points do exist, and there may even be infinitely many of them, but that they nevertheless occur so infrequently that the probability of observing them is zero. Another way of saying this is that an initial condition **almost surely** will not land on a fixed point. This seems like an imprecise or wishy-washy statement, but in mathematics it has a precise meaning: it means that something will occur with a probability of 1.

The main point of this discussion is to again highlight the importance of stability. Stable, attracting orbits are usually all we are concerned about when studying the long-term behavior of a dynamical system. Stable behavior is what is seen in actual experiments, which always involve a little bit of noise or random variation. And stable behavior is almost always what one encounters when studying a dynamical system

[4]You can explore this more in Exercise 7.10 at the end of this chapter.

using a computer or calculator, as we have done in this chapter and will do so frequently in the chapters ahead.

7.5   Other r Values

Thus far in our investigation of the logistic equation we have seen three behaviors. In Fig. 7.4 with $r = 0.8$ there was an attracting fixed point at zero; the population eventually dies off. In Fig. 7.6 with $r = 1.5$, there was an attracting fixed point at $x = \frac{1}{3}$; the population approaches 1/3 and remains there. And in Fig. 7.10 with $r = 3.2$ we found a stable cycle of period two; essentially all orbits are pulled into this cycle of alternating "over supply" and "under supply" of rabbits.

These three behaviors are just the tip of the iceberg. We shall see that the logistic equation exhibits many phenomena beyond simple fixed points and cycles of period 2. You will investigate some of these behaviors in the exercises at the end of this chapter, and we will explore these more fully in Part II of this book.

However, in order to carry out these investigations we will need some tools more powerful than a hand calculator. We will need to make use of a computer to iterate the logistic equation and make time series plots for us. An interface to program that can do this may be found at http://chaos.coa.edu. This program will let you choose an r value, the initial population x_0, and the number of iterates you would like plotted. The program will then generate a time series plot. You can also iterate the logistic equation, or any other function, using a spreadsheet program.

Further Reading

A standard history of ecology, including a full discussion of the development of the logistic model of growth can be found in Kingsland's *Modeling Nature* (1995).

The logistic equation as a model of population growth is a style of model that aims for qualitative as opposed to quantitative insight. The aim is to capture in broad strokes some of the essential features of a population that has some limit to growth. The goal is not necessarily to make quantitative predictions, nor even to be fully realistic or representative. The aim is to produce a style of understanding that is akin to a sketch or caricature as opposed to a detailed photograph. For views on different styles of modeling in the natural and social sciences, see (Levins, 1966), (May, 2002), (May, 2004), (Levins, 2006), and (Epstein, 2008).

Exercises

Exercises 7.1–7.5 concern exponential growth. This is an extremely important phenomenon in science and mathematics. However, it does not occur again in this text, so these exercises are not necessary for subsequent chapters.

(7.1) Exponential growth occurs when the factor by which a quantity increases is constant—i.e., it is multiplied by the same quantity at each time step. What if, instead, the same amount was *added* to the quantity at each time step? Consider the case in which a population of rabbits increases by 10 rabbits each generation.

   (a) What is the function that corresponds to this type of growth?

   (b) What is the long-term behavior of the orbit?

   (c) Let P_0 be the initial number of rabbits. Find a function $P(t)$ that gives the number of rabbits as a function of time.

(7.2) Suppose a population of rabbits starts at 50 in 2010 and doubles every year. What is the rabbit population in 2020? In what year would there be more rabbits than there are people?

(7.3) Suppose that the amount of garbage in a dump doubles every year. After a hundred years the dump is full. The townsfolk notice the dump is getting close to full when it is half-way to capacity. In what year does this occur?

(7.4) Suppose you invest some money in a savings account that earns 4% interest every year.

   (a) Write a function that gives the value of your savings account next year as a function of the amount in the account this year.

   (b) If you invest $5000 in the account, how much money do you have after one year?

   (c) How much do you have after five years?

   (d) How much do you have after fifty years?

(7.5) Suppose you want to have $100,000 for a down payment on a house ten years from now. Today you can deposit money into a savings account that earns 5% interest every year. How much do you need to deposit today so that you have $100,000 ten years from now? How much would you need to deposit of the interest rate was 3%? How much if it was 7%?

(7.6) Consider the logistic equation with $r = 1.5$, as shown in Figs. 7.6 and 7.8.

   (a) Algebraically find all fixed points for this function.

   (b) Estimate the slope of $f(x)$ at $x = \frac{1}{3}$. To do so, trace a straight line that intersects the $f(x)$ curve at $x = \frac{1}{3}$ such that the straight line is momentarily parallel (or tangent) to $f(x)$. Then determine the slope of this straight line.

   (c) What does the slope this line suggest about the stability of the fixed point at $x = \frac{1}{3}$. Is this consistent with what we see in Figs. 7.6 and 7.8?

(7.7) ⋆ Consider the logistic equation $f(x) = rx(1 - x)$ with $r = 2.5$.

   (a) Using a calculator, determine the first three iterates of $x_0 = 0.8$.

   (b) Use the program at http://chaos.coa.edu/ to compute the first three iterates of $x_0 = 0.8$.

(7.8) ⋆⋆ For these exercises use the program at http://chaos.coa.edu/. Consider the logistic equation, $f(x) = rx(1 - x)$. For each of the r values listed below, do the following:

   • Determine the long-term behavior of the system. Does the population die off, reach a fixed point, or reach a periodic point? If the latter, what is the period?

   • Try a few different initial conditions for each parameter values. Remember that your initial conditions should always be between 0 and 1. You should find that the behavior you observe is independent of the initial condition(s) you choose. However, avoid initial conditions that are simple fractions, like $x_0 = 0.5$ or $x_0 = 0.25$.

   • For each r value, make a rough sketch of the time series plot.

Here are the r values to try:

   (a) $r = 0.5$.

   (b) $r = 1.5$.

   (c) $r = 2.8$.

   (d) $r = 3.3$.

(e) $r = 3.5$.

(f) $r = 3.56$.

(g) $r = 3.835$.

(h) $r = 4.0$.

(7.9) Consider the logistic equation, $f(x) = rx(1-x)$.

    (a) Determine an algebraic expression for the non-zero fixed point. (Your answer will depend on r.)

    (b) For what r values is this fixed point positive?

    (c) For what r values is this fixed point less than 1?

(7.10) ♮ Consider the logistic equation with $r = 3.2$, as shown in Fig. 7.10.

    (a) Find the exact value for the fixed point x^* near $x = 0.7$.

    (b) There are two values which, when acted upon by $f(x)$, yield the fixed point. One, of course, is the fixed point itself. But there is another point, denoted x^*_{-1} that lands on the fixed point after just one iteration.

        (i) Sketch the function and illustrate the point x^*_{-1} by graphically iterating and finding an input value that lands on the fixed point.

        (ii) Use algebra to find x^*_{-1} and check that it is consistent with the point you found graphically.

(7.11) ♮ Consider the logistic equation $f(x) = rx(1-x)$ with $r = 3.2$. We have seen that for this parameter value the logistic equation has a cycle of period two. We will use algebra to determine the x values that make up this cycle. To do so, note that if a point x^* is periodic with period two, then it satisfies the following equation:

$$f^{(2)}(x^*) = x^* \, . \tag{7.24}$$

In other words, if we start with x^*, apply f to it twice, we return to x^*.

    (a) Determine an algebraic expression for $f^{(2)}(x)$. It might be easier to not plug in for r until the very end.

    (b) Solve the equation $f^{(2)}(x^*) = x^*$ for x^*. Which of your solutions have period two?

    (c) How do the two periodic points you found compare with those shown in Fig. 7.10?

Newton, Laplace, and Determinism

8

Thus far this book has mainly been concerned with introducing iterated functions. Iterated functions are one of the simplest types of dynamical systems, yet nevertheless show many of the interesting and surprising features found in more complicated dynamical systems. In the next part of the book we will continue our study of iterated functions and will encounter these more interesting dynamical features, including chaos and sensitive dependence on initial conditions, known more colloquially as the butterfly effect. We will be in a better position to understand the significance of these phenomena if we pause for a moment and think about the origins of some of the basic assumptions of science—assumptions that I will suggest are still with us today.

My goal in this chapter is not to give a thorough historical account of the development of science. Rather, I aim to highlight a few of the central, but sometimes unspoken, ideas or assumptions that science makes about the world. In subsequent chapters we will consider the extent to which the study of chaos requires us to reconsider or refine these basic scientific notions. To do so, it is helpful to have an understanding of the character of Newtonian physics and classical mechanics. We thus start by considering the work of Issac Newton.

8.1 Newton and Universal Mechanics

In 1687 Newton published *Principia Mathematica*. The results in this book laid the groundwork for much of physics, if not science itself. There are two general results in the *Principia* that are especially important for the goals of this chapter.

First, Newton laid out a theory of motion—what causes objects to move. This theory is expressed in what are now known as Newton's three laws of motion. Of particular interest to us is Newton's second law, which states that the motion of an object is determined by the forces acting upon it:

$$\vec{F}_{\text{net}} = m\vec{a} . \tag{8.1}$$

The term on the left-hand side of this equation is the *net force*, the total force acting on the object. The arrow on top of the F tells us that force is a vector, a quantity that has a direction in addition to a size.

It is not a surprise that direction enters into this equation; if I give you a push, how you move as a result depends on the direction as well as the strength of my push. On the right-hand side of Eq. (8.1), m is the object's mass. And $\vec{a}$ is the object's acceleration, the rate of change of the object's velocity. The acceleration is also a vector; as was the case with force, direction matters for acceleration.

Newton's second law, Eq. (8.1), tells us why objects move the way they do—it is because of the forces that act on them. Motion is deterministic in the same sense that the iterated functions we have been studying are deterministic. Newton's second law is a differential equation, a relationship between a quantity and its rates of change. So it is a somewhat different mathematical entity than iterated functions. But the determinism is the same. Equation (8.1) lets us determine the future position of an object given its current position and a specification of the forces acting on it. Similarly, for our iterated functions we can specify the future values of the function—its itinerary—given knowledge of its current value and the details of the function that is acting on it.

The second, and in many ways most revolutionary, feature of Newton's *Principia* is that it put forth the idea that the laws of physics are universal. In other words, the same rules that can be used to determine the motion of objects in Cambridge, England, can also be used in Oxford, Paris, or New York. Moreover, the same rules or laws also apply to celestial bodies, such as the moon or the sun. Prior to this time scientists[1] believed that the motion of the moon and the planets would follow different laws than earthly objects. Specifically, Newton put forth this universality when he formulated his law of gravity. Newton showed that the effect of gravity on a falling apple on earth and the gravitational pull of the earth on the moon, causing it to revolve around us, both can be explained or described with the same rule or law. This is now known as Newton's universal law of gravitation.

The results in the *Principia*, Newton's laws of motion and the law of universal gravitation, launched the study of mechanics.[2] These results, together with the invention of calculus—a body of mathematical theory and techniques for understanding quantities that change continuously, such as the position of an object as it moves through space—are still studied today in essentially unchanged form. The mechanics and calculus now learned by physics and engineering students has the same basic structure as it did in the 1700s. This area of physics, known as classical mechanics, has been tremendously successful.

Classical mechanics does not hold in all physical situations. For very small objects, classical mechanics has been replaced by quantum mechanics. For very fast objects, those moving at a fraction of the speed of light, classical mechanics has been replaced by the theory of special relativity. And for very, very large objects, the size of a star or a galaxy, Newton's law of gravity must be modified by general relativity. Yet for describing the motion of objects of everyday experience, Newtonian mechanics endures.

[1] More properly, natural philosophers. The term *scientist* did not assume its modern meaning and gain acceptance until the early 1800s. For an interesting history of the word "scientist", see the article by Sydney Ross (1962).

[2] It was not solely Newton who is responsible for mechanics. The story is, of course, considerably more complex. Copernicus, Galileo, Descartes, Hooke, and Kepler all did crucial work that Newton drew upon in the *Principia*. Nevertheless, the publication of the *Principia* is often taken to mark a turning point in intellectual history and the crowning achievement of the scientific revolution.

8.2 The Enlightenment and Optimism

In addition to launching the field of physics known as classical mechanics, Newton's *Principia* also in many ways was a crowning achievement of the scientific revolution—the emergence of science and the scientific method in the 1500s and 1600s. During this time the geocentric view[3] of the world was replaced with the heliocentric, and careful observation and mathematical analysis became an important part of science. More generally, the scientific method emerged and solidified, wherein knowledge is objectively generated via repeated observation or experiment, building on previous results, and expressed within a logical or mathematical framework. Newton's *Principia* tied together the previous work of Copernicus, Kepler, Galileo, and others, into a unified and powerful theory.

The scientific revolution was followed by the Enlightenment,[4] an era spanning the 1700s, characterized by advances in science, a belief in reason and logic over authority and doctrine, and expanding democracy and individual rights. During this time, influenced by the ideas put forth by Newton, a scientific view of the world solidified. The universe according to Newton is one determined by laws or rules. Objects move because of the forces that act on them, and we can use Newton's second law, Eq. (8.1), to deduce the future positions of objects. In a Newtonian framework the universe is mechanistic, material, and mathematical.

The world is mechanistic because Newton's laws explain motion; a change in motion is caused by a force. In this point of view one pictures, only somewhat metaphorically, objects as being controlled by gears or levers; objects interact, collide, and exert forces on each other, but always obeying the same laws. There is no need for divine intervention, nor is there any need for chance—just let the universe go, and it will do its thing and evolve forward in time according to universal physical laws.

The Newtonian universe is material in the sense that the world was viewed as being made up of stuff—tangible, real objects. It was argued that even forces like gravity that appear to act across empty stretches of space are conveyed by tiny particles, or corpuscles. Moreover, since the universe is material, its behavior can be predicted or understood. Things are they way they are for a reason or a cause. The Newtonian world is mathematical, in that it was viewed that the regularities or laws that describe or govern the world are mathematical in nature. As Galileo puts it, "Philosophy is written in this grand book of the universe, which stands continually open to our gaze. But the book cannot be understood unless one first learns to comprehend the language and to read the alphabet in which it is composed. It is written in the language of mathematics... " (quoted in Godfrey-Smith (2003, pp. 10–11)). This view is certainly still dominant in the physical sciences, and is increasingly influential in the biological sciences as well.

Not all situations will be amenable to laws as simple as Newton's law of gravity. And the laws may not be exact; even Newton's law

[3]In a geocentric model of the universe the earth is seen as being at the center of the world, and all other celestial objects, such as the sun, moon, and other planets, revolve around the earth. In a heliocentric model, the sun is at the center and the planets revolve around the sun.

[4]Some historians argue that there is not a clear boundary between these two periods, and that suggesting that there is boundary oversimplifies a complex history. I am sympathetic toward this view, but nevertheless I think it is a useful way to refer to the period of intellectual history that followed Newton.

of gravity applied to objects on earth has tiny inaccuracies that are corrected for by Einstein's general theory of relativity. But the point is that there *are* laws; there is a fundamental orderliness to things. The prevailing view during the Enlightenment—and perhaps today as well—is that it is the job of science to figure out these laws. Even if we do not currently understand it, the world is understandable. Phenomena occur for a reason, and there are laws of nature that have the potential to be expressed in the language of mathematics. The Newtonian world held the promise that things could be understood.

8.3   Causality and Laplace's Demon

The Newtonian universe is one of cause and effect. Objects move for a reason, according to universal laws. Such a world is said to be deterministic; the present state is determined by the past state. And if we know the current position of an object, and if we know the laws of physics—the forces that the object is subject to—then we can determine the future position of the object. Physics is thus predictive; it allows one to make definite statements about events that have not yet happened. This is the promise of physics in the Enlightenment. If we measure accurately enough, and understand the laws and forces of nature, then the future becomes predictable.

However, extrapolating these ideas one quickly arrives at bit of a puzzle. If someone knew the initial position of all the objects in the universe and the forces and rules that apply to these objects, then this person could predict the entire future. This idea is succinctly put forth in a famous passage from Pierre-Simon Laplace, a French mathematician in the late 1700s and early 1800s. He writes:

> We may regard the present state of the universe as the effect of its past and the cause of its future. An intellect which at a certain moment would know all forces that set nature in motion, and all positions of all items of which nature is composed, if this intellect were also vast enough to submit these data to analysis, it would embrace in a single formula the movements of the greatest bodies of the universe and those of the tiniest atom; for such an intellect nothing would be uncertain and the future just like the past would be present before its eyes. (Laplace, 2009).

Such an intellect is now often referred to as **Laplace's demon**, although Laplace himself did not use the term.

One of the issues raised by Laplace is that of free will. If the universe really is deterministic, if the world is fundamentally material and the objects of the world obey fixed, deterministic laws, then in a sense the future has already been written—it is an inevitable consequence of the way things are today. Nevertheless, we perceive that we as individuals are capable of making real choices—should I take a break from writing this paragraph and eat a cookie, or should I continue on? But perhaps this choice is an illusion. Depending on one's disposition, such matters can be fun to think about, or can be a journey down an abyss of doubt and despair. However, questions of free will are not the main reason I bring up Laplace's idea of determinism.

Instead, what is interesting and important about the Laplace passage above is that it spells out quite compactly one of the aspirations of science: to become as close as possible to the intellect of Laplace's demon. Doing so requires three things. First, one needs exact measurements of the current state of affairs. Second, one needs to know the laws or rules that the world obeys. And third, one needs sufficient computational power so as to be able to calculate the future behavior.

Of course, the above is not exactly possible. Certainly the first and third items can never be fully achieved; there will always be some uncertainty in our measurements, and even modern, fast computers are limited in what they can calculate. Nevertheless, I think that many scientists—and non-scientists, for that matter—believe in a modified version of the vision of determinism articulated by Laplace. Namely, measurement, knowledge of the laws of nature, and computation enable one to make good predictions about the future. These predictions are not exact, but they can be made better with more careful measurement, more thorough explication of the laws of nature, and with more computational effort.

8.4 Science Today

The optimism of the Enlightenment can seem quaint and even naive. Today we tend to have a more tempered view of the power of reason and science to explain the world. Some scientific problems—finding a cure for AIDS, predicting the path of a hurricane, or even forecasting next week's weather—still seem out of reach. And the great scientific advances of the twentieth century did not prevent the carnage of that century's genocides and wars. Here in the twenty-first century, despite overwhelming scientific evidence of human-caused climate change, there is little action to prevent climate change from occurring.

Today attitudes toward science vary widely. For that matter, the nature of science varies widely, too. Genetics, astronomy, particle physics, geology, and ornithology are all generally viewed as science, but the nature of the work done by scientists in these fields is very diverse. Nevertheless, I believe that the basic scientific impulse described in the previous section is still with most scientists and non-scientists. If we just knew a little more, or could measure things more accurately and had more data, or if our computers were more powerful, then our ability to predict and understand and explain would be greater.

For example, weather forecasts of more than a few days are notoriously unreliable. At issue is not our understanding of physics, for there is little doubt that we know the fundamental physical laws that describe energy and air and moisture in the atmosphere. So presumably what is needed are better measurements of the current atmospheric conditions, and faster computers. With these advances, the thinking goes, it should be possible to get closer to Laplace's demon and improve our weather predictions. We shall see that the phenomenon of chaos does not undermine this basic premise, but it does suggest severe limitations on our practical ability to make accurate long-term weather predictions.

An additional aspect of our Newtonian legacy is a belief that basic laws of nature are likely to be mathematical and simple. Newton's laws of motion, together with the universal law of gravitation, describe in a concise and powerful way a breathtaking array of phenomena. The laws of electricity and magnetism, developed in the 1800s, are similarly concise. Just four equations—Maxwell's equations—together with one additional rule known as the Lorentz force law—accurately describe all electromagnetic phenomena at scales larger than that of a few molecules.

Of course, just because the fundamental laws of nature are simple does not mean that the world is simple or predictable. Indeed, complexity and randomness appear to be all around us. We might ask, then, how this complexity and randomness arise. One assumption is that complicated behavior arises when a system is large or complicated. For example, the turbulent flow of water results from the fact that there are a vast number of molecules in water, each of which is more or less free to move independently. In this view, complexity and randomness arise from the fact that there are an enormous number of constituents in water. It is a complicated system, and so one might expect complicated behavior. Conversely, a simple system is likely to have simple behavior. Indeed, we have seen this in previous chapters; simple iterated functions have simple behavior. Almost all the orbits we have encountered have either tended toward a fixed point, or flown off to infinity.

8.5 A Look Ahead

What does chaos have to say about the Newtonian world? How could randomness arise in a deterministic universe? Does chaos oblige us to revise our mechanistic and deterministic view? Is Newton wrong? Hardly. We will explore these questions, and more, in the next part of the book where we will encounter the phenomenon of chaos. We will see that simple, deterministic systems hold some interesting surprises. Simple deterministic dynamical systems can produce behavior that is apparently random. Such dynamical systems can also produce remarkably complex and interesting patterns. Simple iterated systems of the sort we have been studying are not doomed to repetitiveness or bland predictability, but rather can produce surprise and continual novelty.

Further Reading

This chapter contains just the barest sketch of the origins of science and the scientific method. For further reading, I recommend the first chapter of Peter Godfrey-Smith's book on the philosophy of science (Godfrey-Smith, 2003). The essay on causal determinism (Hoefer, 2010) in the *Stanford Encyclopedia of Philosophy* is an excellent introduction to the implications of Newtonian determinism. For a discussion of determinism and the origins of science in relation to chaos, I recommend the first two chapters of Ian Stewart's *Does God Play Dice?* (2002).

Part II

Chaos

Chaos and the Logistic Equation

In the previous chapter I introduced the logistic equation:

$$f(x) = rx(1 - x),$$ (9.1)

where x represents a population, expressed as a fraction of the maximum possible population. Hence, x is always between 0 and 1. The variable r is a parameter. Thus far we have analyzed orbits for a few different values of r. In this chapter we will look at this issue more systematically and in much more detail. What happens as we change r, and how can we summarize and visualize these changes?

Have you done Exercise 7.8 in Chapter 7? If not, I strongly suggest that you go back and give it a try. This exercise leads you to discover some of what I will show you in this chapter, and it will be much more fun if you get to experience this discovery for yourself.

9.1 Periodic Behavior

We begin by considering a handful of different r values. For each, I will plot the function and show graphically the orbit of a typical initial condition. I will also illustrate the long-term behavior of the orbit with a type of diagram that is similar to a phase line.

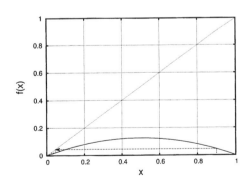

Fig. 9.1 A graph of the logistic equation, $f(x) = rx(1 - x)$, for $r = 0.5$. Also shown are the graphical iterates of the initial condition $x_0 = 0.9$. The corresponding time series plot is shown in Fig. 9.1, and the final-state diagram is shown in Fig. 9.3.

I will start with $r = 0.5$. Figure 9.1 shows the orbit of $x_0 = 0.9$ for the logistic equation with $r = 0.5$. The population decreases quickly and approaches zero. There is an attracting fixed point at $x = 0$. Hence, all initial conditions will approach 0. If we wait a long time, eventually the

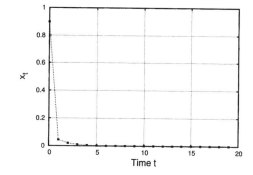

Fig. 9.2 The time series plot for the graphical iteration shown in Fig. 9.1. The corresponding final-state diagram is shown in Fig. 9.3.

Fig. 9.3 The final-state diagram for the logistic equation, $f(x) = rx(1-x)$, for $r = 0.5$. Compare with Figs. 9.1 and 9.2.

population will be at 0. Strictly speaking it will always be a little bit above zero, but this difference will quickly become imperceptible.

This is illustrated in Fig. 9.3, which can be thought of as a final-state diagram. I have drawn the possible values for x as a line segment; recall that x is always between 0 and 1. I have then drawn a dot at $x = 0$ to indicate that this is the final state of the system. This is similar to a phase line, however, on this sort of diagram we do not draw any arrows. All we do is indicate the final state or states of the orbit. Such a diagram is a succinct summary of the fate of the orbits for a particular r value. These final-state diagrams will be the key to developing a diagram that will let us see, all at once, all the different behaviors of the logistic equation. This will be the topic of Chapter 11. For now, though, let us continue with our exploration of the logistic equation, one r value at a time.

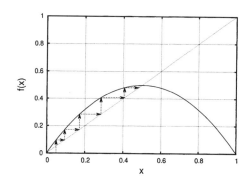

Fig. 9.4 A graph of the logistic equation, $f(x) = rx(1-x)$, for $r = 2.0$. Also shown are the graphical iterates of the initial condition $x_0 = 0.05$. The corresponding time series plot is shown in the Fig. 9.5, and the corresponding final-state diagram is shown in Fig. 9.6.

For our next r value, we try $r = 2.0$. The results are shown in Figs. 9.4–9.6. We can see that the orbit grows fairly quickly and approaches a fixed point at $x = 0.5$. We indicate this with a single dot at $x = 0.5$ on the final-state diagram of Fig. 9.6.

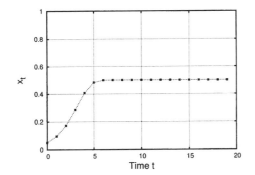

Fig. 9.5 The time series plot for the orbit shown in Fig. 9.4. The corresponding final-state diagram is shown in Fig. 9.6.

Fig. 9.6 The final-state diagram for the logistic equation, $f(x) = rx(1-x)$, for $r = 2.0$. Compare with Figs. 9.4 and 9.5.

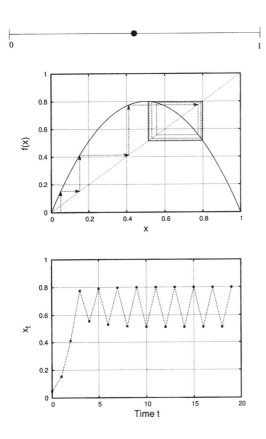

Fig. 9.7 A graph of the logistic equation, $f(x) = rx(1-x)$, for $r = 3.2$. Also shown are the graphical iterates of the initial condition $x_0 = 0.05$. The corresponding time series plot is shown in Fig. 9.8, and the corresponding final-state diagram is shown in Fig. 9.9. The orbit approaches a cycle of period 2.

Fig. 9.8 The time series plot for the orbit shown in Fig. 9.7. The corresponding final-state diagram is shown in Fig. 9.9. The long-term behavior of the orbit is periodic with a periodicity of 2.

Fig. 9.9 The final-state diagram for the logistic equation, $f(x) = rx(1-x)$, for $r = 3.2$. The orbit is pulled toward a period-2 cycle. Compare with Figs. 9.7 and 9.8.

Next, we try $r = 3.2$. The results are shown in Figs. 9.7–9.9. This time, rather than approaching a single value—i.e., an attracting fixed

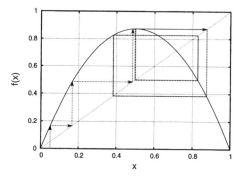

Fig. 9.10 A graph of the logistic equation, $f(x) = rx(1 - x)$, for $r = 3.5$. Also shown are the graphical iterates of the initial condition $x_0 = 0.05$. The corresponding time series plot is shown in Fig. 9.11, and the corresponding final-state diagram is shown in Fig. 9.12.

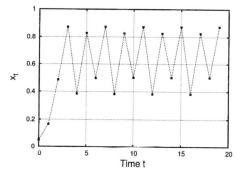

Fig. 9.11 The time series plot for the orbit of Fig. 9.10. The corresponding final-state diagram is shown in Fig. 9.12. The orbit is drawn to a cycle of period 4.

Fig. 9.12 The final-state diagram for the logistic equation, $f(x) = rx(1 - x)$, for $r = 3.5$. The orbit is pulled toward a cycle of period 4. Compare with Figs. 9.10 and 9.11.

point—the orbit approaches a cycle of period 2. This can most clearly be seen in the time series plot, Fig. 9.8. This period-2 behavior should not be much of a surprise, as we considered this case in the previous chapter. On the graphical iteration in Fig. 9.7, the period-2 behavior manifests itself as the square pattern in the dashed line of the graphical orbit. The population oscillates between $x \approx 0.52$ and $x \approx 0.8$. This is indicated on the final-state diagram of Fig. 9.9 as *two* dots, one at $x \approx 0.52$ and one at $x \approx 0.8$.

Now for something new: let us try $r = 3.5$. The results of iterating the logistic equation with this r value are shown in Fig. 9.10. Looking carefully at the time series plot in Fig. 9.11 we can see that the behavior is periodic, but this time the period is 4; it takes four iterations for the population to cycle back. In the long run the orbit values are approximately 0.50, 0.87, 0.38, 0.83, repeating every four iterates. Accordingly, the final-state diagram in Fig. 9.12 now has four dots, indicating that in the long run the system will oscillate among four values.

Continuing our survey of the behavior of the logistic equation for different r values, we next consider $r = 3.56$. The results for this r value

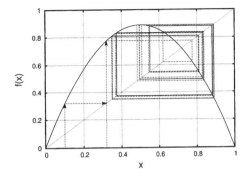

Fig. 9.13 A graph of the logistic equation, $f(x) = rx(1 - x)$, for $r = 3.56$. Also shown are the graphical iterates of the initial condition $x_0 = 0.05$. The corresponding time series plot is shown in Fig. 9.14, and the corresponding final-state diagram is shown in Fig. 9.15.

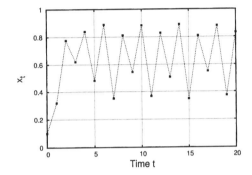

Fig. 9.14 The time series graph corresponding to the orbit shown in Fig. 9.13. The corresponding final-state diagram is shown in Fig. 9.15. The long-term behavior is periodic with a period of 8.

Fig. 9.15 The final-state diagram for the logistic equation, $f(x) = rx(1-x)$, for $r = 3.56$. The long-term behavior of the orbit is periodic with a period of 8. Compare with Figs. 9.13 and 9.14.

are shown in Figs. 9.13–9.15. The long-term behavior of the orbit is now periodic with period 8. It is a little bit hard to see the period-8 behavior. It looks a lot like period 4. But if you look carefully at the time series plot in Fig. 9.13, you will see that it actually takes 8 cycles for the orbit to repeat itself. In Fig. 9.15, we represent this period-8 behavior with 8 dots.

For our final example of this section we consider $r = 3.84$. The results for this parameter value are shown in Figs. 9.16—9.18. Here, the orbit eventually becomes periodic with a period of 3. It takes a fairly long time for the period-3 behavior to become evident. In the time series plot, Fig. 9.17, the period-3 behavior is not seen until $t = 22$. From this point on, however, the period-3 behavior is fairly clear. The orbit cycles from 0.15 to 0.49 to 0.96, repeating every three time steps.

Summarizing, we have seen that the logistic equation is capable of a number of different periodic behaviors. Depending on the r value, we have found periodicities of 1, 2, 3, 4, and 8. (Note that period 1 is the same as a fixed point.) This little equation, $f(x) = rx(1 - x)$, can do quite a bit. What else is the logistic equation capable of? How are the r

Fig. 9.16 A graph of the logistic equation, $f(x) = rx(1 - x)$, for 3.84. Also shown are the graphical iterates of the initial condition $x_0 = 0.1$. The corresponding time series plot is shown in Fig. 9.17, and the corresponding final-state diagram is shown in Fig. 9.18.

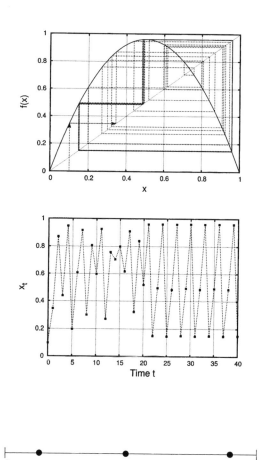

Fig. 9.17 The time series plot corresponding to the orbit shown in Fig. 9.16, and the final-state diagram is in Fig. 9.18. The long-term behavior of the orbit is periodic with period 3. The corresponding final-state diagram is shown in Fig. 9.18.

Fig. 9.18 The final-state diagram for the logistic equation, $f(x) = rx(1 - x)$, for 3.84. The long-term behavior of the orbit is periodic with a period of 3. Compare with Figs. 9.16 and 9.17.

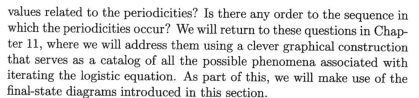

values related to the periodicities? Is there any order to the sequence in which the periodicities occur? We will return to these questions in Chapter 11, where we will address them using a clever graphical construction that serves as a catalog of all the possible phenomena associated with iterating the logistic equation. As part of this, we will make use of the final-state diagrams introduced in this section.

For the rest of this chapter we will investigate a new, non-periodic sort of behavior. To do so, we examine the orbits of the logistic equation for $r = 4$.

9.2 Aperiodic Behavior

In Fig. 9.19 I have shown the effects of graphically iterating the logistic equation with $r = 4$ for the initial condition $x_0 = 1$. I have not put any arrows on the graphical iterates, as there is not space on the diagram. The main point to observe is that there does not seem to be a periodic attractor—or if there is, it is taking the orbit a very long time to get pulled into it. A more helpful view is the corresponding time series plot,

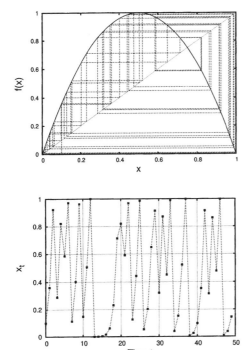

Fig. 9.19 A graph of the logistic equation, $f(x) = rx(1 - x)$, for $r = 4.0$. Also shown are the graphical iterates of the initial condition $x_0 = 0.1$. The corresponding time series plot is shown in Fig. 9.20. The orbit does not appear to be periodic.

Fig. 9.20 The time series plot corresponding to the orbit shown in Fig. 9.19. The orbit does not appear to be periodic.

shown in Fig. 9.20, in which it appears as if the orbit is not periodic. This is in contrast to the other r values we have looked at, where we were able to discern the periodic behavior. What is going on in Fig. 9.20?

Perhaps the orbit really is periodic, but it takes a while for the periodic behavior to set in. This does seem plausible. For example, in Fig. 9.17 we saw that the behavior was periodic, but it took around twenty iterates for the orbit to reach the period-3 behavior. So to see if the orbit shown in Fig. 9.20 for $r = 4.0$ might be periodic I have again iterated the seed $x_0 = 0.1$. However, this time I have calculated $10,000$ iterates. All of these numbers will not conveniently fit on a graph, so in Fig. 9.21 I have plotted only the last 51.

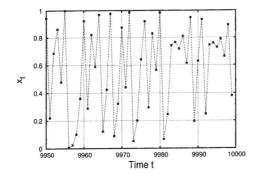

Fig. 9.21 Time series plot for the logistic equation, $f(x) = rx(1 - x)$, for 4.0, for the initial condition $x_0 = 0.1$. Only iterates $9,950$ to $10,000$ are plotted. The orbit does not appear to be periodic.

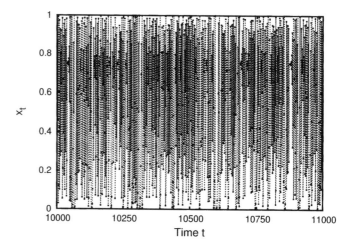

Fig. 9.22 Time series plot for the logistic equation, $f(x) = rx(1 - x)$, for 4.0, for the initial condition $x_0 = 0.1$. Only iterates 10,000 to 11,000 are plotted. The orbit does not appear to be periodic.

Figure 9.21 is somewhat surprising. The orbit *still* does not appear periodic, even after 10,000 iterates. Perhaps it is the case that it is periodic, but that the period is very long. I have been plotting only fifty points at a time. Maybe the period is longer than fifty, in which case we would never be able to see the periodicity on such a short time series plot. So in Fig. 9.22 I have plotted the time series from generation 10,000 to generation 11,000—a total of 1001 generations. It is somewhat hard to see, but the orbit in Fig. 9.22 still does not appear to be periodic; it does not seem to repeat or possess any regular pattern.

In fact, it turns out that the orbit of $x_0 = 0.1$ for $r = 4$ *never repeats*. We could keep iterating forever—literally—and we would never encounter exactly the same number in the itinerary. Such behavior is said to be **aperiodic**. But how can I possibly claim that this is the case? How do I know that the orbit does not repeat after some super large number of iterates? There are a number of ways to respond to these legitimate questions and doubts. First, we could ask our computer program to check if the itinerary ever repeats. For example, we could iterate the equation for 1 million timesteps. After all these iterations, the computer could record the current x value, $x_{1,000,000}$. We could then keep iterating, and at every step check and see if the current x value ever returns to $x_{1,000,000}$. If we did this experiment on a computer, we would indeed find that the orbit never repeats.

[1]In fact, it turns out that in some cases the computer "runs out of numbers" faster than one would guess. What happens is that due to the round-off errors associated with the finite-precision arithmetic that the computer does, the orbit starts to repeat surprisingly quickly in some cases. An interesting discussion of this can be found in Peitgen, Jürgens, and Saupe (1992, pp. 533–5).

But actually this is not quite true. Computers can only keep track of a finite number of digits. So decimals that go on forever—and most numbers between zero and one are decimals that go on forever—are rounded off or truncated in a computer's memory. This means that for a computer there are not an infinite number of numbers between zero and one. So eventually the numbers have to repeat, because the computer will run out of new numbers. Thus, the computer experiment I mentioned in the previous paragraph will not actually work as I claimed.[1] Moreover, even if the computer could store an infinite number of numbers, it could not keep iterating literally forever.

In general, computers cannot provide air-tight evidence or formal proofs of statements involving infinities. But computers can provide very strong evidence in such realms, as is the case here. For the particular case of the logistic equation with $r = 4$, there exist rigorous proofs that the orbit really does never repeat. I am using the word "rigorous" in the sense that mathematicians use it: a proof or demonstration is rigorous if it relies on clear, standard mathematical logic, and does not depend on circumstantial evidence such as that which might be provided by a computer.

The proof of the aperiodicity of the logistic equation with $r = 4$ is beyond the scope of this book; it involves some mathematical techniques that are likely unfamiliar to most readers. Nevertheless, I hope you will take my word for the fact that the aperiodicity of the logistic equation at $r = 4$ has been rigorously established; this statement has been proved without the use of computers.[2]

Finally, there is the matter of representing this chaotic behavior in a final-state diagram, as we have done for the periodic behaviors encountered previously. But this poses a bit of a conundrum; what do we mean by "final states" for a system that repeats forever? In practical terms, one way to think about this is as follows. Imagine constructing the final-state diagram via the following procedure. Start with an initial condition, and iterate the system for a long time—perhaps one thousand time steps. Then iterate for, say, 200 more time steps. The time series plot will bounce up and down, never settling into a periodic process. So we just record on our final-state diagram these 200 points. They will most likely entirely fill up the line segment from 0 to 1. This is illustrated in Fig. 9.23.

[2]Such a proof is a standard part of most junior-level mathematics courses on chaos and dynamical systems. See, e.g., Chapter 15 of Hirsch, Smale, and Devaney (2004).

0   1

Fig. 9.23 The final-state diagram for the logistic equation, $f(x) = rx(1-x)$, for 4.0. The long-term behavior of the orbit is aperiodic.

9.3 Chaos Defined

The orbits of the logistic equation with $r = 4$ are said to be **chaotic**. A dynamical system is chaotic if it possesses all of the following properties:

(1) The dynamical rule is deterministic.

(2) The orbits are aperiodic.

(3) The orbits are bounded.

(4) The dynamical system has *sensitive dependence on initial conditions.*

Let us consider each of these in turn. First, by dynamical rule I mean the rule that determines the orbit of the dynamical system. In this case, the rule is just the function that iterate. A deterministic function is one in which the input determines the output. That is, if you give the function the same input numerous times, the function will always return the same value. The logistic equation, as well as all the other functions we have been working with, is deterministic, as discussed in Section 1.3.

Second, an orbit is aperiodic if it never repeats. The itinerary never retraces its steps; we keep seeing new numbers. The was the case for the logistic equation with $r = 4$. The orbit bounces around forever between zero and 1, and yet it never repeats.

This leads us to the third condition: the orbits must be bounded. This means that the iterates do not fly off to infinity; they stay between an upper limit and a lower limit. For the $r = 4$ logistic equation, these limits are 1 and 0. In contrast, consider for a moment the doubling function, $f(x) = 2x$. Orbits for this equation are not bounded. For example, the orbit of 2 is $2 \longrightarrow 4 \longrightarrow 8 \longrightarrow 16 \longrightarrow 32 \cdots$. This orbit tends toward infinity. This orbit is aperiodic; it clearly will never repeat. But this sort of non-repeating is not very interesting. If an orbit flies off to infinity it is not at all surprising that it does not repeat. In any event, for an orbit to be chaotic it must be bounded. This "fine print" serves to exclude unbounded orbits like those of $f(x) = 2x$ from being considered chaotic.

The fourth, and final, criterion is that the dynamical system display **sensitive dependence on initial conditions** (SDIC). This is a phenomenon that we have not yet encountered and which will be the topic of the next chapter. In brief, though, a system that has SDIC has the property that a very small change in the initial condition will lead to a very large change in the orbit in a relatively short time. sensitive dependence on initial conditions is more colloquially known as **the butterfly effect**.

9.4 Implications of Aperiodic Behavior

Let us pause for a moment and think about what aperiodicity might mean. When we iterate a function we are doing the same thing over and over again. So it is reasonable to expect that the results of repeating this action—i.e., the orbit of the seed—will themselves repeat. However, for the logistic equation this turns out to not be the case. We can keep doing the same thing *forever* and never see the same thing twice. Speaking only somewhat metaphorically, the repetitive sameness of iteration almost paradoxically gives rise to continuous novelty and surprise.

Suppose that instead of iterating a function, you observed something like Fig. 9.21 or 9.22 as the result of an experiment. Perhaps you obtained these measurements by counting rabbits on an island or keeping track of daily stock prices. If you saw such data you might presume that the underlying process governing the phenomena was very complicated. Or, you might assume that there was not any rule governing the situation at all, positing instead that the rabbits or stock returns are a random process—their future behavior is a matter of chance. The people or rabbits are behaving randomly, figuratively tossing coins to determine their next steps, and so the next value in the time series is not determined by the previous value, but instead is determined, at least in part, by chance. Or it could be that the system is being strongly affected by outside noise, such as the weather or other unpredictable external events.

But the study of chaos shows us that a non-repeating and apparently random phenomenon need not be governed by complicated equations, nor is it the case that the system is must be driven by external noise or randomness. Rather, apparent randomness and unpredictability can be generated by a simple, deterministic rule such as the logistic equation.

Note that the phenomenon of aperiodicity and the apparent randomness of the time series do not violate the basic idea of Newtonian determinism, discussed in the previous chapter. The behavior of the orbit in the logistic equation with $r = 4.0$ is most certainly governed by a deterministic rule. What is a surprise, however, is that a simple deterministic rule of this sort can produce such complicated behavior.

For a system to be chaotic in the mathematical sense, it must have sensitive dependence on initial conditions (SDIC), the fourth criterion in the list in the previous section. In the subsequent chapter we will explore SDIC in detail. We will see some visual examples, and will also define SDIC more carefully. Then, in Chapter 11 we will return to our survey of the logistic equation.

Exercises

For Exercises 9.1–9.6 you will need to use a program that can iterate the logistic equation for different values of r and produce a time series plot. This will let you make plots similar to those that are in this chapter. You can find such a program at http://chaos.coa.edu/. Choose the *logistic orbits* option. You could also use a spreadsheet to perform these calculations.

(9.1) Use a calculator to calculate the first four iterates for the seed $x_0 = 0.3$ for the logistic equation with $r = 4.0$. Use the logistic orbits program to do the same thing and verify that they give the same results.

(9.2) Use a calculator to calculate the first four iterates for the seed $x_0 = 0.1$ for the logistic equation with $r = 2.5$. Use the logistic orbits program to do the same thing and verify that they give the same results.

(9.3) Use the logistic orbits program to experiment with different r values. Find at least one additional r

value (besides $r = 4$) that has an aperiodic orbit.

(9.4) Use the logistic orbits program to determine the range of r values for which the long-term behavior of the orbits is periodic with period 2. (Recall that we have seen that $r = 3.2$ is one parameter at which we saw period-2 behavior.)

(9.5) Use the logistic orbits program to determine the long-term behavior of the following r values. For each r value determine as best you can if the orbits are periodic or aperiodic. If they are periodic, state the period:

(a) $r = 2.9$
(b) $r = 3.4$
(c) $r = 3.61$
(d) $r = 3.628$
(e) $r = 3.7$
(f) $r = 3.92$

(9.6) ⋆ Use the logistic orbits program to do the following.

(a) For the logistic equation with $r = 4.0$, make a plot of the first thirty iterates of $x_0 = 0.1$.

(b) Now make a plot of the first thirty iterates of $x_0 = 0.11$.

(c) Do your two time series plot differ significantly? If so, at what iterate does the difference become noticeable?

(d) Now make a plot of the first thirty iterates of $x_0 = 0.1001$ and compare to the time series plot for $x_0 = 0.1$ Do your two time series plot differ significantly? If so, at what iterate does the difference become noticeable?

(9.7) The logistic equation is not used for r values above 4.0. If r is above 4.0 the model of population growth no longer makes sense. To see why, we will consider the logistic equation with $r = 5.0$. This function is plotted in Fig. 9.24.

(a) Choose a few initial conditions and iterate them graphically.

(b) What is the long-term fate of these orbits?

(c) What does this let you conclude about the model for $r = 5.0$? Why does this model not make sense if used to describe the growth of a population?

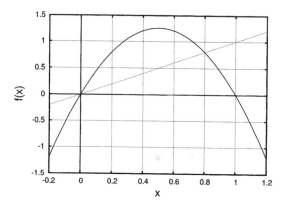

Fig. 9.24 A plot of $f(x) = 5x(1 - x)$, the function for Exercise 9.7.

(9.8) Compute by hand the first several iterates of $x_0 = 0.50$ for the logistic equation with $r = 4.0$. How does the orbit you calculated compare with the orbit shown in Fig. 9.20? What is going on? Explain.

(9.9) Compute by hand the first several iterates of $x_0 = 0.25$ for the logistic equation with $r = 4.0$. How does the orbit you calculated compare with the orbit shown in Fig. 9.20? What is going on? Explain.

The Butterfly Effect

10

In this chapter I discuss in more detail the phenomenon of sensitive dependence on initial conditions (SDIC), the fourth criterion listed in the definition of chaos in the previous chapter. The main idea of SDIC is that small changes in the initial condition can make a large difference in the orbit's behavior. In this chapter we explore the idea of SDIC with considerably more detail and precision.

10.1 Stable Periodic Behavior

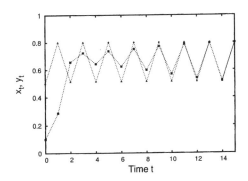

Fig. 10.1 The time series diagram for the logistic equation, $f(x) = rx(1-x)$, for $r = 3.2$. The itineraries of the initial conditions 0.15 and 0.5 are shown. The difference between these two orbits is plotted in Fig. 10.2.

I will begin with an example of a function that does not have SDIC. Let us return to the logistic equation, $f(x) = rx(1-x)$ with $r = 3.2$. We have seen previously that the orbits are periodic for this r value. And I have argued that this behavior is attracting; different orbits get pulled closer to the period-2 orbit. This is illustrated in Fig. 10.1, which shows the time series plots for two different initial conditions, 0.15 and 0.5. I will denote these two initial conditions as x_0 and y_0, and subsequent points in the time series as x_t and y_t. This is fairly standard notation, but could be potentially confusing. The symbol y_0 does not denote a coordinate position on the x-y plane. Rather, it simply denotes a different initial condition. In any event, in Fig. 10.1 we see that the two different orbits are pulled together, and that both are approaching the period-2 attractor.

Figure 10.2 provides a new way to see that orbits are getting pulled toward the attractor. This figure plots the difference between the two time series: $x_t - y_t$. When the two orbits are far apart, this quantity is large. And when the orbits are close, $x_t - y_t$ is close to zero. So, in Fig 10.2 we see that $x_t - y_t$ gets closer to zero as time goes on. The quantity $x_t - y_t$ can be positive or negative, depending on which of the two orbits is larger. However, the difference between the two orbits approaches zero. This is an indication that the orbits are getting closer together, as we saw in Fig. 10.1.

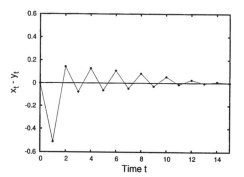

Fig. 10.2 The difference between the two time series plotted in Fig. 10.1.

10.2   Sensitive Dependence on Initial Conditions

Let us repeat this experiment for $r = 4$, where we have seen that the behavior of the orbits is aperiodic. If we start with two nearby initial conditions, will their orbits get drawn closer together? In Fig. 10.3 I have plotted the time series for the two initial conditions $x_0 = 0.4$ and $y_0 = 0.41$. The orbit for the seed $x_0 = 0.4$ is shown with square points connected with a solid line. The orbit of $y_0 = 0.41$ is shown as triangles connected with a dashed line. In the figure we that the two orbits start close together but depart noticeably by $t = 5$. This can also be seen

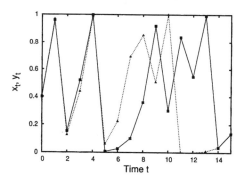

Fig. 10.3 The time series diagram for the logistic equation, $f(x) = rx(1-x)$, for $r = 4.0$. The itineraries of the initial conditions $x_0 = 0.4$ and $y_0 = 0.41$ are shown. The difference between these two orbits is plotted in Fig. 10.4.

in Fig. 10.4, which plots the difference between the two orbits shown in Fig. 10.3. The difference between the two plots is close to zero for the first four or five time steps. But by time $t = 7$, the difference is larger than 0.5.

So the behavior of the two orbits for $r = 4.0$ is the opposite of what we saw for $r = 3.2$. For $r = 3.2$ (Figs. 10.1 and 10.2) orbits are pulled

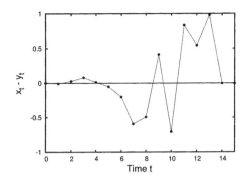

toward an attractor of period 2, whereas for $r = 4.0$ (Figs. 10.3 and 10.4) orbits are pushed apart, and the long-term behavior is aperiodic.

The fact that nearby initial conditions soon end up far apart has some significant implications. Suppose that we are interested in actually using the logistic equation to predict the value of a rabbit population on an island in the years to come. You go to the island, count the rabbits, and determine that the population is 0.41. However, the actual rabbit population is 0.4.[1] Perhaps you accidentally counted a few rabbits twice—rabbits do tend to hop about, and a lot of them look alike. In any event, your measurement error is 2.5%; the value you measure differs from the true value by 2.5%.

This error seems fairly small. However, Fig. 10.4 shows us this error can make a big difference quite quickly. This figure shows a plot of the difference between the actual population, given by the orbit of 0.4, and our predictions using the logistic equation—i.e., the iterates of the seed 0.41. Thus, Fig. 10.4 can be interpreted as a plot of our prediction error: the difference between reality and our model. It would appear that our ability to accurately predict the population is rather limited. After just seven generations our prediction is off by around 0.5, 50% of the maximum number of rabbits.

Being able to predict only seven generations into the future is perhaps a little disappointing. This disappointment is all the more pronounced when we note that in this admittedly artificial example, we assume that we know exactly the dynamics of the situation. That is, in this scenario our model of the population dynamics, $f(x) = 4x(1 - x)$, is exactly the rule that governs the real population. In a more realistic setting, our rule for the population dynamics likely would only be approximate. There also may be external influences not accounted for in the model that will introduce further errors. Here, however, we see prediction errors that are due entirely to our initial measurement errors. It is striking that our ability to predict into the future is so limited, without even taking into account complications such as noise or external influences.

Given that measurement inaccuracy is the root of our trouble, we could return to the island and count the rabbits again, this time taking extra care to count every rabbit once and only once. Let us imagine that we do an excellent job of counting, but that we are not quite perfect. We

[1]Remember that we measure the population as a fraction of the maximum possible number of rabbits. So a population of 0.4 means that the population is at 40% of its maximum, not that there are 0.4 rabbits.

measure 0.400001 while the exact value is still 0.4. This is a tiny error—just 0.00025%. We anticipate that this will improve our predictions. But by how much?

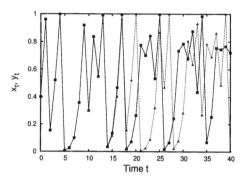

Fig. 10.5 The time series diagram for the logistic equation, $f(x) = rx(1-x)$, for $r = 4.0$. The itineraries of the initial conditions $x_0 = 0.4$ and $y_0 = 0.400001$ are shown. The difference between these two orbits is plotted in Fig. 10.6.

In Fig. 10.5 I have plotted the time series for the initial conditions 0.4 and 0.400001. As in Fig. 10.3, the orbit of the seed $x_0 = 0.4$ is shown with square points connected with a solid line. The orbit of $y_0 = 0.41$ is shown as triangles connected with a dashed line. The two orbits are almost exactly on top of each other for around fifteen generations; the squares hide the orbit plotted with triangles. However, after $t = 18$ the two orbits depart, and the behaviors are quite different. This can be seen more clearly in Fig. 10.6, which shows the difference between the orbits. Recall that we can interpret this difference as our prediction error. Thus, Fig. 10.6 tells us that our predictions will now be reasonably accurate for around eighteen generations.

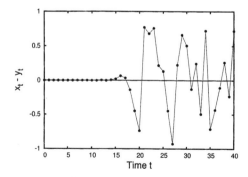

Fig. 10.6 The difference between the two time series plotted in Fig. 10.5.

At first, it might seem like this has been a success. By increasing our measurement accuracy we extended our predictions from six years to eighteen. We have improved our forecasting range by three times. However, in order to do so we had to improve our measurement accuracy by 10,000 times. If we roughly equate measurement accuracy with how hard we would need to work to make the measurement, then what this tells us is that we have increased our work by 10,000 times but only increased our prediction range by three times. Viewed in this light, this is rather disappointing. An analogy may drive this home: imagine working 10, 000 times harder at a job but only earning three times as much money.

Moreover, it is almost impossible to measure anything to the accuracy of 0.00025%, at least using readily available measuring tools. Even with specialized equipment, it is difficult to perform measurements to this degree of accuracy. Doing so would entail, for example, measuring the height of a typical human to within around 0.00004 or 4×10^{-5} meters. This is roughly the size of a large bacterium. Even if we could somehow perform such an accurate measurement, it does not even seem meaningful to speak of height as being specified this exactly. We shrink and expand a little over the course of a day, and a single strand of hair has a thickness of around 0.0001 meters.

The basic lesson is that systems that have SDIC are impossible to accurately predict for anything other than the short-term. Increasing the accuracy with which we the measure the initial condition helps, but eventually the small inevitable small inaccuracies in our initial measurement will cause our prediction to be way off. Thus, systems that are chaotic are, as a practical matter, unpredictable for anything beyond a very short time horizon. Note that this unpredictability occurs despite the fact that the equation governing the orbit is completely deterministic. For this reason, chaotic systems are a **deterministic source of randomness.**[2]

The phenomenon of sensitive dependence on initial conditions is more colloquially know as the **butterfly effect**. The idea is that the weather is a chaotic system, and hence a small disturbance, such as the flapping of a butterfly's wings, could lead to large changes in the path of the orbit. In the context of the weather, this changed orbit could make a difference in the path of a tornado. For a fascinating, short history of the origin of the term "butterfly effect", see (Hilborn, 2004).

In everyday life we are perhaps used to the notion that there are some circumstances in which small perturbations in a system can lead to large changes in its behavior. But this is not necessarily sensitive dependence on initial conditions in the sense it is used by mathematicians and physicists studying chaos. For a system to have SDIC, essentially *every* initial condition[3] must have the property that a small perturbation leads to a large change in the long-term behavior. In a sense, it is as if every iterate is like a marble perched unstably on the top of an upturned bowl. At every step, a small change in the orbit will lead to large changes further down the line.

10.3 SDIC Defined

In the previous section I discussed SDIC at some length and illustrated this phenomenon with a variety of plots. However, this has largely been a qualitative discussion. In this section my goal is to define the notion of sensitive dependence on initial conditions more precisely and quantitatively. I will begin by stating a standard mathematical definition for

[2]In light of SDIC, you may at this point wonder what "randomness" really means. This is discussed in Chapter 14.

[3] Why do I say "essentially every" initial condition instead of "every" initial condition? The reason is that even in a chaotic system with aperiodic orbits, there are still periodic orbits. These orbits are unstable, and so we would not typically observe them, as discussed in Section 7.4. Because of the presence of these unstable periodic orbits, it is not the case that *every* pair of orbits will diverge. However, because these periodic orbits are very rare, it is vanishingly unlikely that we will observe them. A more precise definition of SDIC that avoids the difficulties associated with unstable periodic points is given in the next section.

SDIC; subsequently I will interpret the statement in a more intuitive way.

Let f be a function, and let x_0 and y_0 be two possible initial conditions for f. Then f has *sensitive dependence on initial conditions* if there is some number δ such that for *any* x_0 there is a y_0 that is not more that ϵ away from x_0, where the initial condition y_0 has the property that there is some integer n such that, after n iterates, the orbit of y_0 is more than δ away from the orbit of x_o. That is, $|x_n - y_n| > \delta$.

This is quite an abstract statement. It will become clearer with a concrete example. The basic idea is this. Suppose you choose some error threshold δ. For the sake of concreteness, let us use $\delta = 0.6$. Then you choose some initial condition, perhaps $x_0 = 0.4$. Lastly, you choose an initial error ϵ—let us imagine you choose $\epsilon = 0.05$. Then, you challenge me to do the following: find some point that is no further than ϵ away from x_0 such that its orbit eventually gets δ away from the orbit of x_0. In this example, my challenge would be to find an initial condition y_0 between 0.35 and 0.45 such that its orbit eventually gets 0.6 away from the orbit of $x_0 = 0.4$.

For a parameter value for the logistic equation, such as $r = 4$, it is not difficult to find an initial condition y_0 that meets the criteria. A little bit of experimenting determines that $y_0 = 0.44$ does the trick. This is illustrated in Table 10.1, where I list the orbit for 0.4 and 0.44, and also the distance between these two orbits. One sees that by the fifth iterate the distance between them is larger than 0.6.

The definition for SDIC says that, if a function has SDIC then I will always be able to meet a challenge of this sort. Whatever you choose for x_0, δ, and ϵ, I will be able to find an initial condition y_0 that is not farther than ϵ away from the initial condition x_0, yet nevertheless ends up more than δ away from the orbit of x_0.[4] For a system to be chaotic, it has to mix up and scramble orbits so much that nearby any initial condition there are other initial conditions that get far away from it.

Here is another way to see how a chaotic system pulls nearby orbits apart. In Fig. 10.7 I have plotted 1000 different orbits for the logistic equation with $r = 4.0$. All the initial conditions are initially in a very small interval; they are between 0.395 and 0.405. Upon iterating, the small interval grows. Due to sensitive dependence on initial conditions, the orbits spread out. This continues, and quite soon the orbits, initially confined to a small region 0.01 wide, now can be found all along the full interval. On Fig. 10.7 I have also plotted the time series for the orbit that started exactly at 0.40. One can see that the orbits get mixed up very quickly. The nearby orbits do not stay nearby for long.

In contrast, let us look at what happens for a non-chaotic system. In Fig. 10.8 I have again plotted 1000 orbits, this time for the parameter value $r = 3.2$, for which the logistic equation has an attracting cycle of period 2; all orbits oscillate between roughly 0.80 and 0.51. Initially, the orbits are spread out along the entire interval. But fairly quickly they get pulled toward the attractor, and by the fortieth iterate all orbits are essentially on the period-2 cycle. This dynamical system does not

Table 10.1 The orbit of two nearby initial conditions, $x_0 = 0.400$ and $y_0 = 0.440$, for the logistic equation with $r = 4.0$. The orbits eventually get further than $\delta = 0.6$ apart.

n	x_n	y_n	$|y_n - x_n|$
0	0.400	0.440	0.040
1	0.960	0.986	0.027
2	0.154	0.057	0.097
3	0.520	0.214	0.306
4	0.998	0.673	0.325
5	0.006	0.880	0.874

[4]There is an additional condition placed on δ. The error threshold δ can not be so large than in order for it to be exceeded the orbit would have to leave the region that bounds x. For the logistic equation, x is always between 0 and 1, so δ needs to be less than 1.

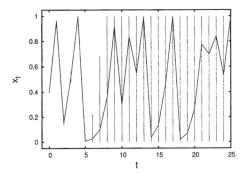

Fig. 10.7 One thousand different orbits for the logistic equation with $r = 4.0$. The 1000 different initial conditions are uniformly spaced from 0.395 to 0.405. I have also included the time series plot for the orbit that starts exactly at 0.400. The orbits are initially bunched together but they very quickly spread apart. By the eighth orbit, the initial conditions, which were originally confined to an interval just 0.01 wide, have expanded to fill the entire interval. (Based on Fig. 9 from Smith (2007).)

show sensitive dependence on initial conditions. Orbits are pulled closer together, rather than further apart.

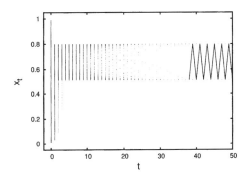

Fig. 10.8 One thousand different orbits for the logistic equation with $r = 3.2$. The 1000 different initial conditions are uniformly spaced from 0 to 1, and subsequently are pulled toward the period-2 attractor. By the fortieth iterate, all orbits are on one of the two period-2 points. I have included the time series plot of just one orbit for $t > 38$ to illustrate the period-2 nature of the long-term behavior. (Based on Fig. 9 from Smith (2007).)

Sensitive dependence on initial conditions is usually not difficult to observe with a computer; one just tries iterating nearby orbits and looks to see if they get pushed apart or pulled together. Rigorously proving that a dynamical system has SDIC, however, can be quite difficult. It has been proven that the logistic equation with $r = 4.0$ has SDIC, but this has not been proven for other chaotic r values. The proof of SDIC for $r = 4.0$ is a fairly standard topic in more advanced dynamical systems courses, but is beyond the scope of this text.

10.4 Lyapunov Exponents

If a dynamical system has SDIC, we know that two orbits will eventually get far apart. However, we do not know how fast they diverge. How long do we have to wait for the two orbits to get far apart? Five iterations? Ten? One hundred? In the definition of SDIC, there is nothing specified about how large n has to be before $|x_n - y_n|$ is larger than our threshold δ. A dynamical system either has SDIC or it does not; it is a binary distinction. In this section I discuss a way of measuring how fast nearby orbits are pulled apart. This will provide us with a measure of the degree to which a system has sensitive dependence on initial conditions.[5]

[5]This section is more technical than most others in this book. It can be skipped if the reader so desires.

Let x_0 and y_0 denote two initial conditions that start off close together. So $|x_0 - y_0|$ is small. We are interested in the absolute value,[6] because we do not care whether x_0 is larger than y_0 or vice versa. All we are interested in is how far apart they are. Let us call this initial separation D_0:

$$D_0 = |x_0 - y_0| .\tag{10.1}$$

The separation after t iterations we will call $D(t)$:

$$D(t) = |x_t - y_t| .\tag{10.2}$$

If a system has SDIC, we expect that $D(t)$ will increase, since the orbits get pushed apart, as in Fig. 10.7. But how fast are they pushed apart? To answer this question we would like to know how $|x_t - y_t|$ changes with t, the number of iteratations.

It turns out that for many systems the behavior of $|x_n - y_n|$ can be well described by an exponential function of the following form:

$$D(t) \approx D_0 2^{\lambda t} ,\tag{10.3}$$

for small t. Here $D(t)$ is the difference between the two orbits, as defined in Eq. (10.1), and λ is a quantity known as the **Lyapunov exponent**.[7] If λ is greater than zero, then the quantity $2^{\lambda t}$ gets larger as t gets larger, and hence the two orbits are being pushed apart. However, if λ is less than zero, then $2^{\lambda t}$ gets smaller as t gets larger. So if $\lambda > 0$, orbits are pushed apart, and the function has sensitive dependence on initial conditions. The larger the Lyapunov exponent λ, the faster the orbits are pulled apart, and the greater the sensitivity on initial conditions.

Equation (10.3) holds only for small t. The reason for this is that the distance between the two orbits cannot grow forever, since the orbits are bounded. Also, note that the relationship in Eq. (10.3) is approximate. On average, orbits get pushed apart according to Eq. (10.3), but for any single orbit it is just an approximation.

As a concrete example, suppose $\lambda = 1$. Then Eq. (10.3) becomes

$$D(t) \approx D_0 2^t .\tag{10.4}$$

In other words, the distance between the two orbits approximately doubles every time step. We would then expect nearby orbits to be pushed apart quite rapidly. In just eight time steps the initial distance between the two orbits would increase by roughly a factor of 256, since $2^8 = 256$.

This phenomenon is illustrated in Fig. 10.9. Here I have plotted the distance between two different orbits for the logistic equation with $r = 4.0$. The initial conditions I used are $x_0 = 0.3$ and $y_0 = 0.300001$. So the initial D is small: 0.000001. As expected, the distance $D(t)$ grows. The logistic equation with $r = 4.0$ has a Lyapunov exponent of $\lambda = 1$. Thus, we expect $D(t)$ to be well described by Eq. (10.4). This is plotted as the dashed line in the figure. We see that this function is indeed a reasonable approximation to $D(t)$, as claimed.

To summarize, the Lyapunov exponent captures the average rate at which the distances between two nearby orbits changes. If the Lyapunov

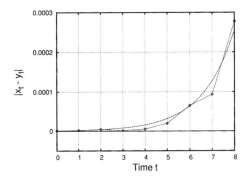

Fig. 10.9 The absolute value of the difference between two time series for $f(x) = rx(1 - x)$ with $r = 4.0$. The dashed curve is $D_0 2^{\lambda t}$ with the Lyapunov exponent $\lambda = 1$.

exponent λ is positive, on average the orbits are pushed apart, and the system has SDIC. The larger λ is, the faster the orbits are pushed apart, and the more unpredictable the orbits are. The Lyapunov exponent is a standard and broadly applicable way of detecting and quantifying sensitive dependence in initial conditions. Lyapunov exponents are used for many different types of dynamical systems—not just the discrete dynamical systems that have been the focus of the book so far.

10.5 Stretching and Folding: Ingredients for Chaos

A chaotic system has sensitive dependence on initial conditions and has orbits that are bounded. Both of these properties arise from geometrical features of the dynamical system. In order for the system to have SDIC, the dynamical system must perform some sort of a *stretch*. This stretch has the effect of pulling apart nearby initial conditions, leading to SDIC. In order for the orbits to stay bounded, however, this stretching cannot occur indefinitely. Thus the dynamical system also needs to perform a *fold* that brings orbits back together so they do not grow without bound.

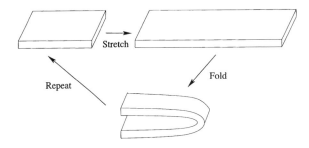

Fig. 10.10 An illustration of stretching and folding. The dough is repeatedly stretched out and folded back upon itself. (Based on Fig. 12.1.2 from Strogatz (2001).)

This stretching and folding can be visualized by picturing the process of kneading dough, as illustrated in Fig. 10.10. One starts with a piece of dough and then stretches it out, perhaps with a rolling pin. After the dough is roughly twice the original length, it is folded back on itself, and one starts the process again. The process is thus iterated: the same procedure is used again and again, and the output from one step is used as the input for the next step.

Imagine two points in the dough that are initially close to each other: perhaps two adjacent specks of cinnamon. During each stretch, the distance between the points gets larger. During the fold, however, the two points might get closer together. This will occur if the two points are on opposite sides of the midpoint of the dough. So the stretching continually pushes the points apart, and the folding brings them closer if they are far apart and on opposite sides of the midpoint. In this way stretching and folding produce chaotic trajectories; the orbit is bounded and has sensitive dependence on initial conditions.

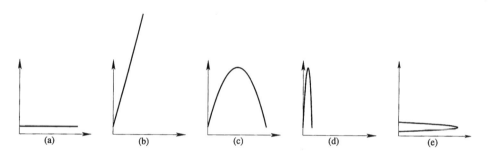

Fig. 10.11 The logistic equation viewed as stretching and folding. Part (a) shows the initial conditions as a horizontal line. This is then stretched (b) and folded, yielding (c). To iterate the process, the output values on the vertical axis are used as input values on the horizontal axis. This is illustrated in parts (d) and (e).

The logistic equation geometrically has the effect of stretching and folding. This is illustrated in Fig. 10.11. The initial conditions—or the dough—are shown in part (a) as the horizontal line. The initial conditions are then stretched (b) and folded, to produce the familiar logistic equation in part (c). This equation is then iterated. Geometrically, this means that the output of the function—the values on the vertical axis— are used as input values, on the horizontal axis, for the next step. This process is illustrated in parts (d) and (e) of Fig. 10.11.

I introduced the logistic equation in Chapter 7 as a generic model of limited population growth and argued that it captured some general features of a population that initially grows exponentially, but which has some feature which prevents the population from growing continually. The logistic equation performs stretching and folding, a property common to all chaotic systems. And so the logistic equation is not just a generic population equation, it is a generic equation for chaos.[8] The stretching and folding illustrated in Fig. 10.10 are the basic geometric ingredients for chaos. All chaotic systems can be viewed as performing these two basic operations.

[8] In fact, we will see in Chapter 12 that some properties of logistic equation's transition to chaos are the same for almost all chaotic systems.

10.6 Chaotic Numerics: The Shadowing Lemma

In this and the previous chapter we have primarily used computers to iterate equations so that we may study their long-term dynamical behaviors. This is a fairly typical situation. Computational work of this sort is very common in the study of dynamical systems. Computers are generally viewed as being trustworthy, at least when it comes to simple things like arithmetic. So reliance on computers for the sort of repetitive arithmetic associated with iterating functions is not cause for concern.

Or is it? As we have seen in this chapter, a chaotic system has sensitive dependence on initial conditions; a small change in the value of an orbit can lead to very large changes down the road. A computer uses a finite amount of memory when it stores a number. This is not an issue when it is trying to store small integers, like 6 or 13 or 1969. Decimals are another matter. It is not hard to store 0.45 or 6.789. But what if the computer tries to record the result of dividing 1 by 11? This number is

$$\frac{1}{11} = 0.0909090909090909090909090909090\ldots. \tag{10.5}$$

The decimal keeps repeating forever. The repeating block is usually indicated as follows:

$$\frac{1}{11} = 0.0\overline{90}. \tag{10.6}$$

But a computer cannot store all of these digits. So it will have to settle for an approximate answer and will round off. In the computer's memory the result of 1 divided by 11 may look like:

$$\frac{1}{11} = 0.09090909090909091. \tag{10.7}$$

Equations (10.5) and (10.7) seem essentially identical. If the numbers in question referred to some physical quantity, they would be completely indistinguishable. However, when iterating a system with sensitive dependence on initial conditions, tiny differences in an initial condition get magnified quickly. The numbers on the right-hand side of Eqs. (10.5) and (10.7) would differ by around 0.1 after around 53 iterations if iterated with the logistic equation with $r = 4$.

Suppose I am doing a numerical experiment in which I want to know about the long-term behavior of the orbit of the initial condition $\frac{1}{11}$. My computer might use the slightly inexact Eq. (10.7) for the value of this number. For the first 50 iterates this inexactness is not a big deal. The orbit the computer calculates is very very close to the true orbit—the orbit that would result if the exact value for $\frac{1}{11}$ was used.

But after 55 iterates or so, the orbit the computer is calculating is no longer a good approximation to the true orbit of $\frac{1}{11}$. So what *is* the computer calculating at this point? It would be reasonable to assume that it is giving us rubbish: a completely fictional orbit. This is true, if what we are interested in is predicting the exact orbit of of $\frac{1}{11}$. By the time we get to, say, the sixtieth iterate, the computer orbit has almost no relation to the true orbit.

Suppose that I want to use results of computer experiments such as this—using a computer to iterate a number of different initial conditions for a function—to conclude that that function has sensitive dependence on initial conditions. This would seem to be highly problematic. At some point, perhaps quite quickly, the round-off errors made by the computer lead to a "fictional" orbit, one unrelated to the chosen initial condition. It seems that this might doom efforts to use computers to explore chaotic systems.

Amazingly, however, all is not lost. It turns out that even though the orbit loses relation to the chosen initial condition, the orbit nevertheless is very close to the orbit for some *other* initial condition. That is, there is some other true orbit that is very well approximated by the computer-generated, fictional orbit. This phenomenon is known as **shadowing**. Even though the computed orbit is not a good approximation of the true orbit for the exact initial condition, nevertheless, the computed orbit shadows—i.e., follows closely—some other exact orbit.

Fig. 10.12 Illustration of the effect of round-off error on a computed orbit. The true, exact orbit is shown as the squares connected with dashed lines. The computed orbit is the circles connected with solid lines. Round-off errors and SDIC lead to the computed orbit diverging from the true orbit. (Based on Fig. 10.45 of Peitgen, Jürgens, and Saupe (1992, p. 578).)

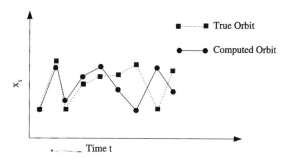

This state of affairs is illustrated in Figs. 10.12 and 10.13. In both figures the orbit calculated by the computer is shown as circular points connected by a solid line. In Fig. 10.12 the dashed line with squares is the true, exact orbit; the solid line with circles is the orbit given to us by the computer. Round-off errors and sensitive dependence on initial conditions cause the computed orbit to depart from the exact orbit.

However, the computed orbit is a good approximation for some other orbit, just not for the initial condition chosen. This is illustrated in Fig. 10.13. There is some exact orbit that is very close to the computed orbit. In the figure, the computed orbit is again shown as circles, while the other, exact orbit is shown as triangles. So the conclusion is that although round-off errors and SDIC lead the computed orbit to depart from the original true orbit, the computed orbit nevertheless approximates some other true orbit. So the computed orbit is not bogus; it still tells us something real about the dynamical system. This result is known as the *shadowing lemma*[9]

This seems like magic, or wishful thinking, or both. It is beyond the scope of this book to prove this result, but there are a few things I can say to help make this seem plausible. First, an analogy. Suppose you

[9]A *lemma* in mathematics is a result that is used as an intermediate step to prove or demonstrate some central or important theorem.

are commissioned to draw a portrait of someone. Unfortunately you are not very good at drawing, and so the portrait is inaccurate. Perhaps the eyes are somewhat the wrong color, the nose is too small, ears too large, and so on. The result is that your portrait does not resemble the true image you were trying to capture. However, the shadowing lemma says that the portrait you created is an accurate representation of someone else, and hence does capture something true about humanity.

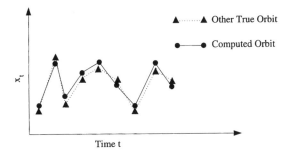

Fig. 10.13 Illustration of shadowing. The computed orbit, shown as circles connected by solid lines, is a close approximation to some other true orbit— the true, exact orbit of some nearby initial condition. This other true orbit is shown as triangles connected by dotted lines. (Based on Fig. 10.45 of Peitgen, Jürgens, and Saupe (1992, p. 578).)

In a chaotic dynamical system the orbits are bounded—in the case of the logistic equation the iterates stay between 0 and 1. So the errors that occur due to round-off cannot be too crazy. The computed orbit still represents a possible orbit, just not the one that corresponds to the exact initial condition you thought you were starting with. Orbits for a chaotic system are aperiodic; they never repeat. Thus, if one watches an aperiodic orbit unfold, eventually a portion of a chaotic itinerary will appear that is very similar to the computed orbit. This is not a proof of the shadowing lemma by any means, but I hope it helps to make it seem a little more reasonable.

In summary, sensitive dependence on initial conditions means that very small changes in an initial condition—even a small change resulting from round-off error inside a computer—will grow and become large. The result is that long-term prediction of a chaotic system is impossible. Even if we could know the *exact* initial condition of a chaotic system, a computer likely would not be able to accurately determine the orbit because small round-off errors in the computer would be magnified by SDIC. However, loosely speaking, the aperiodicity of the system ensures that the computed orbit still corresponds to a true orbit of the system. After all, in an aperiodic system almost any orbit is possible. We will return to these ideas in Chapter 13, when we look in more detail at regularities in chaotic behavior.

Exercises

For Exercises 10.5–10.8 you will need a program that can compute orbits of the logistic equation. You can use the program at http://chaos.coa.edu/; you can also use a spreadsheet to perform the calculation.

(10.1) In Table 10.2 are shown the first eight iterates for two different initial conditions. The function is the logistic equation with $r = 3.8$. Make a plot of the difference between x_t and y_t, as was done in Fig. 10.4.

Table 10.2 Orbits for two different initial conditions. Data for Exercise 10.1.

t	x_t	y_t
0	0.60	0.61
1	0.91	0.90
2	0.30	0.33
3	0.81	0.84
4	0.59	0.51
5	0.92	0.95
6	0.29	0.18
7	0.79	0.56
8	0.63	0.93

(10.2) In Table 10.3 are shown the first ten iterates for two different initial conditions. The function is the logistic equation with $r = 2.5$. Make a plot of the difference between x_t and y_t, as was done in Fig. 10.4. How would you describe the fate of the orbits? Do you think this function shows SDIC?

(10.3) In this exercise and the next we will examine the function $f(x) = 2x$. Iterating this function yields a dynamical system that is not chaotic, since the orbits are not bounded. However, the system does possess SDIC. In this exercise you will consider a few particular cases and use the definition in the second paragraph of Section 10.3.

  (a) Let $x_0 = 2.0$. Compute the first ten iterates of x_0.

  (b) Let $x_0 = 2.0$, $\epsilon = 1.0$, and $\delta = 4.0$. Find an initial condition y_0 that is within ϵ of x_0 and which has the property that eventually its orbit is a distance δ away from the orbit of x_0.

(c) Let $x_0 = 2.0$, $\epsilon = 0.50$, and $\delta = 2.0$. Find an initial condition y_0 that is within ϵ of x_0 and which has the property that eventually its orbit is a distance δ away from the orbit of x_0.

(d) Let $x_0 = 2.0$, $\epsilon = .1$, and $\delta = 1.0$. Find an initial condition y_0 that is within ϵ of x_0 and which has the property that eventually its orbit is a distance δ away from the orbit of x_0.

Table 10.3 Orbits for two different initial conditions. Data for Exercise 10.2.

t	x_t	y_t
0	0.500	0.700
1	0.625	0.525
2	0.586	0.623
3	0.607	0.587
4	0.597	0.606
5	0.602	0.597
6	0.599	0.601
7	0.600	0.599

(10.4) ♯ This is a continuation of Exercise 10.3. Again consider the function $f(x) = 2x$.

  (a) Suppose we have two different initial conditions, x_0 and y_0. Show that after one iteration, the difference between these two initial conditions has doubled. In other words, show that:

$$x_1 - y_1 = 2(x_0 - y_0) . \qquad (10.8)$$

  (b) Use this result to argue that the function has SDIC.

(10.5) Consider the logistic equation with $r = 4.0$. Let $x_0 = 0.2$, $\epsilon = 0.1$, and $\delta = 0.5$.

  (a) Find an initial condition y_0 that is within ϵ of x_0 and which has the property that eventually its orbit is a distance δ away from the orbit of x_0.

  (b) Repeat part (a), but let $\epsilon = 0.01$.

  (c) Repeat part (a), but let $\epsilon = 0.001$.

(10.6) Consider the logistic equation with $r = 4.0$. Let $x_0 = 0.2$, $\epsilon = 0.1$, and $\delta = 0.2$.

(a) Find an initial condition y_0 that is within ϵ of x_0 and which has the property that eventually its orbit is a distance δ away from the orbit of x_0.

(b) Repeat part (a), but let $\delta = 0.4$.

(c) Repeat part (a), but let $\delta = 0.6$.

(10.7) Consider the logistic equation with $r = 3.7$. Let $x_0 = 0.2$, $\epsilon = 0.1$, and $\delta = 0.5$.

(a) Find an initial condition y_0 that is within ϵ of x_0 and which has the property that eventually its orbit is a distance δ away from the orbit of x_0.

(b) Repeat part (a), but let $\epsilon = 0.01$.

(c) Repeat part (a), but let $\epsilon = 0.001$.

(10.8) Consider the logistic equation with $r = 2.8$. Let $x_0 = 0.2$, $\epsilon = 0.1$, and $\delta = 0.5$. Find an initial condition y_0 that is within ϵ of x_0 and which has the property that eventually its orbit is a distance δ away from the orbit of x_0.

(10.9) Suppose a dynamical system has a Lyapunov exponent of 1. Two initial conditions are 0.005 apart. Approximately how far apart would you expect them to be after two iterations? How far apart would you expect them to be after six iterations?

(10.10) Suppose a dynamical system has a Lyapunov exponent of 0.7. Two initial conditions are 0.05 apart. Approximately how far apart would you expect them to be after two iterations? How far apart would you expect them to be after six iterations?

(10.11) Suppose a dynamical system has a Lyapunov exponent of -0.5. Two initial conditions are 0.1 apart. Approximately how far apart would you expect them to be after two iterations? How far apart would you expect them to be after six iterations?

(10.12) ♯ Suppose you are trying to make predictions for a population whose dynamics are described by a function with a Lyapunov exponent of 1. You can only measure the initial value of the population with an accuracy of 0.001. I.e., if you measure an initial population of 0.8, the true population could be as large as 0.8001 or as small as 0.799. Your predictions are only useful to you if the error is less than 0.1. For how many iterates will your prediction be useful? How does you answer change if the Lyapunov exponent is 0.5? How does your answer change if the Lyapunov exponent is -0.25?

The Bifurcation Diagram

In this chapter we return to our study of the logistic equation,

$$f(x) = rx(1-x) \,. \tag{11.1}$$

In Chapter 9 we saw that the long-term behavior of iterates of this equation varied considerably as we changed r. For different r values we found stable fixed points, periodic behavior of periods 2, 3, 4, and 8, and aperiodic behavior. What other behaviors do orbits of the logistic equation exhibit? How do these behaviors change with r, and is there a way to visualize this? These questions will be answered below.

11.1   A Collection of Final-State Diagrams

Our starting point is the final-state diagrams introduced in Chapter 9. Recall that a final-state diagram is, for a given r, the final value(s) of a typical orbit indicated with dots on a number line. For example, if the orbit is periodic with period 2, the final-state diagram consists of two dots, one for each x value in the periodic orbit.

The final-state diagrams of Chapter 9 are collected in Fig. 11.1. Note that to each diagram I have added a label on the left indicating the r value. Next, take Fig. 11.1 and turn it sideways. We rotate it 90 degrees counter-clockwise and plot the r value on the horizontal axis. On the vertical axis we plot the final states.

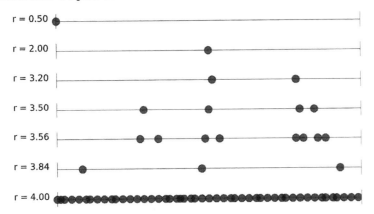

Fig. 11.1 Final-state diagrams for the logistic equation, $f(x) = rx(1-x)$, for various r values.

This is illustrated in Fig. 11.2. Note that I have plotted the r values to scale. That is, unlike in Fig. 11.1, I have put the proper distance between the different final-state diagrams according to the r value. For example, there is very little distance between the $r = 3.5$ and $r = 3.56$ diagrams, and there is a lot of distance on the horizontal axis between $r = 0.5$ and $r = 2.0$.

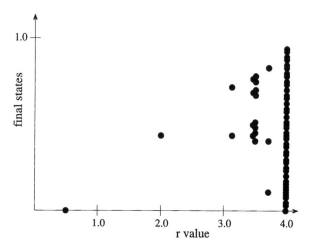

Fig. 11.2 Final-state diagrams for the logistic equation, $f(x) = rx(1-x)$, for various r values. The r values are plotted on the horizontal axis. This type of plot is known as a bifurcation diagram. Compare to Fig. 11.1.

Figure 11.2 hints that there might be a pattern or relationship between the final behaviors and the r value of the logistic equation. In order to see if there is a pattern, we will need more data points—lots more. We will also need to make the individual points smaller, so that they do not overlap each other. I will use a computer to generate many different final-state diagrams for different r values and then make a plot of these final-state diagrams using small points. The result of doing this is shown in Fig. 11.3. This is the same as the previous figure, Fig. 11.2, except in this figure there are many more r values plotted. However, you can see that the handful of r values plotted in Fig. 11.2 can all be found in Fig. 11.3.

At $r = 1$ there is a sudden change in behavior—the rabbits now have a stable, non-zero population, whereas for $r < 1$ the rabbit population died out. This sudden change is an example of a **bifurcation**. A bifurcation is defined as an abrupt, qualitative change in behavior as a system parameter is varied continuously. Another qualitative change in behavior occurs at $r = 3.0$, where the stable population goes from period 1 to period 2. The diagram in Fig. 11.3 is known as a **bifurcation diagram**[1] because it provides a clear way to see bifurcations. The bifurcation diagram is a way of summarizing in one single picture all the possible stable, long-term behaviors of the system.

[1] Some authors use the term **orbit diagram** or **final-state diagram** instead of bifurcation diagram.

Let us continue our examination of Fig. 11.3. We can see that if r is between 3.0 and around 3.4, then the final state is period 2. This is evidenced by the fact that there are two branches on the bifurcation diagram for these values. Accordingly, we found that when $r = 3.2$ the long-term behavior of the orbit is to get pulled to an attracting cycle of period 2. And when $r = 4.0$ we see a solid vertical black line. This corresponds to the aperiodic, chaotic behavior that we explored at some length in Chapter 9.

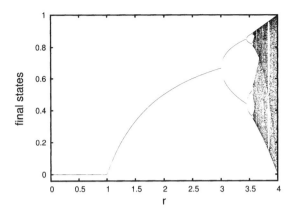

Fig. 11.3 The bifurcation diagram for the logistic equation, $f(x) = rx(1-x)$. The r values are plotted on the horizontal axis. Compare to Fig. 11.2.

The bifurcation diagram of Fig. 11.3 summarizes the behavior of the orbits for *all* r values. We can read off the behavior almost as we can read from a page in a dictionary. To determine the behavior of the orbit for a given r value, first locate that r value on the horizontal axis. Then, draw a vertical line straight up from that r value. If the line you just drew goes through a solid black region of points, then this is an indication that the function is aperiodic for that r value. If the line crosses well-defined "pitchfork tines", then the number of tines the line crosses gives the periodicity of the orbit at that r value.

There appears to be some interesting structure between $r = 3.0$ and $r = 4.0$ in Fig. 11.3, but it is difficult to see because this is a small region of the plot. To get a better view of what is going on in this region, in Fig. 11.4 I have plotted the bifurcation diagram for $3.0 < r < 4.0$ and used a much higher resolution. One can see a remarkably intricate pattern. Note that there appear to be many regions of chaos—these appear as solid or nearly solid vertical regions in the bifurcation diagram. Also, note that the the behavior of the orbits changes frequently as r is increased. In particular, there are chaotic regions that suddenly give way to ordered, period regions. This is another type of bifurcation.

Another striking feature of the bifurcation diagram is the repeated "sideways pitchfork" motif. Each branching corresponds to a doubling of the periodicity. For example, at around $r = 3.45$ there is a branching (or bifurcation) from two to four. This indicates that the orbits change from period 2 to period 4 at this r value. Period 4 then doubles to period 8, and then to 16, and so on. Eventually, at around 3.57, the orbits become chaotic; there are solid vertical lines. We will explore this transition to chaos in more detail in the next chapter. But the chaos does not continue uninterrupted as we increase r. Instead, there are various **periodic windows** that emerge. For example, at around 3.83, a window of period 3 opens up. But every time a periodic window emerges, those periods double, then double again, and then again, and eventually the orbits become chaotic.

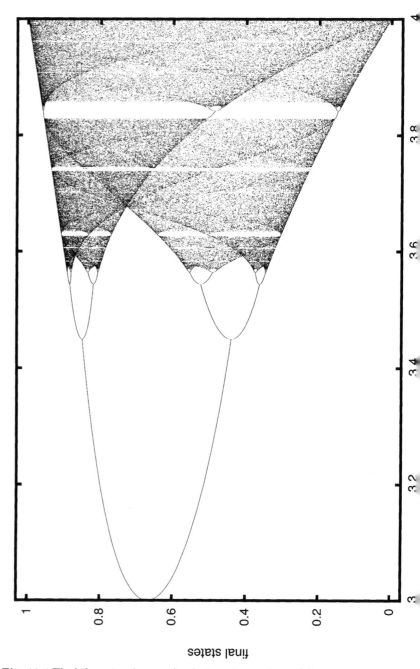

Fig. 11.4 The bifurcation diagram for the logistic equation, $f(x) = rx(1-x)$. This is the same as Fig. 11.3, except the r range is only from 3.0 to 4.0 and the resolution is higher.

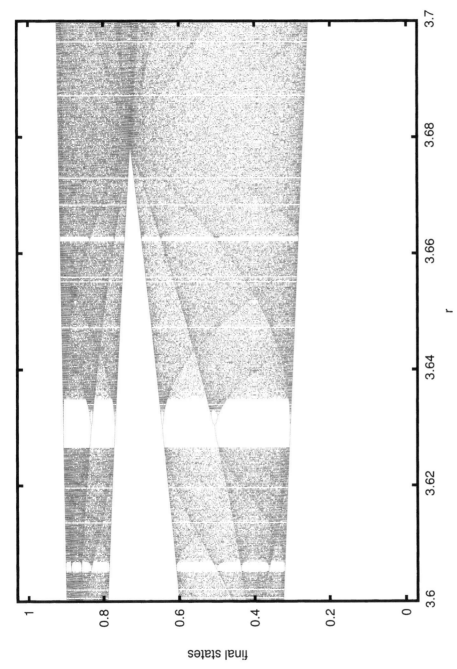

Fig. 11.5 The bifurcation diagram for the logistic equation, $f(x) = rx(1 - x)$, for $r = 3.6$ to 3.7.

To get a better look at this successive period doubling, in Fig. 11.5 I have plotted another close-up of the bifurcation diagram. Again, one sees a remarkably intricate structure. There is a periodic window around $r = 3.63$; the period here is 6. This doubles to period 12, and then 24, and eventually the orbits again are chaotic. There are other, narrower periodic windows interspersed throughout the bifurcation diagram.

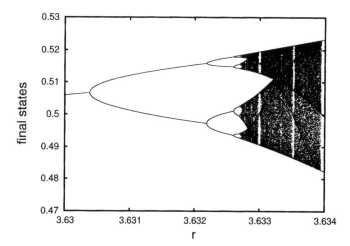

Fig. 11.6 The bifurcation diagram for the logistic equation, $f(x) = rx(1-x)$, for $r = 3.63$ to 3.634. Note the small scale on the vertical axis. This figure looks like a replica of the portion of the bifurcation diagram shown in Fig. 11.4.

Finally, in Fig. 11.6 I have plotted the bifurcation diagram for $r = 3.63$ to 3.634. Note that the r range is very small. I have also shown only the x values from 0.47 to 0.53. We see that the figure looks like a replica of the portion of the bifurcation diagram shown in Fig. 11.4.

11.2   Periodic Windows

As noted above, throughout the bifurcation diagram periodic windows appear. For example, in Fig. 11.5 there is a periodic window of period 6 around $r = 3.64$, and in Fig. 11.4 we can see a window of period three near $r = 3.83$. There are many other periodic windows in the bifurcation diagram. In fact, it can be shown that there is at least one periodic window in any interval of r values that contains a chaotic r value. In other words, there are periodic windows lurking in any chaotic region of bifurcation diagram. There are an infinite number of such periodic windows. Most of them are extremely narrow—too slender to be seen on a bifurcation diagram.

Given that there are so many periodic windows, it is natural to ask if there is any rhyme or reason to the order in which they appear. It turns out that there is a remarkable pattern to the sequencing of the periodic windows in the bifurcation diagram of the logistic equation. There is a way of ordering the integers known as **Sharkovsky**[2] **ordering**, which is as follows:

$$3, 5, 7, 9, \ldots \qquad \text{(odd numbers)}$$

[2] Oleksandr Sharkovsky is a Ukrainian mathematician who has done important work in dynamical systems. There seems to be little agreement on how to spell his name in English; I have seen Charkovsky, Sharkovskii, Sharkovsky, and Sarkovskii. However, on his personal webpage he spells it Sharkovsky.

$2\times3,\,2\times5,\,2\times7,\,2\times9,\,\ldots$ (2 times odd numbers)

$4\times3,\,4\times5,\,4\times7,\,4\times9,\,\ldots$ (4 times odd numbers)

$8\times3,\,8\times5,\,8\times7,\,8\times9,\,\ldots$ (8 times odd numbers)

$$\vdots$$

$\ldots 2^4,\,2^3,\,2^2,\,2,\,1,\,\ldots$ (decreasing powers of 2) .   (11.2)

In Sharkovsky ordering, 3 is the first number, then 5, then 7. After going through all the odd numbers, one then goes through all the odd numbers multiplied by two: 6, 10, 14, and so on. Then the odd numbers multiplied by 4, then 8, then 16. Finally, one counts down from the powers of 2. The largest number in Sharkovsky ordering is $2^0 = 1$. As an example, the following numbers have been placed in Sharkovsky ordering:

$$5,\,37,\,6,\,10,\,22,\,40,\,88,\,48,\,16,\,4,\,1 \,. \qquad (11.3)$$

What does Sharkovsky ordering have to do with the logistic equation? The first appearances of periodic regions or windows in the bifurcation diagram occur in reverse Sharkovsky order. The first period we see as we increase r is 1. This doubles to 2, then 4, 8, 16, and so on. This is exactly Sharkovsky ordering in reverse—i.e., the last line of Eq. (11.2) read backwards.

The bifurcation diagram then shows a transition to chaos around 3.57, followed by chaotic regions punctuated with periodic windows. The order of the first occurrence of window of a certain period is given by the reverse Sharkovsky ordering. For example, suppose as one moves from left to right in the bifurcation diagram one encounters the first instance of a period 14 window. As we left this window and moved further to the right, we would encounter more periodic windows—periods we had already encountered for smaller r values. Eventually we would find a window with a period we had not yet seen. This period would be 10, since this is the next number in reverse Sharkovsky ordering.

The last number in reverse Sharkovsky ordering is 3. Accordingly, moving left to right, the period-3 window is the last new window to appear. I.e., to the left of the period-3 window one could find other periodic windows of all periods except for period 3.

11.3   Bifurcation Diagram Summary

The bifurcation diagram lets us see—all at once—all the different behaviors exhibited by a dynamical system as we vary a parameter. On a bifurcation diagram one can see how these behaviors change as we change the parameter. For the logistic equation we found that there is a particular pattern to these changes. Namely, periods double successively and explode into chaos. Within chaos, periodic windows suddenly emerge. These periods then double successively and again burst into chaos.

The bifurcation diagram for the logistic equation is an object of remarkable complexity. One sees more and more structure as one zooms in. This shows that the logistic equation is capable of a stunning diversity of behaviors. There are chaotic regions and periodic cycles of all possible periodicities. The behavior changes suddenly as the r value is changed. But there is order to how these changes occur—the period doubling occurs at regular intervals, and thus we see the pitchfork motif repeated again and again. It is remarkable that we can get all this from the logistic equation:

$$f(x) = rx(1 - x) \ . \tag{11.4}$$

Iterating this simple quadratic equation produced the data used to make all the figures in this chapter.

In the next chapter we shall look at bifurcation diagrams for other functions. Remarkably, we will find that there are some features of the bifurcation diagram that are the same for broad classes of functions. Even more remarkably, we shall see that these similarities lead to predictions about the transition to chaos in real, physical systems—predictions that have been experimentally verified.

Exercises

For many of these exercises you will need to use a program to make bifurcation diagrams. You will also need to plot the orbits for two different initial conditions to check for sensitive dependence on initial conditions. You can find such programs at http://chaos.coa.edu.

(11.1) By experimenting with the bifurcation diagram program, find r values that yield orbits with the following properties. Once you have found the r value, check that it is behaving as you expect by using the orbit program. There are many possible answers to these exercises. Briefly summarize your findings.

  (a) Period 4

  (b) Period 6 (Hint: Look near period 3.)

  (c) Chaotic behavior for some r not equal to 4. (There are many possible r values to choose from.)

  (d) Period 5 (Hint: Look between 3.7 and 3.8.)

  (e) Periodic behavior of some other period that is not a multiple of 2. (Be sure to state what the period is you have found.)

(11.2) For each r value, do the following.

• Determine the long-term behavior of the orbits. Are the orbits periodic (what period?) or chaotic?

• Does the equation show sensitive dependence on initial conditions? Sketch or print out any graphs you use to draw your conclusions.

  (a) 3.7

  (b) 3.835

  (c) 3.5699456718695445 (do not round off).

(11.3) ⋆ In this exercise you will investigate in more detail the period-doubling route to chaos. This is a preview of what we will do in the next chapter.

  (a) The bifurcation diagram for the logistic equation, shown in Fig. 11.3, shows us that the behavior of the orbits shifts from period 1 to period 2 at $r = 3.0$. Let us call this r value r_1, since it is the value at which the first bifurcation occurs. By zooming in on a bifurcation diagram, locate the r values at which subsequent bifurcations occur. Try to determine these r values to at least a handful of decimal places. If you encounter any difficulties obtaining this accuracy, describe them in your write-up.

(i) Find the r value at which the orbits shift from period 2 to period 4. Call this value r_2.

(ii) Find the r value at which the orbits shift from period 4 to period 8. Call this value r_3.

(iii) Find the r value at which the orbits shift from period 8 to period 16. Call this value r_4.

(b) We are now interested in the ratios between the r values you found above. Define the ratio δ_1 as follows:

$$\delta_1 = \frac{r_2 - r_1}{r_3 - r_2}. \qquad (11.5)$$

Determine the value of δ_1 using the r values you found above.

(c) Now determine the value of δ_2, where

$$\delta_2 = \frac{r_1 - r_2}{r_4 - r_3}. \qquad (11.6)$$

These ratios, δ_1, δ_2, and so on, will play an important role in the next chapter.

(11.4) Place the following numbers in Sharkovsky order:

(a) $1, 2, 3, 4, 5, 6, 7, 8, 9, 10$

(b) $10, 20, 30, 40, 50$

(c) $2, 4, 6, 8, 10, 12$

(d) $125, 126, 127, 128, 129$

Universality

12

In the last three chapters we have seen chaotic behavior: aperiodic, bounded orbits that have sensitive dependence on initial conditions and are generated by a deterministic equation. We have focused exclusively on the logistic equation thus far. It is natural to ask, however, if chaotic behavior can be seen in other functions. Is there something special about the logistic equation, or is chaos a common dynamical behavior? To address this question I will begin by investigating the orbits of a few different functions, and we will see additional examples of chaotic behavior. Looking at these other functions' bifurcation diagrams we will notice some surprising similarities to the logistic equation's bifurcation diagram we investigated at length in the previous chapter. This will then lead to the remarkable phenomenon of universality: there are properties of a type of transition from order to chaos that are the same across large classes of mathematical and—remarkably—physical systems.

12.1 Bifurcation Diagrams for Other Functions

Let us start by considering iterates of the following function:

$$f(x) = rx^2(1-x) \, . \tag{12.1}$$

This looks like the logistic equation, but there is a difference. Note that the function has an x^2 after the r and not an x, as the logistic equation does. If I expand Eq. (12.1), it becomes:

$$f(x) = rx^3 - rx^2 \, . \tag{12.2}$$

We thus see that it is a cubic equation, whereas the logistic equation is a quadratic equation. I will thus refer to Eq. (12.1) as the cubic equation.[1] A plot of this equation is shown in Fig. 12.1 for $r = 5.5$. Note that it is similar to the logistic equation, but the cubic equation is a little lopsided. Its maximum occurs at $x = 2/3$, while the maximum for the logistic equation occurs at $x = 1/2$. Varying r has the effect of increasing or decreasing the height of the function, while its overall shape remains the same.

Let us try iterating the cubic equation. Figure 12.2 shows the time series plot for the initial condition $x_0 = 0.4$ for the cubic equation with $r = 6.0$. The plot appears aperiodic. The iterates stay between approximately 0.5 and 0.9, but within those bounds they appear to bounce

[1] This name is perhaps a little misleading, since there are many different cubic functions, among which Eq. (12.2) is just one possibility. However, it is the only cubic function in this section, so I can refer to it as *the* cubic equation without ambiguity.

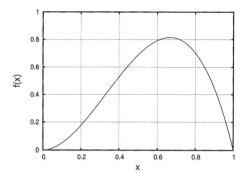

Fig. 12.1 A plot of the cubic function, $f(x) = rx^2(1 - x)$, with $r = 5.5$.

around like we have seen for many parameter values for the logistic equation in the previous chapters. We could try two initial conditions that begin close to each other and plot them both to test for sensitive dependence on initial conditions (SDIC). I have not shown this plot, but the result is not surprising—the orbits do indeed show SDIC. We have thus found chaotic behavior in the cubic equation with $r = 6.0$.

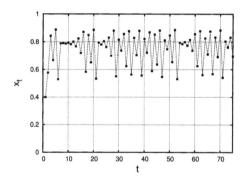

Fig. 12.2 The time series for the cubic function, $f(x) = rx^2(1 - x)$, with $r = 6.0$. The initial condition is $x_0 = 0.4$.

We could try other r values for the cubic equation, testing each one for chaos. But we now have at our disposal a way to visualize the range of behavior of a function all at once: the bifurcation diagram, which was the topic of the previous chapter. Recall that the bifurcation diagram plots on the horizontal axis the parameter value, while on the vertical axis are plotted the final states of the function when iterated.

The bifurcation diagram for the cubic equation, Eq. (12.1), is shown in Fig. 12.3. This result certainly looks familiar—it is strikingly similar to the bifurcation diagram for the logistic equation (see, e.g., Fig. 11.2 in Chapter 11). In fact, you may at first think there has been some error and that I have accidentally used a logistic equation bifurcation diagram instead of the cubic equation bifurcation diagram. There has been no mistake. The two bifurcation diagrams really are that similar.

The bifurcation diagrams are not identical, however, as you will notice if you look closely at the two plots. For the cubic equation, the bifurcation from period 1 to period 2 occurs at around $r = 4.75$. For the logistic equation, this transition occurs at $r = 3.0$. And for the logistic

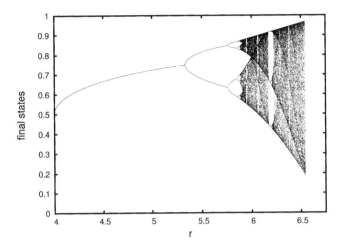

[2]For r greater than 6.75 iterates of the cubic equation are no longer bounded; they tend toward positive or negative infinity, in the same way that orbits of the logistic equation are unbounded if r exceeds 4.0.

Fig. 12.3 The bifurcation diagram for the cubic function, $f(x) = rx^2(1 - x)$.

equation maximal chaos occurs at $r = 4.0$, where the orbits range from 0 to 1. For the cubic equation, maximal chaos occurs at $r = 6.75$, and the orbits range from 0.2 to 1.0.[2] Nevertheless, the similarities between the two bifurcation diagrams are striking. Both have the same general shape, and both show period-doubling bifurcations, as evidenced by the pitchfork shapes that successively split as one moves from left to right on the diagram.

Perhaps this is all a coincidence. After all, the cubic equation is just the logistic equation with one extra x. Let us try another function and see what its bifurcation diagram looks like. We will consider the sine function:

$$f(x) = r \sin\left(\frac{\pi x}{2}\right). \tag{12.3}$$

This function is plotted in Fig. 12.4. Note that input values for this function range from 0 to 2. For this example you do not need to be familiar with sine functions; it suffices to know that the sine function of Eq. (12.3) looks like Fig. 12.4 when plotted.[3]

[3]You might be wondering about the π in the equation. In Eq. (12.3) I am measuring the input x in radians. In radians, the function $\sin(x)$ does a complete up-and-down cycle as x goes from 0 to 2π. I want just the first cycle of the sine function, and I want the cycle to start at 0 and end at 2. Multiplying the argument of the sine function by $\frac{\pi}{2}$ does the trick.

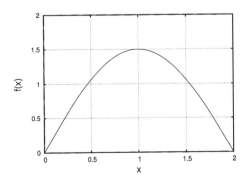

Fig. 12.4 A plot of the sine function, $f(x) = r \sin(\frac{\pi x}{2})$, with $r = 1.5$.

Let us look at the bifurcation diagram for the sine function, Eq. (12.3). Such a plot is shown in Fig. 12.5. Again, we see a striking similarity to the bifurcation diagram for the logistic equation. There are successive period doublings as r is increased. For larger r values there are regions of chaos intermingled with other periodic regions. One can see a period-3 window around 1.85. The sine equation's bifurcation diagram is not identical to the bifurcation diagram for the logistic equation, but the similarities are readily apparent.

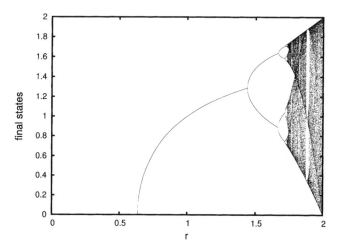

Fig. 12.5 The bifurcation diagram for the sine function, $f(x) = r\sin(\frac{\pi x}{2})$.

I began this chapter by asking whether or not chaotic behavior can be found in other iterated functions. The bifurcation diagrams for the cubic and sine functions, Figs. 12.3 and 12.5 show that the answer to this question is "yes". Both show multiple regions of chaos. That these three functions—the sine, cubic, and logistic equations—behave similarly is perhaps is not that surprising, since the functions themselves are similar; all three equations have a single peak. What is a bit of a surprise, however, is the similarity between the three bifurcation diagrams. In the next section we will focus on a structural feature common to these bifurcation diagrams: the period-doubling route to chaos.

12.2   Universality of Period Doubling

As we have seen, as r is increased for the logistic equation, orbits go from period 1 to 2, then 2 to 4, 4 to 8, and so on. This can be seen in the repeated branchings in the bifurcation diagram. The same motif is apparent in the bifurcation diagrams for the cubic and sine functions. Eventually, as r gets larger, the period doublings give way to chaos. This is referred to as the **period-doubling route to chaos**. Let us look at this process more closely. Figure 12.6 shows a region of the bifurcation diagram for the logistic equation. Of interest are the durations of each of the periodic regimes. Geometrically, this corresponds to the lengths of each of the pitchforks on the bifurcation diagram.

We begin by noting the r values at which the bifurcations occur. The first period-doubling bifurcation occurs at $r = 3$. Here we can see the

Table 12.1 The parameter values at which the first several bifurcations occur in the period-doubling route to chaos in the logistic equation.

Bifurcation	r value
$1 \to 2$	$r_1 = 3.0000$
$2 \to 4$	$r_2 = 3.4500$
$4 \to 8$	$r_3 = 3.5440$
$8 \to 16$	$r_4 = 3.5644$

period shifts from 1 to 2. We will call the r value at which this occurs r_1. The next bifurcation, from period 2 to 4 occurs $r_2 \approx 3.45$. These values, along with the r values for the next two bifurcations, are shown in Table 12.1. Except for r_1, it is impossible to determine accurate values for the bifurcation points just by looking at Fig. 12.6; to get these values I generated additional data and zoomed in on the plot.

I will use Δ to refer to the length of a periodic region.[4] The first region, which is of period 2, I will call Δ_1. The value of Δ_1 can be determined as follows:

[4] Δ is the capital Greek letter "delta".

$$\Delta_1 = r_2 - r_1 \approx 3.45 - 3.00 = 0.45 . \tag{12.4}$$

The quantity Δ_1 is illustrated in Fig. 12.6; it is the length of the first pitchfork.

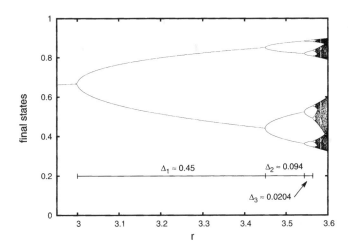

Fig. 12.6 The bifurcation diagram for the logistic equation. Also shown are the quantities Δ_1, Δ_2, and Δ_3, used in the calculation of Feigenbaum's universal number δ.

Let us now consider Δ_2, the length of the period-4 region. Consulting Table 12.1, we find

$$\Delta_2 = r_3 - r_2 \approx 3.544 - 3.45 = 0.094 . \tag{12.5}$$

Similarly, the next Δ is:

$$\Delta_3 = r_4 - r_3 \approx 3.5644 - 3.544 = 0.0204 . \tag{12.6}$$

We can keep zooming in on the bifurcation diagram, and we will keep seeing more and more period-doubling bifurcations. The regions of the higher periods—period 16, 32, 64, and so on—are smaller and smaller. We can see this in Fig. 12.6. The width of the periodic regions gets smaller as the period increases.

We can capture this by looking at the ratio of successive Δ_n's. I will call this ratio δ_n.[5] The quantity δ_n is defined as follows:

[5] δ is the lowercase Greek letter, "delta".

$$\delta_n = \frac{\Delta_n}{\Delta_{n+1}} . \tag{12.7}$$

In words, δ_n is the length of the nth periodic region divided by the length of the next periodic region. Thus, for $n = 1$, we have

$$\delta_1 = \frac{\Delta_1}{\Delta_2} \approx \frac{0.45}{0.094} \approx 4.787 . \tag{12.8}$$

So the length of the first pitchfork is 4.787 times the length of the second pitchfork. Similarly, δ_2 is given by

$$\delta_2 = \frac{\Delta_2}{\Delta_3} \approx \frac{0.094}{0.0204} \approx 4.608 . \tag{12.9}$$

We can keep zooming in on the bifurcation diagram and figure out more and more Δ's, and thus more and more δ's. If we do so, we find that the δ's—the length of a pitchfork divided by the length of the next smallest pitchfork, approaches a fixed value. It turns out that:

$$\delta_n \to 4.66920160... \text{ as } n \text{ gets large} . \tag{12.10}$$

This number is now known as <u>Feigenbaum's constant</u>, after the mathematical physicist Mitchell Feigenbaum, who discovered the number in the late 1970s.[6] What this means is that as we zoom into the bifurcation diagram in the period-doubling region, every successive pitchfork is approximately $\frac{1}{4.67}$ times smaller than the one that preceded it. .

The ratio $\delta = 4.669...$ turns out to be not just a property of the logistic equation. If one were to carry out a similar analysis for the cubic or the sine equation, one would find the same number. The details of the three bifurcation diagrams are not the same, but what *is* the same is the ratio of successive pitchfork lengths as the periods get higher and higher. In fact δ is the same for almost any function that has a period-doubling route to chaos. Recall that all three functions, the logistic, cubic, and sine, have a single maximum. Any function of this sort, so long as the maximum is second-order, will have a δ given by Eq. (12.10). A second-order maximum means that the maximum must appear locally like a small upside down parabola—i.e., it looks like a rescaled and inverted version of $f(x) = x^2$. A plot of an upside-down parabola is shown in Fig. 12.7.

Almost any function that has a maximum will look parabola-like if one looks closely enough. And so almost any function whose bifurcation diagram shows a period-doubling route to chaos will have the same δ. Thus, there is not only something *qualitatively* similar about the three bifurcation diagrams shown in this chapter, there is something *quantitatively* similar: namely, the ratio δ. There are mathematical constraints on how a function can undergo period doubling on its way to chaos. As one approaches the transition, the *only* possible ratio for the lengths of successive periodic regimes is Feigenbaum's δ.

Quantities like δ that are the same across a range of different functions are said in physics parlance to be **universal**. Strictly speaking, the result that δ is approximately 4.669... is true only as n gets large, corresponding to very high periods in the period-doubling sequence. However, in practice usually δ_n is close to 4.669 even for small n. For example, for

[6]For a detailed account of this discovery, see Chapter 10 of Stewart (2002). See also the list of suggested further reading at the end of this chapter.

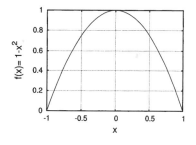

Fig. 12.7 An upside-down parabola, $f(x) = 1 - x^2$. Any function with a single parabola-like maximum that undergoes a period-doubling transition to chaos will have the universal value for δ.

the logistic equation, we found above in Eq. (12.9) that $\delta_2 \approx 4.608$, a little more than 1% below the large-n value. Thus, Feigenbaum's constant is a very good description of period doubling even for small periods.

12.3 Physical Consequences of Universality

Perhaps this is all just a mathematical curiosity. It does seem a little odd, though, that this number $\delta \approx 4.669$ appears in the bifurcation diagram of the logistic, cubic, and sine functions, and almost all other functions as well. However, we are just playing around with mathematical functions that have little to do with the real world; even the logistic equation, which models limited population growth, is at best a crude approximation to a real phenomenon. It turns out, though, that the number 4.669201... is universal enough that it appears in physical systems as well, not just mathematical explorations on a computer or calculator.

One such physical phenomenon is a dripping faucet. The drops' timing changes as one increases the flow rate, the amount of water that flows out of the faucet every minute. In some instances the dripping faucet will undergo a period-doubling transition to chaos. The basic scenario is as follows. For a low flow rate, the drops may be periodic with period 1. The drops would sound something like this:

$$\cdots \; \ast \; ...... \; \ast \; ...... \; \ast \; ...... \; \ast \; ...... \; \ast \; ...... \; \ast \; ...... \; \ast \qquad ; \cdots \qquad (12.11)$$

That is, there is the same amount of time between each drip. The idea in Eq. (12.11) is that "$\ast$" represents a drip splashing, and "......" represents the time interval between drips.

Next, increase the flow rate slightly. In this example the flow rate plays the role of the parameter r; it is the thing that we vary as we conduct our experiment. At some higher flow rate the drops will shift to a different pattern: they will be periodic with period 2:

$$\cdots \; \ast \; .. \; \ast \; ...... \; \ast \; .. \; \ast \; ...... \; \ast \; .. \; \ast \; ...... \; \ast \; .. \; \ast \; ...... \; \ast \; \cdots \qquad (12.12)$$

Now the time between the drops alternates between a short interval and a long interval. Increasing the flow rate a little bit further will result in a shift to period-4 behavior. The drop pattern might now sound like

$$\cdots \; \ast \; .. \; \ast \; ...... \; \ast \; ...... \; \ast \; ...... \; \ast \; .. \; \ast \; ...... \; \ast \; ...... \; \ast \; ...... \; \ast \; \cdots \qquad (12.13)$$

The repeating pattern is three long intervals between drops, followed by a short interval. The sequence has period 4, since it repeats every four drops.

One can record the flow values at which these transitions occur, just as we noted the r values at which the bifurcations occur in the logistic equation. We then calculate δ_1, the ratio of Δ_1 to Δ_2, as we did for the logistic equation in Eq. (12.8) in the previous section. Amazingly, if we carried out this calculation, we would find that δ_1 would be close to the Feigenbaum number, 4.669. Somehow a number that was obtained by iterating simple single-variable functions captures some feature of a complex, multi-dimensional physical phenomenon.

I should pause and note that it is very difficult to conduct an actual experiment of the sort that I just described. There are at least two reasons for this. First, the period-doubling transitions occur over a very small range of flow rates. It is very difficult to control the flow rate precisely enough to move through the periodic behaviors in a controlled way. Second, the actual drip dynamics are very sensitive to mechanical vibrations; people walking in a nearby room can be enough to disturb the experiment, again making it difficult to distinguish one period from another. Because of these difficulties I am not aware of anyone who has actually carried out a calculation of δ for a dripping faucet. However, the dripping faucet has been much studied, and period doubling has been consistently observed.

There are quite a few other physical systems that undergo period doubling and for which experimenters have been able to calculate δ. I will briefly describe one of these experiments. Consider a fluid such as water in a very small box. The bottom and the top of the box are at different temperatures. When the temperature difference is small, the fluid remains motionless, and heat is conducted from the warm bottom of the box to the cooler top.

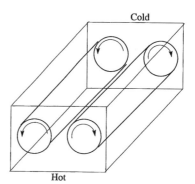

Fig. 12.8 An illustration of convection rolls in a box of fluid. The bottom of the box is at a higher temperature than the top. Hot fluid rises, is cooled at the top, and then falls back to the bottom. The fluid moves in two well-defined cylinders, or rolls.

As the temperature is increased, the fluid begins to move. The warm fluid on the bottom rises, since it is less dense than the cool fluid on the top. Once at the cooler top of the box, the fluid cools off and falls. This type of heat transport, where a hot fluid moves, is known as convection. If the size of the box is right, the rising and falling fluids will form structures known as convection rolls. These are illustrated in Fig. 12.8. The fluid moves in such a way as to create two rotating cylinders inside the box. The warm fluid moves up in the middle of the box and down on the sides. Although fluid is moving in the cylinder, the cylinder itself is a stable structure, much as rapids in a river can give rise to stationary structures even though the water is continuously flowing.

Experiment	Number of Period Doublings Observed	Estimated δ
Convection Rolls (water)	4	4.3 ± 0.8
Convection Rolls (helium)	4	3.5 ± 1.5
Convection Rolls (mercury)	4	4.4 ± 0.1
Electronic Circuit (diode)	4	4.5 ± 0.6
Electronic Circuit (diode)	5	4.3 ± 0.1
Electronic Circuit (transistor)	4	4.7 ± 0.3

Table 12.2 Estimates for δ obtained from experiments on different physical systems. Based on Table 9.1 on p. 29 of Cvitanović (1989).

However, as the temperature is increased further, the convection roll develops a little wiggle. The wiggle appears as a small bend or kink in the cylinder, and the wiggle moves up and down the cylinder. At their onset, these oscillating wiggles have a certain periodicity. As the temperature difference between the top and the bottom of the box is increased, this period doubles. And as the temperature difference increases further, the period doubles again, and so on. One can record the temperature differences at which these period doublings occur and form estimates for the first few δ_n's.

Physicists have carried out this experiment for different fluids. The experiment is tricky; it is much more difficult than I made it seem in the preceding several paragraphs.[7] The results of some of these experiments are summarized in Table 12.2. Also in this table are results other experiments examining period doubling in certain types of electric circuits. In brief, these are circuits that contain some sort of feedback element, resulting in oscillatory behavior. As the driving voltage of these circuits is increased, one sees period doubling.

The results in Table 12.2 show very good agreement with the universal value, $\delta \approx 4.699$. Not all uncertainty ranges include the exact value for δ, but we would not expect them to, since the exact value only holds as the periodicity gets large. Physicists have conducted other experiments looking for period doubling in addition to the few that I have listed here. None of the results contradict the idea of a universal value for δ.

Universality is a stunning result. To see why, it may be useful to step back for a moment and summarize. I introduced the logistic equation several chapters ago as a very simple model of population growth. We found that it displayed a range of different dynamical behaviors. Examining the bifurcation diagram, we noted that it makes transitions from periodic behavior to chaos via a series of period-doubling bifurcations. We then looked at a few other functions and found that they also show period doubling. It turns out that the δ ratio—how much larger a periodic region is compared to the next periodic region—was the same for all of these functions. The value of this ratio is 4.669.

[7] See the Further Reading section at the end of this chapter for references that discuss some of the experimental details.

We can then go out in the world—not mathematics on a computer but real physical objects—perform experiments, and observe period doubling. And these period-doubling transitions have δ's that are consistent with 4.669. We started with a simple quadratic function, and this has led to quantitative predictions about the behavior of dripping faucets, convection rolls in fluids, and oscillating electronic circuits. Amazingly, the logistic equation—a simple, one-dimensional function, contains information about disparate multi-dimensional physical systems like electric circuits and fluid flow.

The phenomenon of universality seems almost magical. How is universality possible? There are two distinct questions that need addressing. First, how is it that all parabola-like one-dimensional functions that undergo the period-doubling route to chaos have the same numerical value for δ? Second, what do phenomena like dripping faucets, convecting fluids, and oscillating circuits have to do with an iterated one-dimensional function?

12.4   Renormalization and Universality

Let us start with the first question: how can the number 4.669 arise from so many different equations? There is a beautiful mathematical theory that explains why this is so, but it is difficult to explain it without some fairly involved and advanced math. However, I will try to qualitatively sketch some of the key elements of this theory. To do so, will follow a line of reasoning similar to that put forth by Ian Stewart on pp. 189–193 of his book *Does God Play Dice?* (2002).

The key idea is to look at how certain features of the system—in this case the logistic equation or whatever function we are iterating—change when the scale is changed. If we have a geometric picture and zoom in, how does the picture change? The central insight is that in some unusual circumstances the shape does not change when magnified.

Here is a simple example that illustrates this point. Consider a curve, such as that shown in Fig. 12.9. Now imagine magnifying a small portion of the curve. The magnified version will look straighter. If you magnify the magnified version, it will look straighter still. Keep zooming in, and eventually it looks like a straight line. No matter how curvey the original

Fig. 12.9 A curve. If you zoom in on any point on the curve, eventually it will look like a straight line.

curve is, it will still look like a straight line if it is magnified enough. There is one exception to this, however. If the curve is not merely curvy but is pointy—if it has a sharp edge—then it will not look like a straight line no matter how many times it is magnified. This is illustrated in Fig. 12.10.

The technical term for this magnification process is **renormalization**. Normalization in this context refers to the process of setting a length scale—choosing which units to use to measure length or determining the scale on a graph. Renormalizing then implies a changing of the length scale. Collectively, the set of all possible magnifications is referred to as the **renormalization group**. One also speaks of renormalization group theory.

Using this terminology, almost any curve will look like a straight line under successive renormalization. The details of the curve do not matter. Only pointy curves like Fig. 12.10 fail to eventually look like a straight line. And a straight line is a special curve; it is the only curve that looks the same under renormalization. If you zoom in on a straight line it still looks like a straight line; its shape is unchanged.

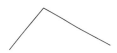

Fig. 12.10 A pointy "curve". The sharp point cannot be made into a straight line, regardless of how many times it is magnified.

Let us now return to iterated functions and period doubling. Imagine doing the same zooming-in trick on the bifurcation diagram near the transition to chaos. Initially we see pitchforks with pitchforks and, when we zoom in, we will again see pitchforks with pitchforks. The image will be unchanged, just as the straight line is unchanged under renormalization.

The bifurcation diagram is not just an arbitrary picture. It arises from an iterated function. We can use the fact that the bifurcation diagram is the same under renormalization to infer properties of the function that creates the bifurcation diagram. When at the transition point to chaos, the iterates are essentially infinitely periodic. The period gets longer and longer and longer before the transition to chaos occurs. We will consider the basic mechanism of period doubling, and then argue that this mechanism has to be the same for all period doublings, since the bifurcation diagram is unchanged by renormalization.

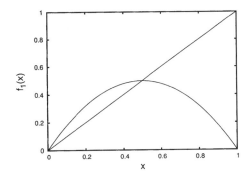

Fig. 12.11 The logistic equation $f_1(x)$ for $r = 2.0$, the super-attractive parameter value for period 1.

Let us think about the transition from period 1 to period 2. In Fig. 12.11 I have plotted the logistic equation for $r = 2.0$. There is an attracting fixed point at $x = 0.5$. The parameter $r = 2.0$ has a special property. For this r value the period-1 behavior is the most attracting; nearby orbits are pulled toward the fixed point at the fastest rate compared to other r values for which orbits have period 1. The attracting point in this case is said to be **super attractive**. The parameter value for the super-attractive period-1 fixed point occurs in the middle of the period-1 region in the bifurcation diagram. I will denote the logistic equation for this special value of r for which the period-1 attractor is super attractive $f_1(x)$. I.e., $f_1(x) = 2x(1-x)$.

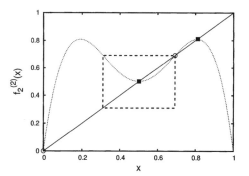

Fig. 12.12 The second iterate of the super-attracting period-2 logistic equation $f_2^{(2)}(x)$. The super-attractive parameter value is for $r \approx 3.236$. There are unstable period-1 points at $x = 0$ and $x \approx 0.69$, indicated with circles. The two period-2 points are indicated with squares.

The parameter for which the period-2 attractor is super attractive turns out to be $r \approx 3.236$. I will call the logistic equation for this special parameter value $f_2(x)$. I am focusing on the super-attractive values because loosely speaking they can be thought of as the most periodic value for a given period. I am interested in comparing period 1 to period 2, and so if I choose the super-attractive values for each period, then we can argue that it is a fair comparison.

In any event, in Fig. 12.12 I have plotted the second iterate of $f_2(x)$. In other words, I have plotted $f_2(f_2(x))$, denoted by $f_2^{(2)}(x)$. Note that we can see four fixed points for $f_2^{(2)}(x)$. Two of the fixed points are actually period 1. These appear as fixed points of $f_2^{(2)}(x)$ because if a point is period 1 it is also period 2.[8] These two period-1 points are 0 and 0.69.

Recall that the point of this exercise is to think about renormalization—how zooming in on the function might change its shape. However, because the bifurcation diagram does not change when one zooms in, we anticipate that the logistic equation will also have this property. We can see this if we look at the small dashed box in the center of Fig. 12.12. Notice that the $f_2^{(2)}(x)$ curve inside the dashed box resembles an upside down version of $f_1(x)$ from Fig. 12.11. If we took the curve inside the box, turned it upside down, and stretched it out to full size, it will very closely resemble $f_1(x)$.

The result of doing this is shown in Fig. 12.13. The original period-1 curve from Fig. 12.11 is the solid curve. The dashed curve that is very close to the solid curve is the stretched and flipped portion of the dashed curve inside the box of Fig. 12.12. This plot shows that rescaling $f_2^{(2)}$ yields a curve that is almost the same as $f_1(x)$. This rescaling is an example of a renormalization; we have zoomed in on one function (and turned it upside down) and arrived at essentially the same function.

We can view this rescaling as a type of operation that takes one function and returns a new function. This is similar to a function, which takes a number as input and returns a number as output. But here, an

[8] If something repeats every time, then it also repeats every other time.

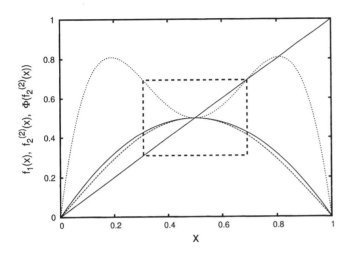

Fig. 12.13 The super-attractive period-1 logistic equation $f_1(x)$, the super-attractive period-2 logistic equation, and $\Phi[f_2^{(2)}(x)]$, the super-attractive period-2 function after it has been rescaled.

entire function is the input and an entire function is the output. Let us denote this function-changing operation by Φ.[9] So, for the situation under discussion, we can write

$$f_1(x) \approx \Phi[f_2^{(2)}(x)] . \tag{12.14}$$

Again, in words, this equation just says that $f_1(x)$, the period-1 function is approximately equal to the period-2 function $f_x^{(2)}$ after it is rescaled ("Phi-ed").

We can make the same comparison between the period-2 and the period-4 function, the period-4 and the period-8 function, and so on. In so doing, we are rescaling multiple times. Each rescaling is governed by an equation similar to Eq. (12.14), and changes the function from the logistic equation of one period to the logistic equation of another period.

Now comes the key step in the argument: this process of successive rescaling is the same as iterating Eq. (12.14). We start with a function and rescale repeatedly. We have seen that sometimes when iterating functions, there is an attracting fixed point. All initial conditions end up at that fixed point.

The same story holds for the rescaling Φ transformation, except instead of initial conditions being pulled toward a fixed *number*, now we have initial conditions being pulled to a fixed *function*. Almost any initial condition will end up at the same fixed function. The initial condition for Φ is also a function; different functions, when renormalized repeatedly, end up at the same function. This attracting function is universal—there is just one such function, just as for many iterated functions there is only one fixed point. Feigenbaum's constant δ can be derived from this universal function.

Moreover, we will arrive at this universal function no matter what equation we start with; the logistic equation, the cubic equation, or the sine equation. Indeed, any equation that is parabola-like will get pulled toward the universal function upon rescaling. This explains the mystery of how all these different functions can have the same value for

[9] Φ is the capital Greek letter "Phi."

δ. The mathematics of determining the universal function is not easy, but conceptually it is no more or less deep than the idea of a function having an attracting fixed point.

To summarize, let us go back to the example that I started with. If one zooms in on, i.e., renormalizes, a curve, eventually it looks like a line, provided that the curve does not have any sharp edges. The line is an attractor; many different curves end up as a line after multiple renormalizations. Similarly, if one renormalizes a function using Φ, one arrives at a universal function, provided that the initial function has a single maximum and that this maximum is parabola like. The universal function is an attractor; many different functions end up as the universal function after multiple rescalings with Φ. From this universal function one can calculate a number of quantities, including δ.

The mathematics of carrying all this out is not easy; it is well beyond the level of this book. Nevertheless, I hope that this discussion sheds some light on universality: how it is that the number 4.669 appears in almost all bifurcation diagrams that exhibit a period-doubling route to chaos.

12.5   How are Higher-Dimensional Phenomena Universal?

In the previous section I attempted to convey how renormalization explains the phenomenon of universality. However, this explanation is limited to one-dimensional functions. That is, renormalization explains why almost all one-dimensional functions that undergo a period-doubling transition to chaos have a universal value for δ. But how is it that the same number δ appears in the analysis of fluid convection rolls and dripping faucets? How can generic, single-variable functions like the logistic equation say anything about multi-dimensional, physical phenomena?

We will be in a better position to answer these questions in Sec. 31.7, after we have considered systems that are modeled by three-dimensional sets of equations. For now, though, a few general comments. I argued in Sec. 10.5 that the key geometric ingredients for chaotic behavior are stretching and folding. The stretching is what produces the sensitive dependence on initial conditions (SDIC), as neighboring orbits are pushed away from each other. The folding is necessary to keep orbits bounded. One-dimensional functions like the logistic equation implement stretching and folding when iterated, as shown in Fig. 10.11.

Higher-dimensional systems, like a dripping faucet or convection rolls, also must stretch and fold as they evolve in time for the same reasons: the stretching leads to SDIC and folding keeps orbits bounded. The stretching and folding occurs in three dimensions, and so the details of the stretching and folding can be more complicated than the simple transformations shown in Fig. 10.10—the dough undergoes a potentially complex kneading process in three dimensions. However, one can take an imaginary slice through this three-dimensional process and the result is a one-dimensional process of the sort analyzed by renormalization. Thus, in a sense one-dimensional systems are embedded in three-dimensional systems. As a result, the universal properties of the period-doubling route to chaos in one-dimensional systems extend to multi-dimensional systems as well.

Further Reading

There are surprisingly few non-technical discussions of the phenomenon of universality. The clearest such account that I have found is Chapter 10 of Stewart (2002). The chapter titled "Universality" in Gleick (1987) is an engaging and exciting account of Mitchell Feigenbaum's discovery of universality. For a discussion of universality in the context of physics, see pp.61–66 of Watts (2004). More technical discussions of universality can be found in Chapter 6 of Smith (1998) and chapter 11 of Peitgen, Jürgens, and Saupe (1992). The edited volume by Cvitanović (1989) collects many of the important early papers on universality in dynamical systems, including the results of experiments measuring δ in physical systems. Rob Shaw's 1984 book *The Dripping Faucet as a Model Chaotic System* (1984), although currently out of print, was a highly influential monograph on the application of ideas and techniques from chaotic dynamics to physical systems.

Exercises

(12.1) What is the long-term behavior of iterates of the cubic equation for $r = 5, 0$, $r = 5.5$, and $r = 6.0$? Refer to Fig. 12.3, the bifurcation diagram for the cubic equation.

(12.2) Find all fixed points for the cubic equation for $r = 5.0$. To do so, use the fixed point equation, $f(x) = x$.

(12.3) Find all fixed points for the cubic equation for $r = 2.0$. To do so, use the fixed point equation, $f(x) = x$.

(12.4) Calculate by hand the first three iterates of $x_0 = 0.4$ for the cubic equation with $r = 6.0$. Do your numbers appear consistent with those plotted in Fig. 12.2?

(12.5) Use Fig. 12.3 to make rough estimates Δ_1 and Δ_2 for the cubic equation.

(12.6) For what r value would you estimate that the logistic equation shows a bifurcation from period 16 to period 32? At what r value would you expect to

see the bifurcation from period 32 to 64? Briefly explain.

(12.7) For this exercise you will need a program that can make high-resolution bifurcation diagrams for the logistic equation, such as those at http://chaos.coa.edu/. Zoom in on the period three window, near $r = 3.83$. You should see a sequence of period-doubling bifurcations, from period 3 to 6 to 12, and so on. Determine the r values at which these bifurcations occur. Then uses these r values to estimate Δ_1, Δ_2, and Δ_3, and δ_1 and δ_2.

(12.8) A function exhibits the period-doubling route to chaos. Suppose the bifurcation from period 1 to period 2 occurs at $r = 4.2$, and the bifurcation from period 2 to period 4 occurs at $r = 4.8$. At what r value would you expect to see a bifurcation from period 4 to period 8? Explain.

(12.9) ♯ Suppose you are conducing an experiment with an oscillating electric circuit. You want to estimate as many δ_n's as you can. You vary the voltage and observe period doublings, as expected. The first period-doubling bifurcation occurs at 3.0 Volts, and the next at 4.5 Volts. The accuracy on your experimental equipment is such that you cannot control the voltage more accurately than 0.0001 Volts. How many period-doubling bifurcations do you expect to be able to observe? Explain.

Statistical Stability of Chaos

13

In Chapter 10 we saw that sensitive dependence on initial conditions places severe limits on our ability to perform long-term prediction of chaotic systems. Two initial conditions that start close to each other very quickly get pushed apart. In this chapter we revisit this general phenomenon and will encounter some other ways to visualize and characterize the behavior of chaotic orbits. In so doing, we will see that although the path of a particular orbit is unpredictable, chaotic systems nevertheless possess statistical regularities.

13.1 Histograms of Periodic Orbits

To examine statistical properties of orbits we will use histograms. A review of histograms can be found in Appendix B. If you have not encountered histograms before you might want to read at least the first part of the appendix before continuing in this chapter. Recall that we have seen in previous chapters that the orbits of the logistic equation $f(x) = rx(1 - x)$ can be periodic or chaotic, depending on the value of the parameter r. Our goal in the following is to find a way to characterize chaotic orbits beyond simply stating that they do not repeat and thus are aperiodic. For example, does an aperiodic orbit spend more time in some regions than others?

Before considering chaotic orbits it will be helpful to look at a non-chaotic example. When $r = 3.4$, the logistic equation has an attracting cycle of period 2. This can be seen in Fig. 13.1, which is a time series plot for the initial condition $x_0 = 0.3$. We see that quite rapidly the orbit settles in to its period-2 behavior, oscillating between 0.45 and 0.84.

Figure 13.2 contains a new type of plot: a histogram. This lets us see the relative frequency of different x-values along the itinerary. The figure shows that the orbit spends an equal amount of time at two values and no time anywhere else.[1] To make this plot, I took a time series of length 1000, discarded the first forty iterates, and then made the histogram. (A time series is just a list of numbers, so to make the histogram I just followed the procedures described in Appendix B.) I used a bin size of 0.01. The histogram shows that all of the numbers in the time series fall in just two bins. This is exactly what we would expect for a period-2

[1] Working out the units and scale for the vertical axis of a histogram such as this is somewhat involved; see Appendix B for a discussion. When interpreting the histograms in this chapter, what is important is the relative heights, not the absolute height of the histogram.

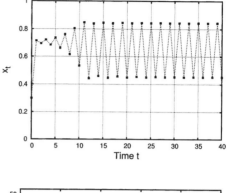

Fig. 13.1 A time series plot of an orbit for the logistic equation with $r = 3.4$. The long-term behavior is periodic with period 2.

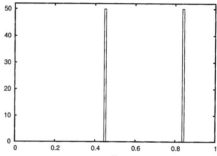

Fig. 13.2 The histogram for the orbit shown in Fig. 13.1. The long-term behavior is periodic with period 2.

time series. For such a sequence a histogram is not really necessary. The distribution of values along the itinerary is quite simple. Once the transient behavior dies away, the system oscillates between roughly 0.44 and 0.84.

13.2   Histograms of Chaotic Orbits

Histograms do not provide new insight into periodic orbits, but they are a powerful tool for examining chaotic orbits. Consider yet again the logistic equation with $r = 4.0$. At this parameter value orbits are chaotic; they have sensitive dependence on initial conditions and they are aperiodic. This aperiodicity can be seen in Fig. 13.3, which is a time series plot for the $r = 4$ logistic equation for the initial condition $x_0 = 0.3$. As expected, the orbit bounces around and does not appear to repeat. However, looking closely at the figure we can see that the orbit spends a lot of time on the "edges"—near 0 or 1—and not as much time in the middle of the interval. A histogram of the orbit shows us that this is indeed the case.

In Fig. 13.4 I have plotted the histogram for the orbit shown in Fig. 13.3. To make this histogram I needed many more than the 100 iterates shown in the time series plot. In order to get a good picture of the distribution of the iterates I used a time series with $100,000$ values and a bin size of 0.01. This histogram tells us that, as anticipated, the

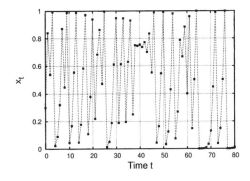

Fig. 13.3 A time series plot of an orbit for the logistic equation with $r = 4.0$. The orbit is aperiodic. The initial condition is $x_0 = 0.3$.

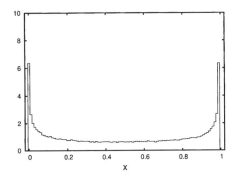

Fig. 13.4 The histogram for the orbit shown in Fig. 13.3.

orbit spends a lot of its time at large values near $x = 1.0$ and small values near zero, and it spends much less time at intermediate values. This is a feature of the orbit that is perhaps not immediately apparent when looking at the time series plot. Another noteworthy feature of this histogram is that it appears to contain no gaps. This means that not only is the orbit aperiodic, but that it roams all over the region between 0 and 1.

We know that the orbits of the logistic equation for $r = 4.0$ possess sensitive dependence on initial conditions. So if I use a different seed, I will get a different orbit. This can be seen in Fig. 13.5, which shows the time series for $r = 4.0$ with an initial condition of $x_0 = 0.31$. The orbit starts similarly to that of Fig. 13.3, but after only ten iterates or so the two time series are quite different. While the particular orbits in Figs. 13.3 and 13.5 are different, they do appear to be similar. Both spend more time near 0 and 1, and both show the frequent arcing upward curves followed by jagged oscillations. Let us take a look at the histogram for the second time series, shown in Fig. 13.6. Remarkably, it looks very similar to the previous histogram in Fig. 13.4. In fact, the two histograms look almost identical. There are, however, slight differences between Figs. 13.4 and 13.6. If I were to plot even more iterates in each of the histograms, the two plots would become more and more similar.[2]

This result is perhaps somewhat surprising. Orbits for the logistic equation with $r = 4.0$ are chaotic, and hence unpredictable due to the butterfly effect. However, they seem to be unpredictable in the same

[2]The histogram of the orbit for the logistic map at $r = 4.0$ appears smooth. You may wonder if it is possible to find a function that approximates this smooth curve. The answer is "yes". One can prove that the curve for the histogram is given by

$$\rho(x) = \frac{1}{\pi \sqrt{x(1-x)}}. \qquad (13.1)$$

This quantity is known as the invariant distribution. The symbol ρ is the Greek letter "rho", and is a standard symbol for invariant distributions. We will not use the formula for $\rho(x)$. I mention it both as an interesting piece of mathematical trivia and as a reminder that there is elegant, rigorous mathematics behind many of the results in dynamical systems. An outline of a derivation of Eq. (13.1) can be found in Peitgen, Jürgens, and Saupe, (1992, pp. 527–9).

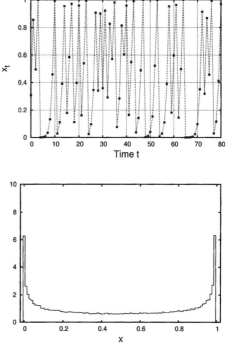

Fig. 13.5 A time series plot of an orbit for the logistic equation with $r = 4.0$ and $x_0 = 0.31$. The orbit follows a different trajectory than the time series of Fig. 13.3.

Fig. 13.6 The histogram for the orbit shown in Fig. 13.5. Note the similarity to Fig. 13.3.

way—different chaotic orbits yield the same histogram. So while the orbits are different, on average they behave the same. In the next section I will further unpack these ideas and introduce some additional terminology.

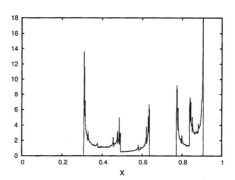

Fig. 13.7 The histogram for an orbit for the logistic equation with $r = 3.625$.

Before doing so, let us look at a histogram for another chaotic orbit. Figure 13.7 shows the histogram for an orbit of the logistic equation for $r = 3.625$, another parameter value for which the orbits are chaotic. The histogram looks quite different from Fig. 13.4, which was obtained from the logistic equation with $r = 4.0$. The orbits do not take all possible values but instead are restricted to two bands: one from roughly 0.3 to 0.65, and the other from 0.75 to 0.9. This is consistent with what we can see on a bifurcation diagram for the logistic equation. Looking back at the bifurcation diagram of Fig. 11.5, one can see that the final states of the orbit for $r = 3.625$ are indeed restricted to two bands.

The histogram gives us direct information that is only seen indirectly in the bifurcation diagram. A vertical slice of a bifurcation diagram gives us a final-state diagram like those shown in Fig. 11.1. The final-state diagram shows what values occur in the long-run for the orbit. A histogram is a final-state diagram with an added vertical dimension that indicates not only what values occur in the time series, but also their relative frequency. In Fig. 13.7 we see that the orbit spends most of its time at the edges of the bands, and there is also a location in the middle of each band that the orbit visits much more frequently than elsewhere. These spikes on the histogram can be seen as darker regions on the bifurcation diagram. Looking again at the bifurcation diagram in Fig. 11.6, one sees dark curves weaving up and down. These curves are regions where there are more points on the bifurcation diagram.

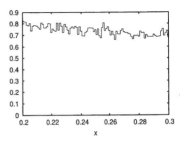

Fig. 13.8 A zoomed-in view of the histogram of Fig. 13.4.

13.3 Ergodicity

Let us consider again the histogram in Fig. 13.4. As noted above, the histogram appears to have no gaps; the orbit visits all regions in the interval between 0 and 1. In Fig. 13.8 I have zoomed in on the histogram so that only the interval from $x = 0.2$ to 0.3 is visible. Again, no gaps are seen in the histogram, indicating that the orbit is visiting all regions in the interval. We can keep zooming in on smaller and smaller portions of the histogram, and we still would not see any gaps.

The technical term for this type of orbit is **ergodic**.[3] An orbit is ergodic if it gets arbitrarily close to any point on the interval in which the orbit roams. (In this case the interval is 0 to 1.) In other words, you can choose *any* point on the interval and eventually the orbit of Fig. 13.3 will come very close to it. The orbit cannot exactly hit every point on the interval—there are far too many points on the interval for the orbit to reach, even if we give it an infinite amount of iterations to do so.

Moreover, the interval also contains an infinite number of unstable periodic points or points that lead to an unstable periodic point. This was discussed in Section 7.4, where I argued that even though there are an infinite number of points leading to an unstable fixed point, there are nevertheless infinitely many more points on the interval that do not lead to the fixed point. As a result, there is zero probability that an initial condition, if chosen at random, will lead to an unstable fixed or periodic point. One says that **almost all** initial conditions of the logistic equation with $r = 4.0$ are aperiodic. The word "almost" is used here in a technical way to indicate that while there are alternatives, they occur so infrequently as to have zero probability. I will use the words "almost all" frequently in the rest of this chapter. The phrase sounds imprecise, but it actually has a rigorous and precise mathematical meaning.[4]

In any event, for the chaotic logistic equation—or any chaotic system for that matter—there are an infinite number of unstable period points

[3]The roots of word ergodic are *ergon* and *odos*, the Greek words for work and path, respectively (Walters, 2000). The term was coined by the physicist Ludwig Boltzmann around 1900 as he was working to derive the average macroscopic properties of matter from their microscopic properties. This field of physics is now known as statistical mechanics. Specifically, Boltzmann used the term to describe a relation between time and spatial averages of physical properties.

[4]These ideas will surface again in Chapter 21 where we will see that there are different types of infinities. The orbit has a countably infinite number of iterates but the interval has an uncountably infinite number of points.

[5]Some initial conditions lie on unstable fixed or periodic points. These are thus not aperiodic and cannot give rise to a histogram such as that of Fig. 13.4. But if one chooses an initial condition at random, with probability 1 the orbit will be aperiodic and a histogram like Fig. 13.4 will result.

that the aperiodic orbits must maneuver round. The aperiodic orbit can never exactly land on these points, since then the orbit would become periodic and would no longer be aperiodic.

On the one hand, ergodicity illustrates just how chaotic the orbit is—during its journey it wanders arbitrarily close to anywhere on the interval. On the other hand, the fact that the orbit wanders and explores every tiny region of the interval leads to a type of predictability. I have argued that both the orbits for $x_0 = 0.3$ and $x_0 = 0.31$ will give rise to the same histogram. This means that the two orbits visit the same regions of the interval equally often. In fact, almost all[5] initial conditions will result in an orbit that has the same histogram. This means that the statistics of each orbit is the same. The specific trajectories that two orbits take will be very different, but each will spend the same fraction of time in each region of the interval. The histograms of Figs. 13.4 or 13.6 are thus predictable and regular features of the logistic equation's chaotic dynamics.

How can this be? How is it that orbits with sensitive dependence on initial conditions lead to the same histogram? Proving this fact is very difficult, but I can say a few things to suggest why this is true. Suppose that we know we have an ergodic orbit—an aperiodic orbit that deftly avoids unstable periodic points while getting arbitrarily close to them. In so doing it builds up the distribution shown in the histogram of Fig. 13.4. Now suppose we have another aperiodic orbit. It, too, will wander all over the interval avoiding unstable periodic points. As it does so, it must come very close to a point that the other orbit has visited. After all, the other orbit is ergodic and so gets arbitrarily close to *any* point on the interval.

So this second aperiodic orbit must end up tracking or shadowing the ergodic orbit for much of the time. The two orbits are aperiodic, but they are also deterministic. So if they get close to each other they have to stay close to each other for at least a little while. Once the second orbit wanders from the first, it will again be near some other point on the first orbit, since the orbit is ergodic. And so it shadows it again for a few iterations. The result is that both aperiodic orbits end up, on average, visiting the same regions of the interval, leading to the same histogram.

There is another way we can think of arriving at the histogram associated with the ergodic orbit. Instead of iterating one initial condition for a very long time, we iterate a vast number of initial conditions all at once. This is illustrated in Fig. 13.9. I began with one million initial conditions chosen at random between 0 and 1. A histogram of these initial conditions is shown in the upper left of the figure. Since the initial conditions were chosen at random, the initial histogram is essentially uniform from 0 to 1. I then iterated all of these initial conditions twice and plotted another histogram, labeled $t = 2$ in the upper right of Fig. 13.9. The next two plots in the figure show a histogram of the million points after five and then fifty iterations. One can see that the histogram changes quite little after just a few iterations.

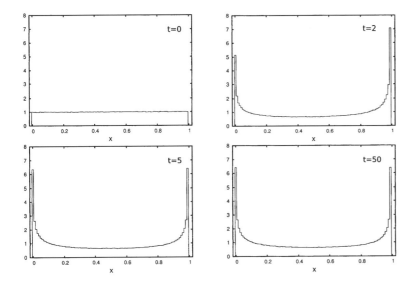

Fig. 13.9 The evolution of one million initial conditions for the logistic equation with $r = 4.0$. The initial conditions are spread uniformly from 0 to 1. After just a few iterates, they are pulled to the invariant density.

Figure 13.9 gives us a somewhat different way of thinking about iterating the logistic equation. Rather than iterating a single point, we iterate a million points at once. We can view the function as acting on the histogram itself, instead of individual orbits. At each step the logistic equation operates on the entire histogram and gives us a new histogram. The histogram on the lower right of Fig. 13.9 is an attracting fixed point of the logistic equation. Any reasonable[6] initial distribution will get pulled to the distribution shown in Fig. 13.4. This distribution is known as the **invariant distribution**, as it is left unchanged by the action of the logistic equation.[7]

This invariant distribution is stable: almost all initial histograms are pulled toward it, and the time series for almost all initial conditions will, when plotted as a histogram, yield the invariant distribution. The histogram is a statistical structure, in the sense that it captures average properties of the orbit. Thus, one can say that the phenomenon of chaos is **statistically stable**, provided that there is an ergodic orbit. Note, however, that the histogram cannot be used to reconstruct an exact orbit, just as a statistical profile of a population cannot be used to faithfully reconstruct a particular individual from the population.

It turns out that for the logistic equation and many other chaotic dynamical systems there can be only one attracting fixed histogram or invariant distribution. (This is why the orbit of almost any initial condition will lead to the same histogram.) This is a difficult statement to prove, but the argument is similar to the reasoning given above that two ergodic orbits have to have the same histogram. Basically, the idea is that there is only room on the interval for one invariant distribution.[8] The logistic equation mixes up orbits, and so any invariant distribution (i.e., histogram) must span the entire interval. In a sense, the mixing up of orbits ensures that all histograms are also mixed together, so there can be only one invariant distribution.

[6] An example of an unreasonable distribution is one where all one million initial conditions lie on an unstable fixed point. Such distributions are vanishingly unlikely.

[7] The invariant distribution is also commonly referred to as the **invariant density**.

[8] This argument is based on James Sethna (2006, pp. 66-7).

The theory behind ergodicity is somewhat difficult, but I hope that the figures in this section have provided some empirical evidence that almost all orbits of the logistic equation for $r = 4.0$ have the same histogram. In the next section I will explore some of the consequences and implications of this remarkable fact.

13.4   Predictable Unpredictability

The key point of Chapter 10 was that sensitive dependence on initial conditions places severe practical limits on our ability to make a long-term prediction for a chaotic orbit. However, the fact that the logistic equation with $r = 4.0$ is ergodic makes a different sort of prediction easy. Suppose we want to know what percent of the time, on average, the orbit spends between 0.1 and 0.2. We can answer this question via the invariant histogram. The answer turns out to be 9.03% regardless of the initial condition. The butterfly effect does not interfere with this sort of prediction. Similarly, we can deduce that the orbit spends, on average, 43.6% of the time between 0.6 and 1.0 and exactly half of its time between 0 and 0.5. So in this sense the chaotic logistic equation is eminently predictable. The chaotic logistic equation is statically predictable even though its orbits are unpredictable.

There is nothing contradictory about the co-existence of statistical predictability and the unpredictability of the butterfly effect. The earth's climate provides an excellent example of this phenomenon. Weather is notoriously unpredictable. Forecasts much beyond a few days are usually quite unreliable. Where I live in the Northeastern United States, it is difficult to predict with any accuracy whether or not it will rain a week from now. However, average, longer-term weather features—i.e. the climate—are quite stable and predictable. It is essentially impossible to predict the amount of rain that will fall 13 days from now, but one can make reasonable predictions about the monthly rainfall. For example, in the last hundred years or so the average rain in August is 2.68 inches. The wettest August saw 8.68 inches of rain and the driest 0.54 inches.[9] Over periods of a century or so the climate—features like average rainfall or average high temperature—are quite stable, even though the weather is generally chaotic and unpredictable.

To summarize, chaotic dynamical systems are unpredictable in the long run because of the butterfly effect. But they are often unpredictable in predictable ways. For the logistic equation with $r = 4.0$ this predictable unpredictability is captured by the invariant histogram, which describes the statistics of the orbits of almost all initial conditions.

[9]The data is for Acadia National Park and was obtained from the US Historical Climatology Network, http://cdiac.ornl.gov/epubs/ndp/ushcn/access.html.

Exercises

(13.1) An orbit of a dynamical system is attracted to a periodic cycle of period 4. The periodic orbit is: . . . 0.3, 0.5, 0.7, 0.9, 0.3, 0.5, 0.7, 0.9 . . .. Sketch a histogram for this orbit.

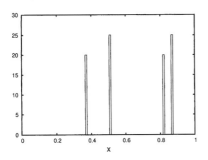

Fig. 13.10 A possible histogram for a period-4 orbit.

(13.2) An orbit of a dynamical systems is attracted to a periodic cycle of period 4. The periodic orbit is: . . . 0.3, 0.9, 0.5, 0.7, 0.3, 0.9, 0.5, 0.7 . . .. Sketch a histogram for this orbit. Compare this histogram with the one you drew for Exercise 13.1.

(13.3) Suppose the orbit of a deterministic dynamical system (not necessarily the logistic equation) is periodic with period 4. Could it have a histogram like that of Fig. 13.10? Why or why not?

(13.4) Figure 13.11 shows histograms of the orbits of the logistic equation with four different parameter values, $r = 3.58$, 3.67, 3.70, and 3.90. By looking at the bifurcation diagram (Fig. 11.4), determine which histogram goes with which parameter value.

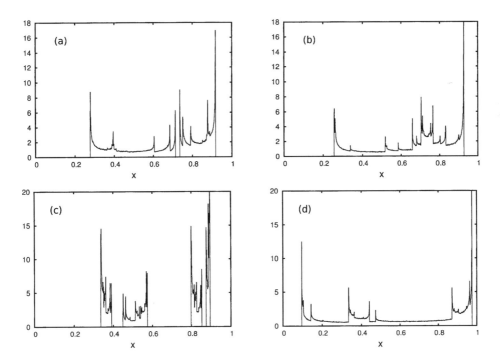

Fig. 13.11 Four different histograms for chaotic orbits of the logistic equation. Each histogram was generated with a different r value.

Determinism, Randomness, and Nonlinearity

14

This chapter, the last before we turn our attention to fractals, contains some additional thoughts on ways to characterize chaotic behavior and think about randomness and determinism. The first few sections introduce a technique known as symbolic dynamics which will allow us to compare the chaotic logistic equation to tossing a coin. Section 14.4 contains some general comments and observations about the distinction between linear and nonlinear functions.

14.1 Symbolic Dynamics

We have seen that the logistic equation with $r = 4.0$ is chaotic; orbits are aperiodic and have sensitive dependence on initial conditions.[1] As a result, long-term prediction is impossible and orbits seemingly move about at random. Is the orbit *really* random? It cannot be—the orbits are generated by a deterministic function, so there is no element of chance involved. However, there is another sense in which the orbits *can* be said to be random. In order to explore this idea, I first need to introduce another tool used in the study of dynamical systems, called **symbolic dynamics**.

The orbits of the logistic equation are a sequence of numbers between 0 and 1. For example, the itinerary for the seed $x_0 = 0.613$ for the logistic equation with $r = 4.0$ begins

$$0.613, 0.949, 0.194, 0.625, 0.937, 0.235, 0.719, 0.809\ldots . \qquad (14.1)$$

There are an infinite number of numbers between 0 and 1, and so there are an infinite number of possible values for iterates.

The idea behind symbolic dynamics is to encode the sequence of numbers in an itinerary into something simpler to analyze but which preserves the key features of the dynamical system. For the logistic equation, the standard encoding is to use the symbol L for any value less than 0.5 and R for any value equal to or larger than 0.5.[2] Thus, the itinerary in Eq. (14.1), in symbolic form, is

$$RRLRRLRR\ldots . \qquad (14.2)$$

We can then study the dynamics of symbol sequences instead of the dynamics of the original itinerary.

---

[1] In this Chapter I focus exclusively on the logistic equation with $r = 4.0$. So I will sometimes refer to it as simply the logistic equation, instead of the cumbersome logistic equation with $r = 4.0$.

[2] The symbols L and R are arbitrary. We could just as well use A and B, or 0 and 1, or ☺ and ☻.

It seems like we are discarding a lot of information when we perform this encoding. After all, the original iterates can take on an infinite number of values, while there are only two symbols, R and L. However, one can show that the two systems—the original iterated logistic equation and the dynamics on the symbol sequences—are in a sense the same. The two systems will have the same number of periodic cycles, and these cycles will have the same stability. If one system is chaotic, the other is too. The technical term for this type of similarity between two dynamical systems is **topological conjugacy**.[3]

The fact that the symbolic representation of Eq. (14.2) and the original itinerary of Eq. (14.1) are equivalent is not obvious. A full proof of this fact is beyond the scope of this book, but I can say a few things to justify it. Let us suppose that we are analyzing symbol sequences—L's and R's—produced by the logistic equation. Perhaps we have seen the sequence LL. We cannot go from this sequence back to the original iterates $x_0 x_1$, since each L could be any number between 0 and 0.5. However, we can say something about this sequence—namely, that the initial condition must have been between 0 and 0.15.

[3]Two spaces are topologically conjugate if there is a function mapping one space to the other that is continuous and invertible. Here, the two spaces are the set of all real numbers between 0 and 1 and the set of all sequences of L's and R's. Since the transformation between the two spaces is continuous—the space does not need to be cut or split apart in any way when being transformed—periodic points are preserved.

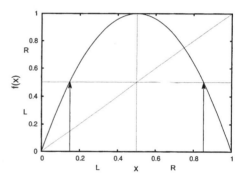

Fig. 14.1 Any initial condition between 0 and 0.15 will end up between 0 and 0.5 when iterated.

To see this, first note that since the first symbol we saw was L, we know that x_0 must have been between 0 and 0.5. Since the second symbol we see is also L, we know that x_1 must also be between 0 and 0.5. Figure 14.1 helps us see that initial conditions between 0 and 0.15 will be between 0 and 0.5 after one iteration. Thus, the sequence LL must have arisen from an initial condition between 0 and 0.15. A similar argument shows that symbols LR have an initial condition between 0.15 and 0.5, RR an initial condition between 0.5 and 0.85, and RL an initial condition between 0.85 and 1.0.

Note that each of the four pairs of two symbols (LL, LR, RL, RR) correspond to a non-overlapping region of possible initial conditions. This is illustrated in Fig. 14.2. Note also that the symbolic labeling

Fig. 14.2 The second generation of the partition that generates symbolic dynamics for the logistic equation.

includes all possible initial conditions. This symbolic labeling is thus said to **partition** the initial conditions. We could next consider all possible blocks of three symbols: LLL, LLR, LRL, and so on. A similar analysis to that of the preceding paragraph would lead to a refinement of the partition of Fig. 14.2. There would be eight different regions, each labeled with a different sequence of L's and R's.

One could continue this process further still by considering sets of four symbols, then five, and so on. At each step the interval for each particular sequence gets smaller and smaller. The result is that as the number of symbols in the sequence of L's and R's grows, the interval of possible initial conditions that leads to that symbol sequence gets smaller. Thus longer and longer sequences of L's and R's correspond to knowing the initial condition more and more accurately; in the limit of an infinite symbol sequence we know the initial condition exactly. A partition that has this property is known as a **generating partition**. All the information about an orbit is contained in its initial condition, since the dynamical system is deterministic. Thus, all the information about an orbit is also contained in the symbol sequence. As a result, the symbol sequences contain the same information as the orbits themselves.

The above exposition is an informal argument and not an exact proof.[4] The main point is that converting orbits of the logistic equation to symbolic sequences results in a dynamical system that contains the same information—has the same periodic and aperiodic orbits—as the original logistic equation. In the next section we will use these symbol sequences to see that the logistic equation can generate randomness.

[4]The argument *can* be made precise. A good, elementary discussion is that in Chapter 3 of Ott, Sauer, and Yorke (1994). See also section 10.7 of Peitgen, Jürgens, and Saupe (1992).

14.2 Chaotic Systems as Sources of Randomness

What do symbol sequences for the logistic equation with $r = 4.0$ look like? Here is a typical sequence:

$$LRRLRRRLLRLRRLLLLLLLRRRRLLRLLRRL\ldots. \qquad (14.3)$$

To make this symbolic orbit I used the initial condition 0.20, calculated the itinerary, and then converted to symbols. There is no obvious pattern. It looks like the L's and R's are equally likely. To test this proposition I need to examine a much larger orbit. I conducted an experiment with a symbolic orbit of one million symbols and found that the fraction of L's was 0.5002 and the fraction of R's was 0.4998. It does indeed seem that the two symbols occur equally often.

What about pairs of symbols? Do these occur equally often, as well? I can test this by doing a similar experiment. I again generated an orbit from the logistic equation with a million symbols and determined the frequency of the four possible adjacent pairs of symbols. The results are shown in Table 14.1. It appears that all four outcomes are equally likely. The four frequencies are not exactly equal, but they are very close, and would get closer if I used more a longer orbit. (The four frequencies do not add up to 1 because the numbers are rounded off.)

Table 14.1 The frequency of occurrence of all four possible pairs of symbols in an orbit of the logistic equation. There were one million symbols in the orbit.

LL	0.2501
LR	0.2502
RL	0.2502
RR	0.2496

The next logical step is to inquire about the frequency of sequences consisting of three symbols. Doing this experiment I find again that all eight possibilities occur with almost exactly the same frequency. I could then examine the frequencies of four symbols, then five, and so on. In each case I would find that all symbol sequences occur equally often.

This is not just a numerical result. It can be proven rigorously—using deductive mathematical logic without relying on computer experiments—that all sequences of L's and R's are equally likely in the symbolic dynamics of the logistic equation with $r = 4.0$. This means that the symbolic dynamics of the logistic equation is as random as a coin toss. A long sequence of tosses from a fair coin—one which is equally likely to come up heads or tails—also has the property that all possible sequences are equally likely.

Here is another way to think about what this result means. Suppose that I gave you two long symbol sequences. One was a symbolic orbit generated by the logistic equation with $r = 4.0$. The other sequence was generated by tossing a coin; if it comes up heads I record L, and if the coin is tails I record R. There is no way to tell which sequence was which. In both sequences all possible occurrences of L's and R's are equally likely .

A coin toss is perhaps the paradigmatic random process. The coin is tossed and it is a matter of chance if one gets heads or tails. Yet we have just seen that the symbolic dynamics for the logistic equation are as random as a sequence generated by a tossed coin. But the logistic equation was generated by a deterministic process—there is no element of chance involved. How can this be?

14.3   Randomness?

To begin to make sense of this we need to think more carefully about randomness. What, exactly, do we mean when we say that something is random? To answer this question it will be helpful to distinguish between two different ideas: the nature of the process that generates a result and the nature of the result itself. Let us begin by considering this first idea: the qualities of the process that generates an orbit. The dynamical systems we have examined so far have all been iterated functions, and all the functions we have studied are deterministic. This means that the output of the function is determined uniquely by the input and nothing else. Giving a deterministic function the same input many times will always result in the same output. Alternatively, one could have a function that incorporates an element of chance, so that the output is not always the same for the same input. In everyday usage we might call such a function random. However, the technical term for such a function is *stochastic*.[5] One speaks of a stochastic function or a stochastic process; these are functions which are non-deterministic. The term "stochastic" is reserved for this sort of indeterminacy, while "random" is used to describe a patternless outcome.

[5]The origin of this word is *stochos*, meaning aim, guess, or target, in ancient Greek.

What about the outcome of a dynamical system? What does it mean to say that an outcome, in this case an itinerary, is random? One way to think about this is that a random outcome is one that has no patterns or regularities. This seems like a subjective statement—perhaps there are regularities present that you can detect but I cannot. However, this notion of randomness can be made objective. Describing how this is done is the topic of the next several paragraphs.

Let us begin with an example of a non-random sequence. Consider a long sequence of L's and R's repeating in a regular pattern such as

$$\ldots LRLRLRLRLRLRLRLR\ldots. \qquad (14.4)$$

Such a sequence is clearly not random. As a result, this sequence can be compactly described. The simple phrase "alternate L and R forever" specifies the entirety of the infinite sequence. We can come up with a compact description of this sequence because it has a regularity or pattern.

This then suggests that we define a random sequence as one which is *incompressible*. A sequence is random if there does not exist a short algorithm[6] that generates the sequence. If there is a short algorithm, then the sequence is compressible. Since the algorithm is shorter than the sequence itself, we have compressed it—found a representation for it that is smaller than the original. For the sequence of Eq. (14.4) the algorithm is simply: print LR and repeat. This algorithm is certainly much shorter than the sequence itself.

In contrast, consider the following sequence which I generated by tossing a coin:[7]

$$HTHHTHTTTHTHHHHHTHHTHTTTTTHH. \qquad (14.5)$$

This sequence is presumably random. I do not see any obvious patterns to exploit, and so the shortest algorithm that reproduces Eq. (14.5) is: print $HTHHTHTTTHTHHHHHTHHTHTTTTTHH$. The algorithm contains a complete copy of the sequence. Thus, the algorithm is not shorter than the sequence, and so we say that the sequence is random.

As a final example, consider the following sequence:

$$11001001000011111101101010100010001000010110100001\ldots. \qquad (14.6)$$

This looks random, but it is not. The sequence is actually the beginning of the number π written in binary, or base-2.[8] As a result, there is indeed a relatively short algorithm for reproducing Eq. (14.6): we need to calculate the digits of π and then convert to base-2. The details of how to do this might be complicated, and it might take a long time to carry out, but that does not matter for this discussion. What does matter is that this algorithm is clearly a lot shorter than a very long sequence of the binary digits of π, and hence we would conclude that Eq. (14.6) is *not* random.

There are some details we need to address in order to make the idea of randomness as incompressibility sufficiently precise. What sort of

[6] "Algorithm" is a technical term for a finite and well-defined set of instructions for carrying out a task. You can think of an algorithm as a computer program, or a recipe, or as some other set of instructions.

[7] Actually, the sequence was generated for me by www.random.org, a website that uses atmospheric noise to generate random numbers.

[8] Binary is discussed in Section 21.5.

device is going to execute the algorithm that reproduces the sequence? Different devices may behave very differently, and the same algorithm could have different lengths when programmed on different machines. The solution to this puzzle lies in an abstraction known as a Universal Turing Machine (UTM). A UTM is a theoretical computing device that is capable of emulating all other computing devices. So the UTM is taken as the standard machine or computer that will execute our algorithms to reproduce a sequence.

There is another question about incompressibility that is more difficult: How can we tell if there is a short algorithm for a sequence? It was not at all obvious that Eq. (14.6) was actually the binary digits of π. And presumably there are lots of lots of other non-obvious patterns. Perhaps *all* sequences have some almost-impossible-to-determine algorithm that generates them, and so there is no such thing as randomness at all. What we need, then, is an algorithm for finding algorithms for compressing sequences. However, it can be proved that such an algorithm cannot exist. At this point it may seem that this business of using incompressibility as a measure of randomness collapses like a house of cards. This certainly does limit the use of this idea. But all is not lost; we can argue that there are not enough algorithms to explain all possible sequences, and so most sequences must be random. This argument proceeds as follows.

There are an infinite number of possible algorithms. There are also an infinite number of possible symbol sequences. However, there are many more sequences than algorithms. In fact, there are infinitely more sequences than there are algorithms that could generate them.[9] Thus, if one chooses an arbitrary infinite symbol sequence it almost surely will be incompressible, and hence random. It thus follows that if one chooses an arbitrary initial condition for the logistic equation and uses it to generate a symbol sequence, it will almost surely be random. Thus, the logistic equation produces a random sequence according to our definition of randomness as incompressibility.

But wait a minute. This still seems odd, since the orbits of the logistic equation are deterministic. Thus, is it not the case that iterating the logistic equation is an algorithm to generate the orbit? If so, then the sequences generated by the logistic equation are not random after all. Iterating the logistic equation is indeed such an algorithm, though there is an important catch. Suppose we wanted to use the logistic equation to generate an infinite symbol sequence. This could be done—the logistic equation with $r = 4.0$ is capable of producing *any* sequence. However, to generate this sequence exactly requires exactly specifying the initial condition, which is a number between 0 and 1. Almost all numbers between 0 and 1 are irrational, meaning that their decimal expansion goes on forever; it neither repeats nor terminates.

Thus, our algorithm for generating the sequence needs an exact initial condition, which is itself an infinite sequence of digits. So now the issue becomes whether or not we can find a short algorithm for the initial condition. It turns out that the vast majority of initial conditions are

[9]The number of algorithms is countably infinite while the number of sequences is uncountably infinite. These two different types of infinity are discussed in Chapter 21.

incompressible. The argument is again based on frequency: there are an infinite number of algorithms but infinitely more numbers between 0 and 1. The upshot is that we cannot use the logistic equation as a short algorithm for generating symbol sequences because such an algorithm requires the exact initial condition, which is itself infinite and incompressible.

So, symbolic orbits generated by the logistic equation really are random, in the algorithmic sense described above.[10] Thus, we have a deterministic dynamical system producing random results. One might expect that random results would require a random generation process, but the study of chaotic dynamical systems shows us that this is not the case. This is one of the key conceptual advances associated with the study of chaos. Before 1970 or so, most scientists would assume that a series of apparently random observations must have arisen by chance, via some sort of stochastic process. Chaos now tells us that this is not a valid assumption to make.

I have two additional remarks on deterministic randomness to conclude this section. First, being able to generate random numbers in a quick and efficient way is surprisingly useful. For example, in computer games we might want something to move randomly across the screen. In strategic interactions, randomness is important because you do not want your adversary to be able to predict your actions. Randomness is essential for cryptography, and it also plays a role in many other mathematical and computational applications. Often a problem is too difficult to solve directly, even with a computer. Instead, one needs to perform a type of random sampling and then average the results. This general process is known as a *Monte Carlo algorithm*. So randomness has many practical and important uses. Chaotic systems similar to the logistic equation are now used to generate random numbers for Monte Carlo algorithms and other applications. In this setting these numbers are often called *pseudo random* to indicate that their source is from a deterministic function and is not generated by a stochastic process.

This leads to my second set of remarks. Given that deterministic systems can produce random behavior, one may wonder if anything is truly random, in the sense of being generated by a random or stochastic process. Is there any chance in the universe? Or is chance just our name for what happens when we cannot predict the outcome of some event due to the butterfly effect? I have repeatedly referenced a coin toss as a truly random event. But the laws of physics that determine the trajectory of the coin are deterministic. So while we describe this as a random process—one usually says that whether the coin is heads or tails is a matter of chance—it cannot be truly stochastic, can it?

One way out of this bind is via quantum mechanics. Unlike the classical mechanics of Newton, discussed in Chapter 8, quantum mechanics is *not* deterministic.[11] Two quantum systems prepared in exactly the same way can behave differently. Randomness at the quantum level—atoms or small molecules—can be amplified by sensitive dependence on initial conditions, leading to everyday phenomena that truly are chance events.

[10]You may object on some level to the notion of algorithmic randomness. If so, there are statistical definitions of randomness based on entropy that are essentially equivalent and lead to the same conclusion.

[11]Almost all physicists agree that non-determinism is intrinsic to quantum mechanics and is not a result of our ignorance or because quantum mechanics is an incomplete theory. Remarkably, one can design experimental tests that prove there can be no hidden variables that render quantum mechanics a deterministic theory. For accessible introductions to these ideas, see Mermin (1992) or Styer (2000). There is an alternative formulation of quantum mechanics, usually called Bohmian mechanics, which *is* deterministic. However, in Bohmian mechanics it is possible for signals to travel faster than the speed of light, and hence it is not consistent with Einstein's theory of special relativity. As a result, the vast majority of physicists prefer the standard formulation of quantum mechanics, but there are some who are so bothered by non-determinism that they opt for the Bohmian approach.

However, there is not agreement whether or not this explains most (or even any) of the randomness we perceive in the world. Understanding different sorts of randomness remains a topic of current scholarship. Work in this area intersects fundamental physics, mathematics, computer science, and philosophy. See the Further Reading section at the end of this chapter for additional references you can turn to if you wish to dig deeper into these issues.

14.4 Linearity, Nonlinearity, and Reductionism

We now shift gears and turn our attention from randomness to nonlinearity. In mathematics there is a fundamental dichotomy between linear and nonlinear systems. Chaos is only possible in nonlinear systems, and so the field of study known as chaos or dynamical systems is also referred to as nonlinear dynamics. Thus, some remarks on linear and nonlinear equations seem in order, especially since the mathematical meanings of these terms are somewhat subtle and counterintuitive. Moreover, these terms are frequently used in non-mathematical contexts as well. In this section I try to unpack some of the different uses of these words.

One often hears the phrase "linear thinking" or reference to a linear approach to a problem. I think in this instance "linear" is being used in a mostly literal sense: linear thinkers like to think sequentially, moving in a clear order from one step to the next. A linear style of thinking might shy away from exploring multiple trains of thought or multiple paths to a problem's solution. Depending on the context, being called a "linear thinker" could be a compliment, or not.

But there is a somewhat more metaphorical and less literal sense in which "linear" and "nonlinear" are used to describe ways of thinking and approaches to problems. In this usage, a linear approach is often associated with reductionism, while nonlinear approaches are associated with less reductive or holistic views. I do not think there is a standard definition of reductionism, but it is generally understood to be the belief that the way to understand a complex object or phenomenon is by understanding the properties of its parts. In this line of thinking, the way to learn about the physical properties of a solid is to understand the properties of the atoms in the solid. Or the way to understand the behavior of a group of people is to learn as much as possible about the psychology of the individuals in the group.

Reductionism is sometimes described in overly simple terms: that the only way to understand an object is to take it apart and see what it is made of, denying the importance of the interactions between an object's constituent parts and failing to recognize that an object's properties may depend on its context. It is easy to make a reductive approach seem almost buffoonish, and in many circles being called "reductionist" is not meant as a compliment.

Reductionism in its extreme is surely a bad idea, but the same can be said about almost any "ism" in science, as a commitment to one explanatory framework precludes other types of understanding and inevitably narrows one's point of view. While reductionism should be approached with caution, I nevertheless think there is much good that can be said about reductive approaches. The reductive approach that has characterized much of modern physical science has been unarguably successful. By the standard criteria of science—the ability to predict the results of repeatable experiments—reductive approaches have been a stunning success. By learning about atoms and molecules and DNA and cells, we have learned a great deal about how some aspects of the world work. So, at least in this sense, reductive approaches work extremely well.

But more generally, I think that all knowledge is reductive. In order to try to understand our world, it is inevitable that we need to choose some portion of the world to focus on that is smaller than the world itself. I do not know any way around this. We cannot study the whole; all we can meaningfully analyze are parts. So I would argue that all knowledge is reductive in at least some sense. To me, the question is not whether or not to be reductive, but what sort of reduction to do and how to do it. There are different levels at which one can understand a phenomenon, and I think a pluralistic approach is best.

Different people will attack problems in different ways, and different problems call out for different approaches. There is, however, some risk in reductive approaches that I think we should be wary of. By necessity we have to study parts and not wholes. But it is dangerous to then imply that the part *is* the whole. For example, there is nothing wrong with studying an organism's genes; but I do think there is something wrong with then implying that an organism *is* its genes or that all traits or behaviors are purely genetic. There is also a tendency in many circles to privilege reductive knowledge. Studying genes and elementary particle physics is often viewed as more prestigious than studying ecology or the physics of macrophenomena, like weather or the properties of materials.

In any event, let us now turn our attention more directly to linear and nonlinear functions and how these might be related to reductionism. I will do so via an example. Suppose you are going shopping at the local grocery store. Your only task is to get ice cream for a gathering that will happen later that evening. Let us imagine that ice cream costs $3 a pint. You put a bunch of pints of ice cream in your shopping cart and head to the checkout line. How much will the ice cream cost? Three dollars times the number of pints you have.

This relationship is graphed in Fig. 14.3. Not surprisingly, the function is linear. This expresses the fact that no matter how many pints of ice cream you buy, the cost per pint is the same. So, imagine that we want to "understand" your cart of groceries. In this particular context, "understand" means "figure out the cost of". We can understand the cart very easily. In fact, we can understand the cart by considering each pint of ice cream one at a time. We figure out the total cost by figuring out the cost of each pint, and then adding them up to get the total cost.

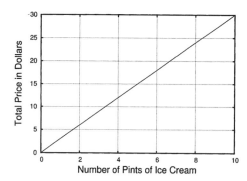

Fig. 14.3 The price of ice cream as a function of the number of pints you buy. In this case the price is a linear function of the amount of ice cream purchased.

A consequence of this is as follows. Suppose you have ten pints in your cart, and you are at the store with three friends. Each of you could check out in a different check-out line. One of you could take two pints, the other three, the other five. You would end up paying just as much if you checked out all at once. No matter how you divided up the pints, your final price would be the same. The whole price of the cart of groceries is simply the sum of its parts.

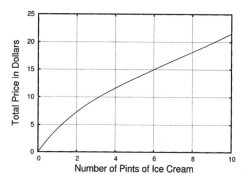

Fig. 14.4 The price of ice cream as a function of the number of pints you buy. In this case the price is a nonlinear function of the amount of ice cream purchased.

Now, suppose that the total price is given by the nonlinear function in Fig. 14.4 as opposed to the linear function of Fig. 14.3. This corresponds to a situation in which there was a volume discount. In this scenario, two pints will cost around $7, while ten pints of ice cream will only cost around $21. So if you get ten pints, you are paying $21/10 =$2.10 for each pint. But if you get two pints, you pay $3.50 per pint.

Imagine again that our task is to understand your grocery cart. Simply grabbing the pints one by one, determining their prices, and then adding up all the prices to get the total price of the goods in your cart will not work. If you and your friends were to check out separately, you would not get the same price as if you checked out all at once.

One way to describe this state of affairs is to say that there is an *interaction* between the pints of ice cream. The pints do not interact literally, in that they are not bouncing off each other. But the presence of one pint has the effect of changing the price of other pints. This is what leads to the lower price when you get a lot of ice cream. Ignoring this interaction and trying to understand the system merely as the sum of its parts will lead to the wrong conclusion about the price of the cart full of ice cream. It is in this sense that there is a connection between nonlinear functions and interactions. A nonlinear relationship means that one cannot break up the system into its components; there are interactions between components that mean that whatever feature of the system the function captures cannot be re-expressed as the sum of its parts.

Here is another way to say this. Let $(x, f(x))$ correspond to x-y pairs for a function. For the linear ice cream example of Fig. 14.3, two such pairs are $(3, 9)$ and $(5, 15)$. For a linear function we can add these two pairs together to get a new pair that is also on the line:

$$(3, 9) + (5, 15) = (8, 24) . \tag{14.7}$$

The point $(8, 24)$ is on the line graphed in Fig. 14.3. However, this does *not* work for nonlinear functions. If you add two points that are on the graph in Fig. 14.4, the resultant point is not on the graph. (You might wish to check this for yourself using the figure.) This is another way of seeing that the whole cannot be decomposed as the sum of its parts.

Generalizing this example leads to a more general definition of linearity. A function is linear if it has the property that if A is a solution and B is also a solution, then $A + B$ is also a solution. This means that if you have two different solutions and add them together, you have a third solution. And to this third solution one could add, say, the first solution, getting a fourth solution, and so on. In addition, for a function to be linear it must be the case that if A is a solution then cA is as well, where c is any number.[12] Combining these two conditions, we say that a function is linear if

$$A \text{ and } B \text{ solutions} \Rightarrow A + cB \text{ is a solution, for all } c , \tag{14.8}$$

where $\Rightarrow$ should be read "implies that".

It is Eq. (14.8) that defines linearity. This definition applies to all functions, not just the one-dimensional functions that we have studied so far. Confusingly, not all functions that are lines fit the definition of a linear function. If the function has a y-intercept that is not zero, it will not satisfy the criteria of Eq. (14.8). You will encounter an example of this in the exercises for this chapter.

Equation 14.8 tells us that linear systems are made up of regular "building blocks"—entities that make up the whole, and whose properties are unaffected by the presence of other building blocks. In the ice cream example, the building block is a single pint of ice cream, and the property is the price per pint. For our nonlinear example, Fig. 14.4, the properties of the pints, their price, changes as more pints are added. This means the nonlinear system is not decomposable in the same way that the linear system is. Thus, modeling a system with a nonlinear function is generally a less reductive approach. In contrast, if one models a system with a linear function, one is implicitly assuming a sort of decomposability.

[12] For example, the point $(3, 9)$ is a solution to the ice cream pricing equation, graphed in Fig. 14.3. Then 2 times this solution is also a solution; the point $(6, 18)$ is also on the line in Fig. 14.3. The same will hold true for any value of c, even if c is not an integer.

This decomposability lends a certain simplicity to a linear system. There exist a host of solution methods for such equations that involve solving the equation for several simple building blocks, and then building up full solutions as a combination of these building blocks. These methods are not always easy, but it is still usually much, much easier to solve a linear system than a nonlinear one. There are general methods for solving classes of linear equations, but few general techniques work for nonlinear equations. But it is not just the solution methods for linear systems that are simple; the dynamics themselves are simpler as well. In fact, in order for a system to be chaotic it must be nonlinear.[13]

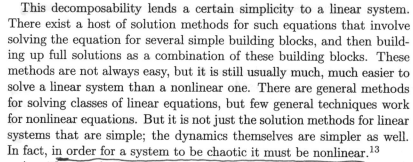

---

[13]An exception to this occurs in infinite dimensional systems, which are far beyond the scope of this book.

As with several other claims I have made in this chapter I cannot prove this assertion, but I can argue why this is so. For the one-dimensional iterated functions we have focused on so far, one can see geometrically that a nonlinear function is needed for chaos. A linear function can have at most one fixed point, and depending on its stability, either all initial conditions will get pulled toward it or pushed away from it. So there is no possibility for a chaotic orbit. What about other types of dynamical systems, such as those in two and three dimensions, which will be the focus of Part V of this book? Here the situation is not as straightforward. There are a number of mathematical subtleties and different cases to consider. However, the basic intuition is that if a system is linear its solutions are made up of building blocks that do not interact. This decomposability or modularity limits the complexity of the solutions such that it is impossible to have orbits that are both aperiodic and have sensitive dependence on initial conditions.

14.5   Summary and a Look Ahead

In this part of the book we have encountered chaos: a dynamical system is chaotic if it is deterministic, has bounded, aperiodic orbits, and sensitive dependence on initial conditions (SDIC). For such a system, long-term prediction is not possible. A small inaccuracy in the measurement of the system's initial condition is quickly magnified under iteration, and so only short-term prediction is feasible. The orbits of a chaotic dynamical system appear random, even though they are generated by a deterministic rule.

Yet systems that are chaotic in the mathematical sense are not structureless. By examining the bifurcation diagram, we have seen regularities in how the logistic equation shuttles back and forth between periodic and chaotic orbits. There is a pattern to the way that the equation's behavior changes as the parameter r is increased. Remarkably, some features of this pattern are universal. Several properties of the period-doubling transition to chaos are the same for different equations and even different physical systems. Not only is there "order in chaos", but some features of that order are the same for different chaotic systems.

Chaotic systems also possess statistical regularities. Two orbits with slightly different initial conditions will follow very different trajectories.

However, histograms constructed from these two trajectories will be very similar, and thus their average properties will also be very similar. A chaotic dynamical system thus combines elements of order and disorder, predictability and unpredictability.

In Chapter 8 I discussed the Newtonian worldview and the hope of Laplacian determinism. Chaos does not supplant Newton's laws, but it does oblige us to rethink at least two of the assumptions that are often a part of the Newtonian view. First, sensitive dependence on initial conditions (SDIC) complicates the Laplacian notion that increasingly accurate measurements will lead to increasingly accurate predictions. It is still strictly true that better measurements lead to better predictions, but a system with SDIC will never be predictable in the long run. Second, chaos shows us that it is not always the case that a simple equation will have simple dynamical behavior. In particular, a deterministic system can be unpredictable and appear random. Randomness and order are not mutually exclusive.

In the next part of the book I will introduce fractals. At first, we will leave dynamical systems to the side as we learn what fractals are and how to characterize them using dimensions. After several chapters, however, we will see that fractals and dynamical systems are linked. Intricate fractals often arise from simple, iterated processes.

Further Reading

As one might imagine, there is a vast literature on different notions of randomness, the extent to which randomness is compatible with determinism, computability and uncomputablilty, and a host of related issues. *Chance and Chaos* by David Ruelle (1993) is an accessible and engaging introduction to these issues. Ruelle writes lucidly about some potentially vexing philosophical and mathematical ideas. For a focused discussion of the philosophical implications of the butterfly effect and what chaos has to say about different sorts of determinism, I recommend Chapters 2 and 3 of Kellert (1993). A very clear, not-too-technical discussion of different types of infinities, Universal Turing Machines, computability, and randomness can be found in Chapters 2 and 3 of Gary Flake's book *The Computational Beauty of Nature* (1999). Chapter 14 of Flake's book relates these ideas to chaotic dynamical systems in a clear and compelling way. I also recommend the essay on causal determinism in the *Stanford Encyclopedia of Philosophy* (Hoefer, 2010); it covers much more territory than just the relationships between determinism and chaotic dynamics. Joe Ford's article "How Random is a Coin Toss?" (1983) is a succinct, although in places fairly technical, overview of randomness in physics. Brian Hayes' article "Randomness as a Resource" (2001) is an entertaining and fascinating overview of the uses of randomness in algorithms and the challenges associated with generating good random numbers.

Exercises

(14.1) Convert the following itinerary into its symbolic representation: 0.33, 0.8844, 0.408947, 0.966837, 0.128253, 0.447215, 0.988855, 0.0440827, 0.168558, 0.560584, 0.985318, 0.0578647.

(14.2) ♯ In this exercise you will derive the fact that any initial condition between 0 and approximately 0.15 will stay between 0 and 0.5 when iterated by the logistic equation with $r = 4.0$: $f(x) = 4x(1-x)$. This was shown graphically in Fig. 14.1. Show this using algebra by finding the x value(s) such that $f(x) = 0.5$. To do so you will need to use the quadratic formula; see Appendix A.2.

(14.3) Starting with Fig. 14.2, sketch the partition for three symbol sequences. The lengths of the intervals do not need to be exact.

(14.4) Suppose a fair coin is tossed many times. You are interested in the frequency with which you see particular sequences of six tosses. What fraction of the time would you expect to see each of the following?

    (a) *HTTHHT*

    (b) *HHHHHH*

    (c) *HHTTHH*

    (d) *HTHTHT*

(14.5) Suppose you observe a very long symbol sequence produced by the logistic equation with $r = 4.0$. You are interested in the frequency with which you see particular sequences of six symbols. What fraction of the time would you expect to see each of the following?

    (a) *LRRLLR*

    (b) *LLLLLL*

    (c) *LLRRLL*

    (d) *LRLRLR*

(14.6) Produce a real-life example of a linear function. Why is it linear?

(14.7) Produce a real-life example of a nonlinear function. What is it about the situation that makes in nonlinear?

(14.8) Consider the nonlinear ice cream pricing, shown in Fig. 14.4.

    (a) How much do four pints of ice cream cost? What is the cost per pint in this case?

    (b) How much do eight pints of ice cream cost? What is the cost per pint in this case?

(14.9) Suppose the ice cream situation is the same as it was in Fig. 14.3, but now there is a $1 fee that you have to pay when you check out at the cashier. The fee is $1 regardless of how much ice cream you buy.

    (a) Graph the price of ice cream as a function of the number of pints.

    (b) Is the function you just plotted linear in the sense of this chapter?

(14.10) Is the function $f(x) = x^2$ linear? To check, determine two (x, y) pairs that satisfy the function. Then add these points together and see if this new point satisfied the function.

(14.11) A one-dimensional linear function $g(x)$ has a solution $(2, 4)$. Find two other solutions to this function. Sketch $g(x)$.

Part III

Fractals

Introducing Fractals

15

We will now put aside chaos and dynamical systems and focus instead on a different topic: fractals. This will not be a complete departure from what has come before; we will see that iteration plays a key role in the generation of fractals. Gradually over the next several chapters we will see that fractals and the sorts dynamical systems we have studied in the first two parts of this book are closely related.

15.1 Shapes

Fig. 15.1 Three shapes from ordinary geometry: a circle, a line segment, and a rectangle.

Figure 15.1 shows three familiar shapes from geometry: a circle, a line segment, and a rectangle. These geometric forms are abstractions of shapes that we encounter in the physical world: a round coffee mug, a clothesline, and a tabletop. But the world we live in is much richer than this. There are many shapes—the branches of a tree, the bumps of a mountain range, the meander of a river—that do not resemble the shapes of Fig 15.1. Consider the images shown in Fig. 15.2. These objects from the natural and physical world are very different from the simple circles and lines of ordinary geometry. Sure, we could describe a winding river or branching trees as a collection of line segments arranged in a particular way. But it seems that this would be missing the essence of the shape that we are trying to describe.

The desire to better describe forms such as trees and winding rivers leads us to a different sort of shape. We construct an initial example by an iterative process, shown in Fig. 15.3. At $n = 0$ we start with a small square. We can think of this as a seed, playing the role that the initial condition x_0 does for an iterated function. To get to the shape at step $n = 1$, we make four copies of the shape at $n = 0$, and place one copy at each of the corners. We then repeat, or iterate, this process. To get to step $n + 1$ we take the shape at n, make four copies of it, and place one copy at each of the corners. The result, as one applies this rule over and over and over, is an intricate (and very large) structure that resembles a snowflake. This snowflake shape is known as a **fractal**. I will describe

Fig. 15.2 Four naturally occurring objects that are well viewed as fractals. Clockwise from top left: tree silhouette (Bruce Thompson, licensed under Creative Commons CC BY-NC-SA 2.0); romanesco broccoli (Licensed under Creative Commons CC0, photo courtesy PDPhoto.org); Baltoro glacier and the Karakoram mountains (Guilhem Vellut, licensed under Creative Commons Attribution-Share Alike 2.0 Generic license); and the Ganges river delta (NASA Earth Observatory).

fractals more fully below, but for now the main thing to note is how different it is than a simple square or circle.

For our next example of a fractal, consider Fig. 15.4. Again, this fractal is constructed via iteration. We start at step $n = 0$ with a line segment. We then remove the middle third of that line segment to obtain the shape labeled $n = 1$. Repeating this step, we get the shape at $n = 2$. That is, we remove the middle third of each line segment at $n = 1$. We keep on doing this, removing at every step the middle third of every line segment. If we carry this process on forever, the result is an infinite number of very, very, very tiny line segments.[1] The collection of these minuscule line segments is known as the Cantor set.

---

[1]Strictly speaking, in the limit that n goes to infinity, the tiny line segments approach points.

15.2 Self-Similarity

The Cantor set and the snowflake fractal are both self-similar. What this means is that a small portion of them looks like the whole. For

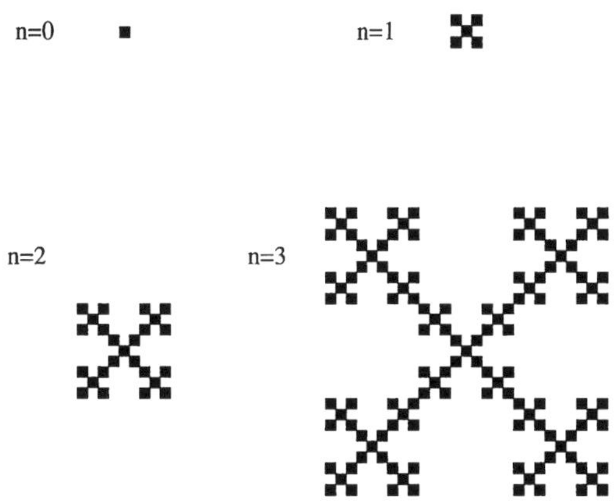

Fig. 15.3 The first several stages in the construction of a "snowflake" fractal. One starts at $n = 0$ with a seed shape, in this case a small square. One builds the next shape by placing a copy of the previous shape on each of the four corners of the current shape.

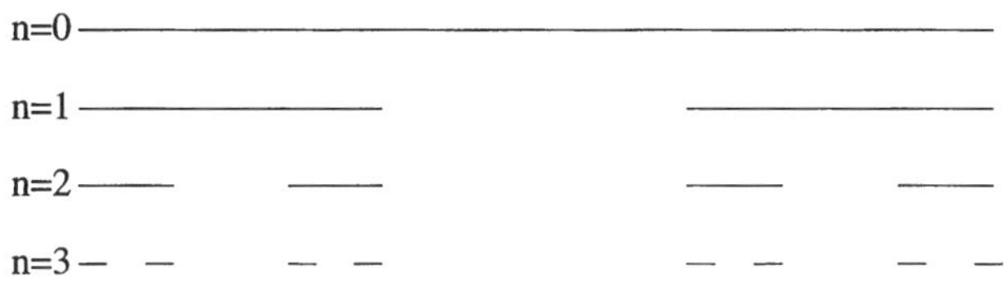

Fig. 15.4 The first several stages in the construction of the middle-thirds Cantor set. One starts with a line segment. At every successive step, one removes the middle third of every line segment in the previous step.

example, imagine breaking off an arm of the snowflake. What you have looks like a miniature copy of the entire snowflake. You could magnify the small portion and you will get a shape that looks like the whole snowflake. You could break a smaller arm off of the arm you already broke off, and again you would have something that is a miniature copy of the whole snowflake. The same property is true of the Cantor set, although this might be a little bit harder to see. If you take a small portion of the set you will have something that is a miniature copy of the entire set.

The natural fractals shown in Fig. 15.2 are also self-similar. Here the self-similarity is not exact; small portions of each picture are not identical to the large picture, but rather bear a close resemblance. For example, a branch of a tree looks similar, but not identical, to the full tree.

This property is called **self-similarity**; an object is self-similar if it contains replicas of itself of many different sizes. Fractals are self-similar geometric objects. As a counter-example, consider a person. People are not fractals. If you break off a person's arm, what you have in your hands will look like an arm, and not a small copy of the person. So a person is not self-similar, and hence is not a fractal.

Note that there are some non-fractals that are nevertheless self-similar. An example is a line segment. If you break off a portion of a line segment, you have another line segment, which looks just like the original line segment. However, this is not considered a fractal, for reasons that will be discussed in Chapter 16. For now, we can think of fractals as being geometric shapes that are self-similar, but in a "non-trivial" or "non-boring" way. Line segments are self-similar in a boring fashion, and hence are not considered fractals.

Here is another way to think about self-similarity. Suppose one day you woke up and you had significantly changed size. Perhaps you wake up Monday morning and you are only six inches tall. How could you tell? A clear indication would be that you were suddenly smaller than your pillow, your alarm clock would appear gigantic, it might not be safe to jump out of bed, and your cat might eat you. In brief, it would be immediately and stunningly obvious to you that you were very small.

However, suppose you lived in a fractal world—perhaps in an arm of the fractal snowflake of Fig. 15.3. Remember that this fractal gets very, very large as n, the number of times we apply the snowflake-building rule, gets large. If you lived on this snowflake, you would not be able to tell that that you had been shrunk. There are no clues as to scale or size. The entire universe consists of snowflakes within snowflakes within snowflakes. So there is no way to tell how big you are. The only thing you can measure yourself with is a snowflake, and there are snowflakes of all different sizes, so you can never really tell how big you are.

15.3   Typical Size?

As discussed in the preceding paragraphs, for a fractal there is not a typical size. For the fractal snowflake, the snowflake motif is repeated again and again and again at different sizes. So it does not really make sense to talk about an average snowflake size. The Cantor set of Fig. 15.4, after we have iterated the construction process many times, consists of points that are clustered together. But these clusters are clustered into clusters, which in turn are clustered into bigger clusters, and so on.

For ordinary, non-fractal objects we are used to describing them by stating their size: a circle with a radius of two inches, a chair that is 3.5 feet tall, a track on a CD that lasts five minutes. For more complicated situations, we often resort to averages or statements about the size of a typical instance of some entity. For example, we might say that a typical person is 5.5 feet tall, an average cat weighs 9 pounds, and most pints of ice cream cost around $2.50.

But this type of description does not work for fractals; stating the average size of a cluster in the Cantor set or the size of a typical snowflake motif in the snowflake fractal does not really capture the essence of the shape. In fact, we shall see in Chapter 19 that there are some fractals for which the concept of an average size is not well defined. But this begs the question: how *can* we describe fractals if not with averages? The answer to this question will come in the next chapter, where we will see that the geometric notion of dimension can be extended so as to capture some of the structural features of fractals.

15.4   Mathematical vs. Real Fractals

Finally, a few words about mathematical versus real fractals. The fractal snowflake and the Cantor set are abstractions. Strictly speaking they are defined by the result of an infinite iteration process. As such, they have structure on all scales. For example, you could keep zooming in on such a mathematical fractal and you would see the same shape repeated endlessly.

The same is not true for real fractals. Consider, for example, a fern. We can zoom in on the fern, and we will see smaller copies of the fern. But eventually this stops, and the fern no longer looks like little ferns. Instead we start to see individual cells. Or, consider the winding path of a river. We could imagine starting with an aerial view taken by a satellite, and then start zooming in. For quite some time we could see a self-similar shape; meanders on top of meanders, wiggles on top of wiggles. But eventually we will zoom in to the river itself, and we would just see water. Similarly, if we started with the satellite image and zoomed out, eventually the river would disappear from view. So real fractals have some cut-off sizes above and below which an otherwise self-similar object fails to appear self-similar. In contrast, mathematical fractals have no such cut-off; in principle we could keep zooming in forever, and see ever-smaller copies of our original shape.

This does not mean, however, that mathematical fractals are useless for describing the real world. It is important to remember that a fractal, like any geometric object, is just an abstraction or an idealization. For example, consider a perfect, mathematical circle. Such a thing is just an idea, albeit a very useful one. In Fig. 15.1 I have shown a drawing of a circle, made with the graphics program xfig. It is a nice circle, but it is not perfect. Look closely enough—you might need a magnifying glass—and you will see imperfections. And in nature, in the physical world in which we live, there certainly are no perfect circles, just shapes that approximate circles to varying degrees. Nevertheless, it would be hard to argue against the geometric idea, or ideal, of a circle. It certainly is a useful approximation or abstraction of things we encounter in the physical world.

So it is with fractals. In the real world there are no perfect fractals. But many have found that fractals are a tremendously useful and evocative way of capturing the qualities of shapes and processes that are, to varying degrees, self-similar.

Exercises

(15.1) Produce a real-life example of an object or phenomena that is a fractal. Explain why your example fits the criteria for being a fractal.

(15.2) The first two steps ($n = 0$ and $n = 1$) in the construction of a fractal known as the Sierpiński triangle are shown in Fig. 15.5. Sketch the shape for $n = 2$ and $n = 3$.

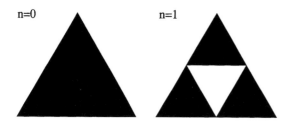

Fig. 15.5 The first two steps in the construction of the Sierpiński triangle.

(15.3) The first several rows of Pascal's triangle are shown in Fig. 15.6. Pascal's triangle is constructed as follows. The top and edges of the triangle are filled with 1's. Other entries in the triangle are obtained by adding the two entries directly above it. For example, the hexagons with the number 4 in them in Fig. 15.6 both have a 1 and a 3 above them.

(a) Continue the triangle so that it extends for fifteen to twenty rows. Use hexagonal graph paper, which you can download at http://incompetech.com/graphpaper/hexagonal/.

(b) Now shade in all the cells that have an odd number in them.

(c) Does the resultant shape look familiar?

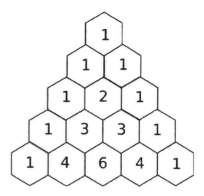

Fig. 15.6 The beginning of Pascal's triangle. The value in each hexagon is equal to the sum of the two hexagons above.

Dimensions

16

In the previous chapter I introduced fractals: self-similar geometric objects. And I discussed how to generate fractals via a geometric iterative process. In this chapter we continue our exploration of fractals by learning how to characterize them by means of the dimension. By defining dimension in terms of the scaling properties of a shape, we will come up with a quantitative way of describing fractals. We begin with a simple geometric exercise that will lead us to the definition of the self-similarity dimension.

16.1 How Many Little Things Fit inside a Big Thing?

Suppose you have a line segment and then magnify it by a factor of 3. The line segment is now three times as long as before. Hence, three of the small line segments fit inside the new, larger line segment. This is illustrated in Fig. 16.1. Now, try the same thing with a square. We start with a small square, and then magnify it by a factor of 3. This means that it is now three times as long and three times as tall as it was before. As you can see in Fig. 16.1, 9 copies of the small square fit inside the bigger square. Finally, let us imagine the same experiment with a cube. We start with a small cube and magnify it by a factor of 3. The cube is now three times as wide, three times as tall, and three times as deep. In this bigger cube, one can fit 27 of the smaller cubes, illustrated in Fig. 16.1.

The results of these experiments are summarized in Table 16.1. In all instances, the magnification factor is 3; it is as if we have put the shape in a photocopier machine and expanded by a factor of 3 in all directions.

Shape	Magnification Factor	Number of small copies that fit within big copy
Line	3	3
Square	3	9
Cube	3	27

Table 16.1 Effects of magnification on different shapes.

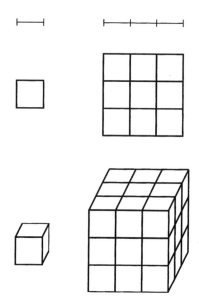

Fig. 16.1 Illustrating the number of small copies of an object that fit inside a big copy.

All lengths in the shape are three times as long as they were previously. E.g., the bigger square in Fig. 16.1 is three times taller and three times wider than the little square.

The question before us, then, is: what property of a shape determines how many small copies of it fit in a bigger copy? The answer to this question is the object's **dimension**. The line, square, and cube all have different dimensions: the line is one-dimensional, the square is two-dimensional, and the cube is three-dimensional. Looking at the table, we see that we can relate the magnification factor, the number of small copies, and the dimension as follows:

$$\text{number of small copies} = (\text{magnification factor})^D \,, \tag{16.1}$$

where D is the dimension.

We may view Eq. (16.1) as defining the dimension D. This might seem like a strange definition, but I hope to have convinced you that it reproduces what we already know about dimension. For example, we expect that a square is two-dimensional. Plugging in $D = 2$ and using the values for the square from Table 16.1, we get

$$9 = 3^2 \,, \tag{16.2}$$

which certainly is true. This definition of dimension is known as the **self-similarity dimension**, because it tells us how many small self-similar pieces of an object fit inside a large piece. There are other other definitions of dimension, but for almost all situations they turn out to be equal. I will usually refer to the dimension D defined in Eq. (16.1) as *the* dimension, unless the context calls for greater specificity. Let us now apply this definition to one of the fractals from Chapter 15.

16.2   The Dimension of the Snowflake

The steps in the construction of the snowflake fractal are illustrated in Fig. 16.2. Our goal is to use Eq. (16.1) to determine the dimension of the snowflake. We begin by focusing on what happens from step $n = 0$ to $n = 1$. The basic shape here is the little square. We can see that there are 5 little squares in the larger shape corresponding to $n = 1$ in the figure. The magnification factor is 3; the new shape is three times as tall and three times as wide as the previous shape.

We can now plug into Eq. (16.1) to obtain

$$5 = 3^D . \tag{16.3}$$

All that remains is to solve this equation for D. We might have been anticipating that the dimension D would equal 2. However, plugging $D = 2$ into Eq. (16.3) does not work; the right-hand side will be $3^2 = 9$, which does not equal 5. We might then try a dimension of 1, but this will not work either. A D of one makes the right-hand side of Eq. (16.3) equal to $3^1 = 3$, which is smaller than 5.

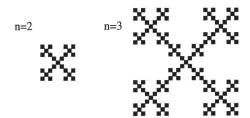

Fig. 16.2 The steps in the construction of the snowflake fractal.

The conclusion, then, is that the dimension must be between 1 and 2! Evidently, the dimension is not an integer. How can this be? Let us put this question on hold for just a moment, and get back to the task of solving Eq. (16.3). We first look for a solution by guessing and checking. Since we know that D is between 1 and 2, let us try 1.5. Doing so, we get

$$3^{1.5} \approx 5.196 . \tag{16.4}$$

(I obtained this number using a calculator.) So 1.5 is too large; we want the right-hand side to be as close to 5 as possible. So let us try $D = 1.4$:

$$3^{1.4} \approx 4.656 . \tag{16.5}$$

Evidently 1.4 is too low. How about $D = 1.45$?

$$3^{1.45} \approx 4.918 . \tag{16.6}$$

A little too low. Let us try $D = 1.46$:

$$3^{1.46} \approx 4.973 . \tag{16.7}$$

We could keep going, trying to make the right-hand side closer and closer to 5. But let us stop here—we are pretty close, and this guessing and checking game can get tedious fairly quickly.

In any event, we have found that the dimension D of the fractal snowflake is:

$$D \approx 1.46 . \tag{16.8}$$

We will consider the meaning of this shortly. But before doing so, let us discuss another way to solve Eq. (16.2). Our task is to solve the following equation for D:

$$3^D = 5 . \tag{16.9}$$

In the preceding paragraphs we found D by guessing and checking. As you might have suspected, there is a quicker way. The key is to use logarithms. A review of logarithms and their basic properties can be found in Appendix A.4.

As a first step, we take the logarithm of both sides of Eq. (16.9):

$$\log(3^D) = \log(5) . \tag{16.10}$$

Using the logarithm property that

$$\log(A^n) = n \log(A) , \tag{16.11}$$

we have

$$D \log(3) = \log(5) . \tag{16.12}$$

Dividing both sides of this equation by $\log(3)$, we obtain

$$D = \frac{\log(5)}{\log(3)} . \tag{16.13}$$

Evaluating this last expression using a calculator, we find that:

$$D \approx 1.46497 . \tag{16.14}$$

[1]Logarithms are incredibly useful. If you had a bad experience with logarithms in the past, as seems to be the case for many students, I encourage you to try again to form a good relationship with them.

Note that this agrees well with the value of 1.46 that we found by guessing and checking. If you do not like logarithms and prefer to solve equations like Eq. (16.9) by guessing and checking, that is OK. Both methods will get you the same answer.[1]

16.3 What does $D \approx 1.46497$ Mean?

So what does a non-integer dimension mean, anyway? There are several ways to think about this. First, a dimension between 1 and 2 means that the shape has some qualities of two-dimensional objects and some of one-dimensional objects. The snowflake is two-dimensional in the sense that it resides in two-dimensional space. It rests on the surface of a piece of paper, which is two-dimensional. However, as the process of building the snowflake proceeds, the shape becomes more and more "edgy", in the sense that its perimeter grows and grows. The shape starts to look like a very long line that is bent and folded to make a snowflake. Thus, the snowflake combines elements of one and two dimensions, befitting an object with a dimension of 1.465.

Another way of giving meaning to a dimension like 1.46497 is as follows. The essential feature of a fractal is that it is self-similar; it is made up of small parts that each resemble the whole, and those small parts are made up of smaller parts that resemble the whole, and so on. So, as suggested in Section 15.3, for a fractal it is not always meaningful to speak of the average or typical size of a component of a fractal. Rather, we want to capture something about what stays the same as we examine the fractal at different length scales.

For the snowflake, what we have seen is that if we increase the length by 3, we get 5 new parts. It is this relationship between 3 and 5 that is constant across scales. This relationship is expressed in the equation

$$3^D = 5 \,. \tag{16.15}$$

Knowing the value of D tells us that if we increase the magnification by 3, we will see 5 times as many pieces. Or, equivalently, every element is made up of 5 smaller elements, each of which looks like the element itself, but scaled down by a factor of 3.

Here is another way to see that the dimension captures something that is constant across scales. In Fig. 16.2, suppose we compare the $n = 0$ and $n = 2$ steps. The shape at $n = 2$ has 25 pieces in it. And each small piece needs to be magnified by 9 to be as wide as the full $n = 2$ shape. So, using these values in Eq. (16.1), we get

$$9^D = 25 \,. \tag{16.16}$$

This equation has the same solution as Eq. (16.3). You can easily check this by plugging in $D = 1.46497$. The point is that D relates the number of small copies of an object to the change in scale, or magnification factor. This relationship is the same for any scale-change we consider. It is in this sense that D captures what stays the same across different scales.

16.4   The Dimension of the Cantor Set

For our next example we return to the Cantor set. The steps in the construction of the Cantor set are shown in Fig. 16.3. The magnification factor is 3, as it was for the snowflake; each line segment must be

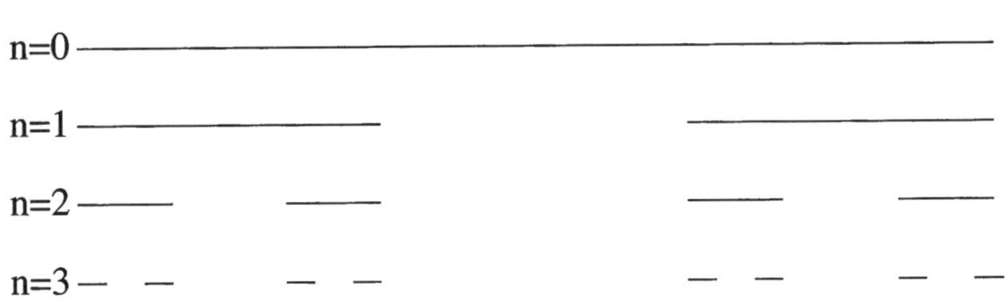

n=0 ——————————————————————

n=1 ——————————   ——————————

n=2 ———   ———   ———   ———

n=3 — —   — —   — —   — —

Fig. 16.3 The steps in the construction of the Cantor set.

stretched to 3 times its length to be as long as the line segments at the previous step. And the number of small copies is 2. At each step there are two small Cantor sets which can be scaled up to reproduce the original. Thus, the dimension equation for the Cantor set is:

$$3^D = 2 . \tag{16.17}$$

To solve for D we take the logarithm of both sides:

$$\log(3^D) = \log(2) . \tag{16.18}$$

Simplifying and solving for D, we obtain:

$$D\log(3) = \log(2) , \tag{16.19}$$

$$D = \frac{\log(2)}{\log(3)} \approx 0.6309 . \tag{16.20}$$

The dimension between 1 and 0 indicates that the Cantor set is in some regards line-like (one-dimensional) and in some regards point-like (zero-dimensional). The Cantor set is constructed from one-dimensional lines, but after so many line segments have been removed, what we are left with is a collection of points. Or, one can think of the Cantor set as being made up of so many points that, even though the points are disconnected, the Cantor set is in some ways line-like. In either case, we see that the Cantor set is between zero and one dimensions.

16.5   The Dimension of the Sierpiński Triangle

As a final example, let us determine the self-similarity dimension of the Sierpiński triangle, a fractal whose construction is illustrated in Fig. 16.4. Looking at the $n = 1$ stage, we can see that there are 3 small copies of the triangle inside the full shape. How big is each small copy compared to the large one? The magnification factor for this fractal is 2. The

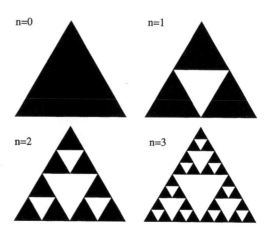

Fig. 16.4 The steps in the construction of the Sierpiński triangle.

easiest way to see this is to look at one of the sides of one of the small triangles. We can see that this side of the small triangle is exactly half the length of the side of the full triangle. So we would need to magnify it by a factor of 2 for it to be as large as the big triangle. Thus, our dimension equation is:

$$2^D = 3 . \tag{16.21}$$

Solving for D, we obtain

$$D = \frac{\log(3)}{\log(2)} \approx 1.585 . \tag{16.22}$$

16.6 Fractals, Defined Again

We are now in a position to give a somewhat more precise definition of a fractal. First, we need to introduce a different sort of dimension: the **topological dimension**. The topological dimension of an object is our intuitive notion of dimension. The topological dimension of a point is 0, of a line is 1, of a plane is 2, and of a cube is 3. Consider the Cantor set. It is made up of points, and thus its topological dimension is zero. It starts off as line segments, but in the limit that an infinite number of line segments are removed, all that are left are points. What about the topological dimension of the Sierpiński triangle? In the limit that we remove more and more triangles, we are left with a structure that is made up of tiny little line segments.[2] Thus, the topological dimension of the Sierpiński triangle is 1.

We can now state one definition of a fractal: A fractal is a geometrical object whose self-similarity dimension is greater than its topological dimension. The Cantor set has a self-similarity dimension of 0.6309 and a topological dimension of 0. And the Sierpiński triangle has a self-similarity dimension of 1.585 and a topological dimension of 1. So, as expected, both the Cantor set and the Sierpiński triangle are fractals by this definition.

This definition for a fractal is fairly standard. However, there is not universal agreement on the definition. Many simply define fractals to be any shape or object that displays self-similarity. Kenneth Falconer, the author of one of the more widely used and influential textbooks on fractals, argues that the "definition of a 'fractal' should be regarded in the same way as a biologist regards the definition of 'life' " (Falconer, 2003, p. xxv). Rather than a rigid definition, Falconer puts forth a number of qualities which most fractals have.

(1) They exhibit self-similarity across a range of scales. This self-similarity can be exact or approximate.

(2) They are not well described by usual geometric forms like circles, cones, and lines.

(3) The self-similarity dimension, or some similar dimension, is larger than the topological dimension.

[2]It might appear that the Sierpiński triangle also turns into a dust of points, much like the Cantor set. However, in the Sierpiński triangle, line segments remain. This can be seen most clearly by looking at the boundaries of the triangle; clearly these are line segments and not points.

Finally, I should mention that there are a number of variants on the self-similarity dimension. These variants, such as the capacity dimension, Hausdorff dimension, and the box-counting dimension, are defined differently, but capture the same idea. All measure how the bulk or volume of the shape scales. One of these variants, the box-counting dimension, is the subject of Chapter 18.

Exercises

(16.1) Consider the fractal snowflake of Fig. 16.2. Determine the magnification factor and the number of small pieces in the big shape when going from $n = 0$ to $n = 3$. Use these numbers and Eq. (16.1) to determine the dimension.

(16.2) Explain why it makes sense for the Sierpiński triangle to have a dimension between 1 and 2. In what ways is it one-dimensional and in what ways is it two-dimensional?

(16.3) Consider a Cantor set constructed by removing the middle fifth of each line segment at each step.

    (a) Sketch the first several steps in the construction of this Cantor set.

    (b) Determine the dimension of the middle-fifths Cantor set.

(16.4) Determine the dimension of the Koch curve, shown in Fig. 16.5.

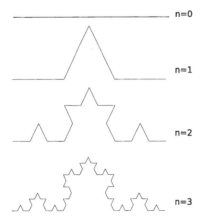

Fig. 16.5 The steps in the construction of the Koch curve.

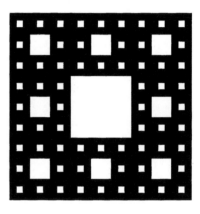

Fig. 16.6 The Sierpiński carpet.

(16.5) Determine the dimension of the Sierpiński carpet, shown in Fig. 16.6.

(16.6) Determine the dimension of the Menger sponge, shown in Fig. 16.7.

(16.7) Consider a four-dimensional "cube". (Such an object is sometimes called a *hypercube*.) If the hypercube were stretched by a factor of three, how many small hypercubes would fit inside the large one?

(16.8) ♯ Suppose one makes a fractal by following the same type of removal process that led to the Menger sponge, Fig. 16.7. However, instead of starting with a cube, start with a four-dimensional hypercube. What is the dimension of the resultant fractal?

(16.9) Determine the dimension of the Sierpiński pyramid, shown in Fig. 16.8.

Fig. 16.7 The Menger sponge. (Image source: Amir R. Baserinia, `http://en.wikipedia.org/wiki/File:Menger.png`, licensed under the Creative Commons Attribution-Share Alike 3.0 Unported license.)

Fig. 16.8 The Sierpiński pyramid. (Image source: Solkoll, `http://commons.wikimedia.org/wiki/Image:Sierpinski_pyramid.jpg`, released into the public domain by its author.)

Random Fractals

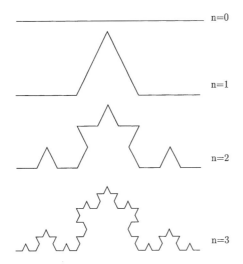

So far we have seen fractals generated by a deterministic procedure: an exact rule is followed at each step of an iterative process. In this chapter we shall see that there are other ways to make fractal shapes. We will consider fractal-generating mechanisms that involve randomness or irregularity. We begin by exploring what happens when we add a little bit of randomness or noise to an otherwise deterministic process.

17.1 The Random Koch Curve

Our starting point is the Koch curve, a classic fractal.[1] The Koch curve, which was the subject of Exercise 1.3, is generated as follows. The seed is a simple line segment. In the first step of iteration, the segment is divided into four smaller segments and arranged as shown in Fig. 17.1. Another way to think of this is that the initial line segment gets bent and stretched upward so it has a triangle in the middle.

[1]The Koch curve is sometimes referred to as the von Koch curve. This is important to know if you are looking for Koch curves in a book's index.

n=0

n=1

n=2

n=3

Fig. 17.1 The first several steps in the generation of the Koch curve.

This process is then iterated. In the next step, moving from $n = 1$ to $n = 2$ in Fig. 17.1, again each line segment is replaced by four smaller line segments. This procedure is repeated again as one goes from $n = 2$ to $n = 3$ in the figure. The end result is shown in Fig. 17.2. In principle, the iteration should be carried out to an infinite number of generations. In the figure I have shown the result after only six iterations; additional generations yield features that are too small to see.

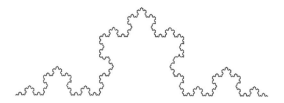

Fig. 17.2 The Koch curve after $n = 6$ steps.

The Koch curve is exactly self-similar. It is made up of small parts that are exact small replicas of the full shape. If one took, say, the left third of the picture and zoomed in by a factor of 3, the resultant image would be identical to the full Koch curve. The Koch curve can be thought of as being similar to a coastline; like a real coastline, the Koch curve has inlets which have inlets on them which have inlets on them, and so on.

However, Fig. 17.2 clearly is too symmetric to bear more than a passing resemblance to a real coastline. The Koch curve is far too regular. The inlets in real coastlines are similar to each other but are not identical. However, we can modify the generation process shown in Fig. 17.1 to produce a shape that is much more realistic. To do so, all we need to do is add a bit of randomness to the iteration. Namely, each time we replace a line segment with a bent segment, half of the time we bend the line up, and half of the time we bend it down. This is illustrated in Fig. 17.3. Repeated application of this iteration rule produces a shape

Fig. 17.3 Illustration of the basic step in the generation of a random Koch curve. At each step every line segment is replaced with a bent line segment. Half of the time the line is bent up; the other half it is bent down. Based on Fig. 9.1 of Peitgen, Jürgens, and Saupe (1992).

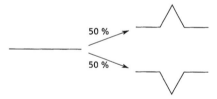

known as a random Koch curve. Note that this generation rule is not deterministic. There is randomness in the rule: it does not yield the same result every time it is applied. Thus, the random Koch curve is not unique; the generation process can produce many different outcomes.

Figure 17.4 shows first several iterations using the random generation rule illustrated in Fig. 17.3. When doing the iteration I randomly chose each direction for the bend in each line segment. For example, for the $n = 2$ step I chose, reading left to right, bends that go up, up, down, and then up.[2] The sixth step of this generation process is the top shape in Fig. 17.5. Note the similarity between this curve and the $n = 3$ step shown in Fig. 17.4. In Fig. 17.5 I have also shown two other random Koch curves. The three curves in this figure were all generated with the same random rule shown in Fig. 17.3. The curves are different because different random choices for the up or down bends were made.

Unlike the fractals in the previous chapter, the random Koch curves of Fig. 17.5 are not exactly self-similar. A small copy of the curve, when

[2]Actually I did not choose these directions; I wrote a program that generated these figures and the program randomly selects for me.

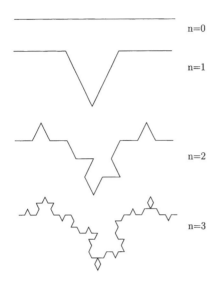

n=0

n=1

n=2

n=3

Fig. 17.4 One possible sequence of shapes obtained when generating a random Koch curve. The end result of this sequence is the random Koch curve shown at the top of Fig. 17.5.

magnified, closely resembles the full curve, but it is not an exact replica. This property of inexact self-similarity is sometimes called **statistical**

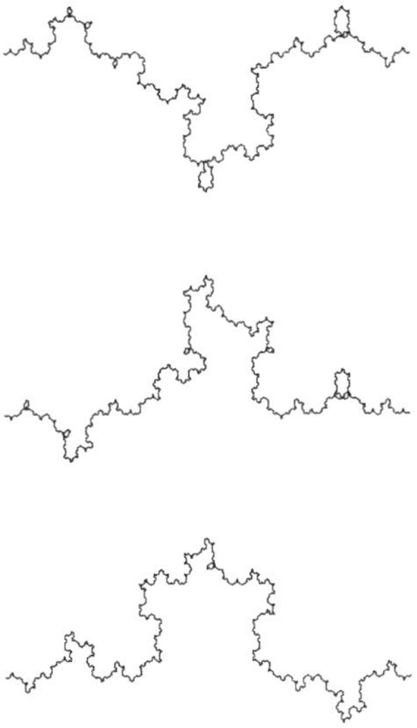

Fig. 17.5 Three different random Koch curves.

self-similarity. The idea is that such a fractal is made not of exact copies of itself, but of smaller parts that have the same statistical properties as the whole.

While the three curves of Fig. 17.5 are different, they do share some qualities. Speaking loosely, they seem to have the same bumpiness or crinkliness. One way of capturing this geometric similarity is via the dimension. The self-similarity dimension of Chapter 16 cannot be used here, since the random Koch fractal is not exactly self-similar. However, there is another related definition of the dimension that can be applied to random fractals such as this.[3] The result is that the random Koch curve

[3]In particular, one can determine the exact value of the Hausdorff dimension for the random Koch curve. Doing so is beyond the scope of this book; it requires some fairly advanced probability theory (Falconer, 2003, Chapter 15).

[4]This happens to be the same dimension as that of the non-random Koch curve of Fig. 17.2. However, it is not usually the case that a random fractal has the same dimension as its deterministic counterpart.

Fig. 17.6 The coastline north of Ragged Point, California, USA. (Photo courtesy of Prof. K.H. Solomon.)

Fig. 17.7 The basic step in the construction of an irregular Sierpiński triangle. To make this figure and all the other irregular fractals in this section I used the Fractal Curve Demo program by Christian Desrosiers, available free at http://profs.etsmtl.ca/cdesrosiers/software.html.

is $\log(4)/\log(3) \approx 1.262$.[4] In Chapter 18 I will introduce another type of dimension that will let us characterize the fractal nature of irregular but statistically self-similar shapes such as the random Koch curves of Fig. 17.5.

Recall that our motivation for investigating the random Koch curve was to find a fractal that does a better job of approximating the shape of a coastline than the Koch curve of Fig. 17.2. Looking at the random Koch curves shown in Fig. 17.5, I would say that we have met this goal. These shapes again do not look exactly like a coastline (see Fig. 17.6) but they certainly are more coastline-like than the exact Koch curve. The random Koch curves resemble other things in addition to a coastline: a crack in the pavement, a rock outcropping, or a torn piece of paper. Adding a little bit of randomness to a regular fractal produces shapes that are strongly reminiscent of many objects that are found in the physical and natural world.

The random Koch curves of Fig. 17.5 are, in a sense, complicated shapes. They bend and twist in irregular ways. At first blush these shapes might seem very difficult to describe, as doing so would require specifying all the details of the curve as it zigs and zags. However, we have seen that this apparently complicated shape is generated by a simple rule. The rule does involve randomness—it is not a deterministic process—but it is simple nonetheless. This illustrates a general result: fractals are surprisingly simple to generate. Even random fractals, which can bear a striking resemblance to physical and natural objects, can be generated by very simple processes.

It is not hard to imagine variations on the rule we used to generate the random Koch curve. For example, the probabilities of the up and down bends need not be equal as they were in Fig. 17.3. E.g., we could chose the upward bend to occur 75% of the time instead of 50%. We could also make the bend appear at different locations on each line segment; it does not have to always occur in the middle. Or we could add some randomness to the bend itself and make some of the bends larger than others. There are many options to explore. We can also add randomness to other classic fractals, such as the Sierpiński triangle and carpet, the snowflake fractal, and the Menger sponge. The possibilities are vast, and the results are often surprisingly beautiful and interesting shapes.

17.2 Irregular Fractals

Another variation on the regular fractals of the previous two chapters is to use an asymmetric or irregular generation rule. Such a fractal is not random, since it is produced with a deterministic rule. Nevertheless, the resultant shape has a somewhat random or disordered feel to it. Here is one example. Figure 17.7 illustrates the basic step in the generation of an irregular Sierpiński triangle. At each step every triangle has a tilted triangle removed from its center. (Compare this with the generation of the regular Sierpiński triangle shown in Fig. 16.4.) When this process is

Fig. 17.8 An irregular Sierpiński triangle.

iterated, the shape of Fig. 17.8 results. I do not know if this shape particularly resembles anything found in nature, but I think it is interesting nevertheless.

One can also generate an irregular version of the Sierpiński carpet. The basic step is shown in Fig. 17.9. Rather than removing a small square from the center of each square, in this version of the Sierpiński carpet one removes a rectangle that is slightly off center. The result of iterating this process is shown in Fig. 17.10. The image has an appealing art-deco feel to it.

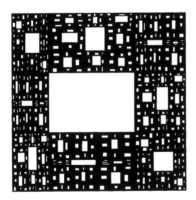

Fig. 17.9 The basic step in the construction of an irregular Sierpiński carpet.

Fig. 17.10 An irregular Sierpiński carpet.

There are many, many different ways one can alter basic fractals to produce interesting and aesthetically appealing shapes. One can also develop different fractal-generating rules with the goal of producing images that are as close as possible to some real phenomena, such as a coastline, a snowflake, lightning, or a tree. The random Koch curve and the irregular Sierpiński triangle and carpets are just the tip of the fractal iceberg. See the section on further reading at the end of this chapter if you would like to explore more.

Finally, note that the two fractals shown above, the irregular Sierpiński triangle and carpet, are not random. While their asymmetry gives them a random appearance, they are generated by a deterministic rule.[5] Nev-

[5]In this sense these fractals do not properly belong in a chapter titled Random Fractals, but they seem to fit here better than elsewhere.

ertheless, some authors refer to shapes like Figs. 17.8 or 17.10 as random fractals, and not irregular.

17.3   Fractal Landscapes

The ideas of the previous two sections—random and irregular fractals—can be extended and refined to produce images that bear a stunning resemblance to real landscapes. Such images are sometimes called fractal forgeries. You have almost surely seen these before, as fractal techniques are routinely used to generate artificial landscapes and scenery used in movies and video games.

Fig. 17.11 A fractal landscape generated by the program Fracplanet by Tim Day.   Reproduced with permission from `http://www.bottlenose.demon.co.uk/share/fracplanet/`.

An example of a fractal landscape is shown in Fig. 17.11. This image was generated by the computer program Fracplanet, written by Tim Day. A procedure similar to that of the random Koch curve, but in two dimensions instead of one, was used to make the mountain ranges. There is a separate algorithm to make clouds. The mountains are shaded to look three-dimensional and rivers are added as well. Fractal landscape-generating programs have become very complicated and sophisticated, and they produce breathtaking results. Often the landscapes include trees and other vegetation and look as real as any natural landscape. The success of these algorithms shows us yet again how simple iterated procedures can produce complicated and intricate results. In the Further Reading section at the end of this chapter I have included links to a few programs that you can download and use to generate your own fractal landscapes.

17.4   The Chaos Game

We now leave landscapes behind and explore another way that a random procedure can generate a fractal. To do so we will iterate a rule and examine the long-term behavior of the orbit. I illustrate this rule with an example. Consider the triangle shown in Fig. 17.12. Choose a point somewhere in the middle of the triangle. I chose the point labeled 0 in the figure. We now decide at random to move toward corner A, B, or C. To do so, I will imagine that we roll a three-sided die.[6] Suppose I roll such a die and that the outcome is A. I then move from my starting point directly toward corner A, but I would go only halfway. This point is labeled 1 on Fig. 17.12. Continuing with this example, suppose I then

[6]Such a die does not exist, however, so in practice one would roll an ordinary six-sided die and move toward A if one got a 1 or a 2, toward B for 3 and 4, and toward C otherwise.

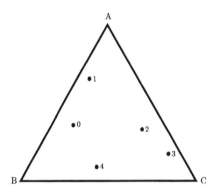

Fig. 17.12 Steps in the chaos game.

roll a C. The point then moves halfway toward corner C, resulting in the point labeled 2. If I then roll a C again, I again move halfway to corner C, resulting in point 3. If I next roll B, I would move halfway to corner B, yielding point 4.

The overall procedure is straightforward. Start somewhere inside the triangle. Randomly choose to move halfway toward corner A, B, or C. Repeat. This procedure was introduced by Michael Barnsley, who called this process **the chaos game**. This is a dynamical system, very much like the iterated functions that we have studied at length in previous chapters. However, unlike those iterated functions, the chaos game dynamical system is not deterministic. Rather, it is a stochastic dynamical system; an element of chance is incorporated in each step. Applying the rule different times to the same point will not always yield the same result, since the result depends on which corner one moves to.

What do you think is the long-term fate of the orbit in the chaos game? Will the orbit get stuck in the middle, caught between the pull of the three corners? Will it move around completely at random, journeying all over the triangle? Is there an attractor of any sort? Let us investigate and find out. The leftmost image in Fig. 17.13 shows a plot of 100 iterates from the chaos game. I have not included the first ten iterates

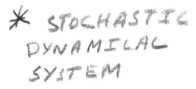

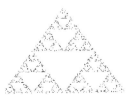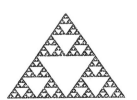

Fig. 17.13 Results of playing the chaos game. Left to right, the figures show 100, 1000, and 100,000 points. In each plot I have not shown the first ten iterates.

Strictly speaking, the point could be in the triangle or on its boundary. To simplify the subsequent discussion I will simply talk about points being in certain triangles, when more properly I should say that the points are in the triangle or on the triangle's boundary.

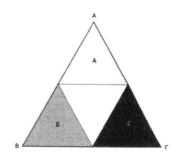

Fig. 17.14 Analyzing the chaos game. After one iteration where one rolls a C, the orbit will be in the dark shaded triangle, regardless of the initial condition.

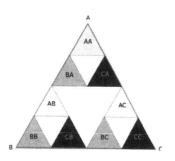

Fig. 17.15 Analyzing the chaos game. If the first roll is C and the second is A, the orbit will end up in the dark triangle labeled CA.

since we are interested in the long-term behavior of the orbit. The middle image of Fig. 17.13 shows 1000 iterates and the right image has 100,000 iterates. Out of nowhere a Sierpiński triangle has appeared. As the figure suggests, the points bounce around randomly depending on which corners are selected when the die is tossed. The orbit does not trace out the Sierpiński triangle in any direct way. Rather, it emerges and slowly comes into focus as more and more points are plotted.

How can this be? How does this random process lead to a regular fractal? Let us dissect the chaos game one move at a time. Recall that we start by choosing an initial condition anywhere inside the triangle. Suppose that in the first move we roll a C. After this move, the point must be in the dark shaded triangle in the lower left of Fig. 17.14.[7] To convince yourself that this is true, you might want to try choosing a few initial points in the big triangle, move halfway to point C, and you will see that the point is always in the shaded triangle, as I have claimed. Similarly, if your initial roll was B, the point must be in the shaded triangle on the lower left. And if the initial roll was A, then the point is in the light triangle on the top.

Let us think about the next iteration. Again, assume that the first roll was a C, so that we know we are in the dark triangle in the lower right of Fig. 17.14. If our next roll is an A, where does the point end up? It must be somewhere in the small dark triangle labeled CA in Fig. 17.15. To see this it may help to choose a few points in triangle C, move each point halfway to point A, and you will see that all of the points end up in triangle CA.

We can also think about what happens if all the points in triangle C simultaneously move half of the way to corner A. This is illustrated in Fig. 17.16. The lower left corner of triangle C moves halfway to corner A and ends up as the lower left corner of triangle CA. The top corner or triangle C moves halfway to point A and becomes the top corner of triangle CA, and so on. Again, the conclusion is that any point in triangle C will move to triangle CA if one rolls an A.

Returning to Fig. 17.15, we can analyze other moves in addition to CA. For example, if the first two chaos game moves were A and then A, the second iterate must be in the small light triangle labeled AA. And if the first two moves were B and then C, the second orbit will be in the gray triangle labeled BC. Note that after one iteration the orbit cannot be in the middle triangle, no matter what initial condition we choose. (The middle triangle is the unlabeled, white triangle in the center of Fig. 17.14.) And after two iterations there can be no points in the large white triangle nor in any of the small white triangles in Fig. 17.15. Each iteration of the chaos game excludes the orbits from the middle of the triangles that were present at the previous step. Thus, in the long run, the orbits are restricted to lie on the Sierpiński triangle. This explains the results of the chaos game: successive iterates eventually "fill up" the Sierpiński triangle, as seen in Fig. 17.13.

17.5 The Role of Randomness

Let us step back for a moment and think about what the chaos game tells us. The chaos game—the procedure shown in Fig. 17.12—is a random dynamical system. There is an element of chance in every iteration, since we need to roll a die in order to figure out where the point goes. Remarkably, the resultant shape does not appear random at all; it is a very precise and symmetric Sierpiński triangle. The image looks just like the Sierpiński triangle generated by the deterministic geometric method of successively removing triangles, as in Fig. 16.4. The chaos game shows that a random dynamical system can give a deterministic result.

This is another example of the surprises to be found in simple dynamical systems. We have seen that the logistic equation, when iterated, can yield results that appear random and are unpredictable due to sensitive dependence on initial conditions. The observed phenomenon—apparent randomness—is a result of determinism. The chaos game turns this around. Here, the observed phenomenon—an exact fractal—is the result of a random dynamical system.

The sequence of moves in the chaos game needs to be random in order for the Sierpiński triangle to emerge. For example, suppose that we chose the moves in the chaos game sequentially instead of randomly. E.g., an A move is always followed by a B move, then a C move, and then an A move again, so that the sequence of moves is . . . ABCABC . . . . The result of playing this modified chaos game is shown in Fig. 17.17. The Sierpiński triangle has disappeared; the long-term behavior is simply a period-3 cycle. This demonstrates that randomness is an essential part of the chaos game; randomness is necessary to produce the Sierpiński triangle. Even though in Fig. 17.17 all the moves (A, B, and C) appear equally often, they are not in random order. The randomness is needed for the orbit to wander all over the triangle, avoiding the forbidden regions while visiting different locations of the Sierpiński triangle. The order in which the orbit visits the Sierpiński triangle is random, but the fact that it eventually produces the Sierpiński triangle is not a matter of chance.

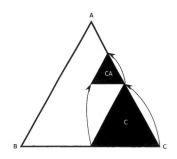

Fig. 17.16 The triangle C is transformed to the smaller triangle CA if the chaos game move is toward corner A.

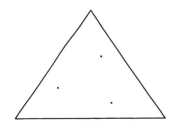

Fig. 17.17 The long-term behavior of orbits under a variant of the chaos game when the moves are chosen sequentially, . . .ABCABC. . . . The orbit is pulled to a period-3 attractor instead of the triangle.

17.6 The Collage Theorem

The chaos game may seem like a special trick that can produce Sierpiński triangle but little else. However, the procedure of using randomness to make fractals turns out to be quite general. For example, if one plays the chaos game with four points arranged in a square, not surprisingly one obtains the Sierpiński carpet, shown in Fig. 16.6. Slightly more complicated versions of the chaos game can yield many other fractal shapes. In fact, it turns out that almost *any* shape—fractal or non-fractal—can be generated by a version of the chaos game. A generalized version of the chaos game turns out to be remarkably powerful and flexible.

This generalized chaos game is constructed as follows. Thus far each move in the game takes a single point and moves it to another location. Which location it gets moved to is determined by a random rule. We could also imagine a chaos game that uses slightly more complicated rules. It is easiest to think of these rules in terms of their effects on shapes rather than single points. To that end, consider operations that not only move a shape but also transform it—stretch, shrink, shear, and/or rotate it. The technical term for these sorts of transformation is affine. An **affine transformation** is any geometric transformation that keeps parallel lines parallel. For example, an affine transformation might take a square, rotate it, shrink it by 20%, and then move it up and to the right, as shown in Fig. 17.18. This action changes the shape, but it does not change the fact that the opposite sides of the shape are parallel, just as they were before the transformation.

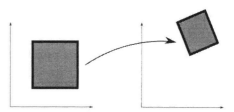

Fig. 17.18 An example of an affine transformation. The square is shrunk, rotated, and moved up and to the right.

In the original chaos game of the previous section, at each step the particular move is chosen at random. In the generalized chaos game, each move is a different affine transformation. For example, there may be four different affine transformations in a particular chaos game. One starts with a random initial point. Then roll a four-sided die to determine which affine transformation is chosen and carry out the transformation. Iterate, rolling the four-sided die again at each step.

Using affine transformations in the chaos game yields impressive results. For just one example, consider the fern shown in Fig. 17.19. This image was made via a chaos game with just four affine transformations. Small variations on these four transformations yields a menagerie of different fern shapes. Other sets of affine transformations yield a variety of other fractals.

In the mid-1980s Michael Barnsley proved a remarkable result. Given essentially *any* shape, one can make a chaos game that will generate it. This result is now known as the **collage theorem**, since in finding the particular affine transformations needed to reproduce an image, one forms a collage in which the full shape is covered with several smaller shapes. These smaller shapes then define the affine transformations that will generate the full shape. The collage theorem is quite general; it applies to essentially all shapes, not just fractals. There is much more to explore about chaos game variants and the collage theorem. See the Further Reading section at the end of this chapter.

I conclude this section by discussing two important implications of the collage theorem. First, the collage theorem has potential applications to image compression. Suppose one has a complex and detailed image

Fig. 17.19 A fractal fern produced by a chaos game with four affine transformations. (Image source: DSP-user, http://en.wikipedia.org/wiki/File: Barnsley_fern_plotted_with_VisSim. PNG. licensed under Creative Commons Attribution-Share Alike 3.0 Unported license.)

stored as a large file. To save space (or time if one needs to send the image from one computer to another), one could determine what chaos game will generate the image. One can then discard the large image and store only the rules for the chaos game. If one needs to generate the image, all that needs to be done is to play the chaos game and display the result. Thus, the specification of the chaos game is a representation of the image that is much smaller than the original file.

The second implication of the collage theorem and the chaos game is somewhat more abstract and concerns the relationship between randomness and order. When we encounter an intricate and precise shape such as the Sierpiński triangle or the fern of Fig. 17.19, I think it is natural to assume that an exact and precise procedure must have been used to generate them. But the chaos game shows us that exact and precise fractal shapes can be generated by a random process. Yet again, we see that iteration—in this case of a random dynamical system—can produce surprising results. A complex and predictable image results from a simple and random process. We tend to think of randomness and order as being opposites, but the chaos game unites these two phenomena.

Further Reading

Chapter 9 of Peitgen, Jürgens, and Saupe (1992) is a good overview of the different ways that randomness can be used to make random or irregular fractals. Chapter 15 of Stewart (2002) is a clear, non-technical overview of fractals and discusses random and irregular fractals.

There are a number of programs that you can use to experiment with different fractals and make fractal forgeries. The program Terragen™ is capable of producing incredibly detailed and realistic landscapes. It is free for non-commercial use and can be downloaded at `http://www.planetside.co.uk/`. The program Fracplanet, which can be downloaded at `http://www.bottlenose.demon.co.uk/share/fracplanet/index.htm`, makes fractal landscapes and entire fractal planets. Additional software can be found via an internet search.

For more about the chaos game and producing fractals through affine transformations, see Chapters 5 and 6 of Peitgen, Jürgens, and Saupe (1992) or Chapter 7 of Flake (1999). Both of these references use more mathematics than does this book, but nevertheless should be fairly accessible. A thorough, but quite technical reference on the chaos game and fractals produced by iterated functions is Barnsley's *Fractals Everywhere*, (2000). A less technical discussion of fractals and chaos games can be found in Chapters 1 and 2 of Peak and Frame (1994). Lessons 1–7 of the workbook by Choate, Devaney, and Foster (2000b) are an excellent general, elementary introduction to fractals and chaos games, and contain worksheets and suggestions for teaching and classroom activities.

Exercises

(17.1) Suppose that a random Koch curve is generated according to the rule shown in Fig. 17.3.

(a) What is the probability that after $n = 2$ iterations a random Koch curve looks like the non-random Koch curve, shown in Fig. 17.1.?

(b) What is the probability that after $n = 3$ iterations a random Koch curve looks like the non-random Koch curve, shown in Fig. 17.1.?

(c) As the number of generations in the construction of a random Koch curve gets larger and larger, what is the probability that the Random Curve looks like the exact curve?

(17.2) By hand, construct a random Koch curve up to the $n = 3$ generation. Toss a coin to determine if each bend is up or down. Repeat this construction, re-tossing the coin to generate a different random Koch curve. Compare the two shapes.

(17.3) In Section 17.4 the initial condition for the chaos game was always inside the triangle. What would happen if the initial condition was outside the triangle?

(17.4) Consider a modified chaos game in which the moves are chosen randomly, but only B or C moves are chosen. What is the long-term behavior of the orbit?

(17.5) Consider a modified chaos game in which the moves are not chosen randomly. Instead, B and C moves are alternated. I.e., the sequence of moves is ...BCBCBC.... What is the long-term behavior of the orbit?

(17.6) Suppose one plays the chaos game and chooses an initial condition exactly in the center of the triangle.

(a) What sequence of moves would result in the orbit being inside triangle x in Fig. 17.20?

(b) What sequence of moves would get the orbit into triangle y?

(c) What sequence of moves get the orbit into triangle z?

(17.7) Suppose now that the starting point for the chaos game is point A. That is, you start exactly on the top corner of the triangle.

(a) What sequence of moves will result in the orbit landing inside (and not on the boundary) of the triangle marked r?

(b) What sequence of moves results in the orbit being inside triangle s?

(c) What sequence of moves results in the orbit being inside triangle t?

(17.8) Suppose now that the starting point for the chaos game is point B. That is, you start exactly on the lower left corner of the triangle.

(a) What sequence of moves will result in the orbit landing inside (and not on the boundary) of the triangle marked r?

(b) What sequence of moves results in the orbit being inside triangle s?

(c) What sequence of moves results in the orbit being inside triangle t?

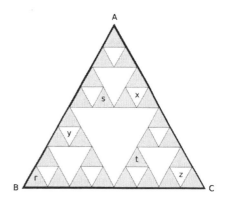

Fig. 17.20 The illustration for Exercises 17.6–17.8.

(17.9) ♮ Consider a two-dimensional version of the chaos game that is set up as follows. Draw two points some distance apart and label them A and B. Choose an initial condition somewhere on the line that connects A and B. Flip a coin to determine the move. If the coin comes up heads, move two-thirds of the way to point A. If it is tails, move two-thirds of the way to point B. Iterate. What is the long-term fate of the orbit? If one were to plot 1000 iterates of this game after discarding the first several moves, what image would appear? Explain.

The Box-Counting Dimension

18

In the previous chapter we saw several examples of fractals that are not exactly self-similar. For example, small parts of the random Koch curve resemble the larger curve, but they are not exact replicas. They are statistically self-similar, or approximately self-similar, but not identical. This is the case for almost all naturally occurring fractals. If one zooms in on a jagged coastline or a winding river, one sees structures that are similar to, but not exact copies of, the larger shape.

In Chapter 16 we analyzed fractals by determining their dimension D. To do so, we used the following relationship:

$$\text{number of small copies} = (\text{magnification factor})^D , \qquad (18.1)$$

where D is the dimension. This equation let us calculate the dimension of a handful of fractal objects, including the Cantor set, the Sierpiński triangle and carpet, and the Koch curve. The definition for D given by Eq. (18.1) relies on there being small pieces of the object that look exactly like small copies of the whole. For this reason, this type of dimension is often known as the self-similarity dimension.

But what about objects that are not exactly self-similar? In this chapter I introduce another way of defining the dimension that will allow us to extend our notion of dimension to objects that are not exactly self-similar. The main idea is that instead of looking at how many small copies of an object are contained in a large copy, we consider instead how the volume or size of the overall shape changes as we change measurement scales. The method we develop will be somewhat tedious, but the type of dimension that results will be more flexible than the self-similarity dimension.

18.1 Covering a Box with Little Boxes

In Chapter 16 we began thinking about dimension by considering how many small boxes fit inside a large box. In this chapter we explore this idea from a different angle: we start with a box and cover it with smaller and smaller boxes. As a starting point, consider a square that has a length and height of 1. We can think of this as a portion of a floor that we want to cover with tiles. Suppose our tiles are in the shape of a square and have a side of length 1/2. We would need to buy four such square tiles in order to completely cover the original 1×1 floor. This is illustrated in part (b) of Fig. 18.1.

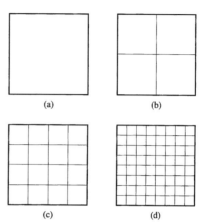

(a)   (b)

(c)   (d)

Fig. 18.1 Covering a square with successively smaller boxes. Figure (a) shows the original square. Figures (b)–(d) show the square covered by successively smaller boxes. The length of the side of the original square is 1. The boxes used to cover the square have sides of length (1/2), (1/4), and (1/8), respectively, in Figs. (b)–(d).

What if we wanted to use smaller tiles to cover the floor? Clearly we would need more tiles. Suppose that each tile is now a square whose side has length 1/4. In this case we need 16 tiles to cover the original square, as can be seen in part (c) of Fig. 18.1. And if we make the tiles smaller still, we would need even more tiles. We will need 64 if the tiles are squares with a side of 1/8, as shown in part (d) of the figure.

You have likely noticed the pattern. If the size of the tiles is halved—i.e., the side of the square tile is half what it was before—the number of tiles needed to cover the shape is squared. When we change from tiles of side 1/2 to side 1/4, the number of tiles needed increases from 4 to 16. Thus, the number of tiles is squared, since $4^2 = 16$. The reason that the tiles are squared, as opposed to being raised to some other power, is that the floor is two-dimensional.

In order to formalize this idea and arrive at an alternative definition of dimension, I need to introduce some notation. Let s denote the length of the side of one of the tiles we used to cover the original square shape. We will denote by $N(s)$ the number of tiles of size s needed to cover the shape.[1] The values of s and $N(s)$ for the example of shown in Fig. 18.1 are contained in Table 18.1. As we have seen, as s gets smaller, $N(s)$ gets larger. The relationship between s and $N(s)$ can be used to determine the dimension. Specifically $N(s)$ and s are related via:

$$N(s) = k \left(\frac{1}{s} \right)^D . \tag{18.2}$$

In the above equation k is some constant—it is a number that does not depend on the box size s. The exponent D in Eq. (18.2) is known as the **box-counting dimension**.

To illustrate the use of Eq. (18.2), consider Fig. 18.1(c). Here $s = 1/4$ and $N(x) = 16$. Plugging these numbers into Eq. (18.2) gives:

$$16 = k \left(\frac{1}{\frac{1}{4}} \right)^D . \tag{18.3}$$

[1] Typically one refers to the tiles as boxes, and hence the dimension defined via this procedure is known as the box-counting dimension.

Table 18.1 Data used to determine the box-counting dimension of a square. The box-counting process is illustrated in Fig. 18.1. The side of the boxes used to cover the circle is s, and $N(s)$ is the number of such boxes needed.

s	$N(s)$
1	1
$\frac{1}{2}$	4
$\frac{1}{4}$	16
$\frac{1}{8}$	64

Simplifying the compound fraction, we get

$$16 = k(4)^D .\qquad(18.4)$$

One solution to this equation is $D = 2$ and $k = 1$. This makes sense, as it tells us that a square is two-dimensional, as we would expect. However, there are also other solutions to this equation. For example, it could be that $D = 1$ and $k = 4$; these values also make make Eq. (18.4) true. However, these values are not consistent with the rest of the data in Table 18.1. The challenge here is that we have two quantities we need to solve for—k and D—and we have a table full of data. We want to take this entire table of data into account when we are figuring out the best values for k and D.

I will discuss a general method for doing just this in Section 18.3. For now, the key observation is that $N(s)$, the number of boxes of side s needed to cover an object, will increase as s decreases. The rate at which $N(s)$ increases is determined by the dimension D, as shown in Eq. (18.2). From a procedural standpoint this definition of dimension is fairly simple: just take the shape and start covering it with boxes. Below, we will see that we can extend this procedure to calculate the dimension of shapes that are much more complex than the simple square considered here.

18.2 Covering a Circle with Little Boxes

Let us repeat the experiment of the previous section for a circle. This is illustrated in Fig. 18.2. In part (a) the circle is covered with boxes of side $s = 0.5$, in part (b) the circle is covered with boxes of $s = 0.25$, and so on. Note that what we are interested in is the number of boxes needed to completely cover the circle. Typically, the boxes on the edge of the shape will extend over the shape itself.

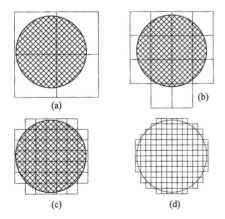

(a)

(b)

(c)

(d)

Fig. 18.2 Covering a circle with successively smaller boxes.

The values of s and $N(s)$ determined using Fig. 18.2 are shown in Table 18.2. Our next task is to use Eq. (18.2) to determine the dimension D. Since a circle is two-dimensional, we anticipate a dimension of 2.

Table 18.2 Data used to determine the box-counting dimension of a circle. The box-counting process is illustrated in Fig. 18.2. The side of the boxes used to cover the circle is s, and $N(s)$ is the number of such boxes needed.

s	$N(s)$
$\frac{1}{2}$	4
$\frac{1}{4}$	14
$\frac{1}{8}$	45
$\frac{1}{16}$	162

Let us begin with the first row in the table, corresponding to $s = 0.5$. Plugging into Eq. (18.2), we get:

$$4 = k \left(\frac{1}{\frac{1}{2}} \right)^D .$$ (18.5)

Simplifying the fraction, this becomes:

$$4 = k 2^D .$$ (18.6)

As was the case with the square in the previous section, we see that this equation has a solution if $D = 2$ and $k = 1$. I am cheating a little, since I know that I want $D = 2$. But this perhaps seems reasonable for this case.

Now let us consider the second row of the table, corresponding to $s = 0.25$. Plugging these numbers into Eq. (18.2) yields:

$$14 = k 4^D .$$ (18.7)

If I take $D = 2$, then k cannot equal one any more. To see this, plug $D = 2$ into the above equation:

$$14 = k 16 ,$$ (18.8)

which gives $k = 7/8$, not 1. But k is a constant—it should be the same for all s. Clearly this is not the case here. How can we fix this? We could change the dimension D. But then we would have that a circle is not two-dimensional, which certainly isn't what we want. What to do?

Equation (18.2) holds in the limit that the boxes get really, really small. The reason for this is that it is only as the boxes get very small that we get an accurate count of the area of the shape whose dimension we are trying to determine. With smaller boxes there is less to "hang over" the edge of the shape. For example, there is less surplus or overhang in Fig. 18.2(d) than in Fig. 18.2(a). So in order to estimate the dimension D using Eq. (18.2), we need to calculate D by investigating what happens as s gets smaller and smaller.

There is an additional potential complication. For a given box size, there is more than one way to completely cover the shape. How many boxes we need might depend on where we start covering. For example, we could lay down our first box in the center of the shape, or we could place the first box so that it is aligned with the top left of the shape. Both techniques will cover the shape, but they might lead to slightly different numbers of boxes. However, as the boxes get smaller and smaller, the details of the covering method matter less and less. Thus, using Eq. (18.2) to determine the dimension D is more accurate as the box size gets smaller and smaller.

18.3   Estimating the Box-Counting Dimension

In the previous sections we have seen that it is not always straightforward to estimate k and D from a table of data. It would be nice to have a way to find the values of k and D that fit the data as well as possible. In this section I will describe a procedure to accomplish this goal. We start with Eq. (18.2):

$$N(s) = k \left(\frac{1}{s}\right)^D . \tag{18.9}$$

Taking the logarithm of both sides of this equation will put it in a more useful form:

$$\log N(s) = \log \left[k \left(\frac{1}{s}\right)^D\right] . \tag{18.10}$$

Simplifying using the properties of logarithms yields:

$$\log N(s) = \log k + D \log \left(\frac{1}{s}\right) . \tag{18.11}$$

(A brief overview of logarithms and their properties can be found in Appendix A.4.)

It might not look like it, but this equation is actually quite useful; this says that the logarithm of $\frac{1}{s}$ and the logarithm of $N(s)$ are linearly related. To help see this, if we let $\log \frac{1}{s} = x$ and $\log N(s) = y$, then we can write

$$y = k + Dx . \tag{18.12}$$

This is just the equation of a line. The y-intercept is k, and the slope is D. Equation (18.12) thus says that the problem of determining the dimension D can be recast as determining the slope of a line. Comparing Eqs. (18.11) and (18.12), we see that D is the slope of the line that results when $\log N(s)$ is plotted versus $\log \frac{1}{s}$.

We already have a table of values of $N(s)$ and s for the circle example. First, let us re-write this table in terms of $\frac{1}{s}$ instead of s. The results of doing so are shown in Table 18.3. Next, take the logarithm of both columns of data to obtain the results shown in Table 18.4. To make this table I used base-10 logarithms. If you use another base, that is fine, but it is important to use the same base throughout a calculation.

Our next step is to plot the data in Table 18.4 and determine the slope of the subsequent line. The data from this table are shown in Fig. 18.3. Note that the points appear to fall along a straight line. The slope of this line is the box-counting dimension D.

To estimate the slope, one draws a line through the data points. The line should be chosen so as to be a best fit; it should come as close as possible to as many of the data points as possible. This notion of a best fit can be made precise; it is a standard result in introductory statistics. Most spreadsheet programs can quickly and unambiguously determine the equation of the line that is the best fit to the data. However, for the exercises at the end of this chapter, it is fine to just simply draw the best line by eye and then estimate the slope of that line.

I used a program to analyze the data shown in Fig. 18.3. The best fit line is drawn in Fig. 18.4. This line has a slope of 1.77 and an

Table 18.3 Data used to determine the box-counting dimension. This is the same data as is in Table 18.2.

$\frac{1}{s}$	$N(s)$
2	4
4	14
8	45
16	162

Table 18.4 Data used to determine the box-counting dimension. This table was obtained by taking the logarithm of the data in Table 18.3.

$\log \frac{1}{s}$	$\log N(s)$
0.301	0.602
0.602	1.146
0.903	1.653
1.204	2.210

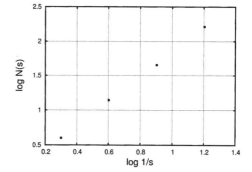

Fig. 18.3 A plot of $\log N(s)$ versus $\log \frac{1}{s}$. The data is from Table 18.4.

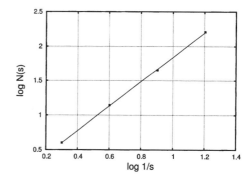

Fig. 18.4 A plot of $\log N(s)$ versus $\log \frac{1}{s}$. The data is from Table 18.4. The line drawn through the points was chosen to be a best fit to the data. The equation of the line is $y = 1.77x + 0.7$.

intercept of 0.07. The slope corresponds to the dimension. This means that we have estimated a dimension of 1.77. To be honest, this is rather disappointing. We have just gone through a lot of work to figure out that the dimension of a circle is 1.77. Clearly this is wrong. A circle has a dimension of 2. What has happened? Recall that this definition for the box-counting dimension D holds only when the size s of the box gets very small. Presumably, if we extended this calculation with a few smaller values of s we would get a more accurate value of D.

This somewhat disappointing result serves as a note of caution. The definition of the box-counting dimension and the procedure for calculating it are conceptually relatively straightforward. However, it is often difficult to get accurate results using this method. Part of the problem is that one needs very small boxes to get good results, but small boxes are difficult to count, and hence are error-prone and/or time-consuming on a computer because there are so many of them. Moreover, using smaller boxes requires a very high-resolution image. Since all images or data sets have finite resolution, eventually it will simply not be possible to use boxes smaller than a certain size.

There are techniques for more accurately determining the box-counting dimension and related quantities than that which I have presented in this chapter. However, these methods are not always easy to apply, and in some cases are the topic of current research. So, any time you encounter an experimental result for the box-counting dimension, you should be at least slightly skeptical. The dimension can be accurately estimated, but it is difficult to do so and requires usually requires a large quantity of data. So take such results with a grain of salt. Nevertheless, it is often possible to get a reasonable estimate of the dimension using the box-counting procedure described above.[2]

[2]For a discussion of several basic experiments concerning fractal dimensions, at a level similar to this book, see Lewis (2002).

18.4 Summary

The box-counting dimension is commonly used to characterize the statistical self-similarity of a wide range of phenomena. It is a powerful and flexible idea that allows us to quantify the properties of objects that are statistically self-similar.

To recapitulate, here are the steps needed to estimate the box-counting dimension D of an object or image.

(1) Count the number of boxes needed to cover the object.

(2) Repeat several times, for successively smaller box sizes s. You should now have a table with a bunch of N values and $\frac{1}{s}$ values.

(3) Take the logarithm of the numbers in your table.

(4) Plot this logarithmic data with $\log(1/s)$ on the horizontal axis.

(5) If the data is approximately linear, determine the slope of the line.

(6) The dimension of the fractal is the slope of the line.

(7) If the data are not approximately linear, you may not have let your box size get small enough, or you may have a complex geometric object that cannot be described with a single dimension.

(8) Any number you get from such a procedure is likely to be inaccurate, so take this number with a grain of salt.

Exercises

For some of these exercises you will need to count boxes. A convenient way to do this is to use graph paper. A good source of graph paper is `http://incompetech.com/graphpaper/`. Here you can print out graph paper with squares of a range of different sizes. Ideally, print the graph paper out on a transparency or a thin sheet of paper and lay it directly on top of the shape.

(18.1) Suppose that we wanted to cover the square shown in Fig. 18.1 with square tiles of side $s = 1/16$. How many tiles are needed? How many tiles would be needed if the tiles' side is 1/32?

(18.2) The Sierpiński triangle has a dimension of approximately 1.585. Suppose that someone covers a Sierpiński triangle with boxes of a certain size and finds that 187 such boxes are needed. Approximately how many boxes would be needed to cover this Sierpiński triangle if the boxes' side is half of what it was before?

(18.3) The Sierpiński carpet has a dimension of approximately 1.893. Suppose someone covers a Sierpiński carpet with boxes of a certain size and finds that 211 such boxes are needed. Approximately how many boxes would you need to cover this Sierpiński carpet if the boxes are now half as large as before?

(18.4) The white cauliflower has a dimension of approximately 2.8 (Kim, 2005). Suppose a certain head of cauliflower is 6 inches tall and weights 2 pounds. How much would a 12-inch tall head of cauliflower weigh?

(18.5) Estimate the box-counting dimension of a "hollow circle". That is, just the circumference of the circle, not the middle. Is the answer close to what you would expect?

(18.6) Estimate the box-counting dimension of a Koch curve, as shown in Fig. 16.5.

(18.7) Estimate the box-counting dimension of the Cantor set. To do so it may be easier to use a series of boxes each of which is a third (rather than half) the size of the previous box.

(18.8) Estimate the box-counting dimension of the shoreline of the coast of Maine.

When do Averages Exist?

It is often not useful to describe fractals in terms of an average size. The reason for this is that fractals are self-similar. Fractals consist of the same basic shape repeated at all length scales, large and small. So stating an average size does not capture what is interesting or noteworthy about the shape.

Moreover, there are some phenomena for which, strictly speaking, an average size simply does not exist. The goal of this chapter is to present a situation in which this is the case: when stating an average property not only is not useful, it is mathematically ill-defined. I think this example will give us additional insight into fractals, and will also let us see that fractals are not just geometric objects—fractals can also be used to describe processes that unfold in time.

Before considering the main example of this chapter, we will need to consider a simpler example. If this next section seems too basic, please have patience; I think it is necessary for the more interesting and subtle example that follows.

19.1 Tossing a Coin

For our initial example, imagine we are playing a simple game. Someone tosses a coin. We will assume that this is a fair coin; the probability of heads is $1/2$ and the probability of tails is $1/2$. If the coin comes up heads, you win one dollar. If the coin comes up tails, you do not win anything. If you play this game over and over, what would your average winnings be?

Since the probability of heads is one half, we expect that half the time you will win a dollar, and half the time you will win nothing. So on average, you will win \$0.50, or 50 cents. We can write this as an equation:

$$\text{Average winnings} = (\text{Probability of heads} \times \$1.00) + (\text{Probability of tails} \times \$0.00) . \quad (19.1)$$

Since the probability of getting heads is $(1/2)$, as is the probability of getting tails, the above equation yields

$$\text{Average winnings} = \left(\frac{1}{2} \times \$1.00\right) + \left(\frac{1}{2} \times \$0.00\right) \quad (19.2)$$

$$= \$0.50 . \quad (19.3)$$

Game	Outcome	Total Winnings	Average Winnings
1	T	0	$\frac{0}{1} = 0.000$
2	H	1	$\frac{1}{2} = 0.500$
3	T	1	$\frac{1}{3} \approx 0.333$
4	H	2	$\frac{1}{2} = 0.500$
5	T	2	$\frac{2}{5} = 0.400$
6	H	3	$\frac{3}{6} = 0.500$
7	H	4	$\frac{4}{7} \approx 0.570$
8	H	5	$\frac{5}{8} = 0.625$
9	T	5	$\frac{5}{9} \approx 0.560$
10	T	5	$\frac{5}{10} = 0.500$

Table 19.1 Sample outcomes for the simple coin-tossing game.

This analysis is probably overkill for this example. But it will be useful to refer to Eq. (19.1) later on when we consider a more complicated example. To simplify notation, in what follows I will omit the dollar sign $ when reporting the winnings from games like this. So, I will write the average winnings for the fair coin-toss game as 0.5, not $0.50.

This average value of 0.5 is what one would expect if one played this game many, many times. If we just play the game ten times, or even a hundred times, it would be very unlikely that our average winning would be exactly one half. The reason for this is that even though the probability of tossing heads is one half, if we toss a coin a number of times we would not expect to observe *exactly* half of the outcomes to be heads. However, as we play the game more and more times, we expect the average to get closer and closer to 0.5.

We illustrate this by analyzing a particular sequence of games. After each game we calculate the average winnings by taking the total winnings and dividing by the total number of games played. Let us suppose that the first toss is tails. Then at this point your total winnings are zero. And this is also your average winning, since you have played once and won nothing. This is shown in Table 19.1.

Suppose that the next toss yields heads. At this point your total winnings are one dollar. You have played the game two times, so the average is 1 divided by 2, or 0.5. Suppose the next toss is also tails. Then your total winnings are still one dollar, but since you have played the game three times your average winnings are 1/3. We could continue playing the game and calculating the average at each step, as shown in the table on the previous page. This type of average is often called the **running average**, since we are actively determining the average while the game is "running". The average jumps around with each successive outcome. But the jumps get smaller the more data points we have—i.e., the more times we play the game.

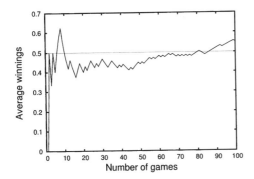

Fig. 19.1 Average winnings for a coin-toss game. Note that the average winning line become less jumpy as the number of coin tosses increases.

This can be seen in a graph of the average winnings plotted as a function of the number of times the game is played. Such a plot is shown in Fig. 19.1. To make this plot I simulated the game 100 times on my computer. Initially the average jumps abruptly. But as we have more and more data points—we play the game more and more—the individual jumps get smaller and smaller. The curve in Fig. 19.1 is getting less and less bumpy. We also see that the average is approaching the anticipated value of 0.5. However, note that even after 100 tosses the average is still a little bit away from 0.5.

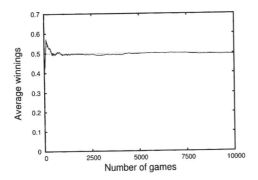

Fig. 19.2 Average winnings for a coin-toss game.

This is not cause for alarm. As noted above, we do not expect the average value to be exactly reached, especially if we have a relatively small number of data points. However, as the number of data points gets larger, we anticipate that the average will jump around less and less as it gets closer to 0.5. This can be seen in Fig. 19.2, in which I have again plotted the average winnings as a function of the number of tosses, but this time I show the results for a total of 10,000 tosses. The average winnings clearly is approaching the predicted value of one half.

Finally, in Fig. 19.3 I have again plotted the average winnings as a function of the number of games played. The range on the horizontal axis extends to one million games played. Note that this graph has a smaller scale on the vertical axis than the previous two. Again we can see that the average approaches 0.5 and that despite an occasional wiggle, the average is getting closer and closer to 0.5 as the number of games played increases.

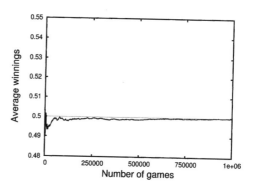

Fig. 19.3 Average winnings for a coin-toss game. Note the small scale on the vertical axis. The last number on the horizontal axis is written in scientific notation: "1e+06" means 1×10^6. So 1e+06 = 1,000,000, or one million.

The main point of this example is to illustrate what it means for something to possess an average. An average is a statistical statement about some aspect of what happens when a situation is repeated again and again. The actual average we calculate will fluctuate as we get more and more data. But the key point is that these fluctuations will, in the long run, get smaller as we get more and more data.

For example, suppose that we were interested in the average height of the students at a university. We could calculate this by measuring the heights of individual students and then averaging. This average would change a little bit as we measure the height of each additional student and add her or him to our data set. But we would expect the average to change less and less as the number of students we have measured grows larger and larger. Indeed, we would be shocked if this was not the case. There might be an unusually large jump in the average at some point—perhaps we meet three unusually tall students and measure their heights one after the other. But we know that it is meaningful to talk of an average height, and it is a pretty straightforward exercise to go out and measure it.

19.2   St. Petersburg Game

We now consider a different example. As with our first example, we will imagine a game in which coins are tossed and money is won. But this time the rules are different. Now, if you toss the coin and get heads, you win \$2. However, if you get tails, you now get to toss again. If your second toss is heads, then you win \$4. If your second toss is tails, you get to toss yet again. If this third toss comes up heads, you get \$8, and if it is tails, you get to toss again, and so on. This entire sequence is considered one round of the game. Unlike the first game, you are now guaranteed to win some money every time you play. This game was introduced by the Swiss mathematician Daniel Bernoulli in the mid-1700s. It sometimes is referred to as the St. Petersburg paradox or the St. Petersburg lottery.[1]

Here is another way to specify the rules. For each round of the game, toss a coin until it comes up heads. Let n be the number of tosses you make until a heads appears. So if your sequence was TTH, n is three.

[1]This is called the St. Petersburg game because Bernoulli published a paper about this process in the *Papers of the Imperial Academy of Sciences in Petersburg*. For further discussion of the St. Petersburg game, see Martin (2008) and references therein. The Wikipedia page on the St. Petersburg paradox is also quite clear and informative.

Game	Outcome	Payoff	Total Winnings	Average Winning
1	TH	4	4	$\frac{4}{1} = 4$
2	TH	4	8	$\frac{8}{2} = 4$
3	TH	4	12	$\frac{12}{3} = 4$
4	H	2	2	$\frac{14}{4} = 3.5$
5	H	2	2	$\frac{16}{5} = 3.2$
6	TTTTH	32	48	$\frac{48}{6} = 8$
7	TTTH	16	64	$\frac{64}{7} \approx 9.14$
8	H	2	66	$\frac{66}{8} = 8.25$
9	TTT	8	74	$\frac{74}{9} \approx 8.22$
10	TH	4	78	$\frac{78}{10} = 7.8$

Table 19.2 Sample outcome for the St. Petersburg game.

Then, the amount of money you get is 2^n. So the payoff for TTH is $2^3 = 8$ dollars.

Clearly this game is going to have a higher average payoff than the previous one, since in every round you win at least \$2. But how much higher? Suppose you had an opportunity to play this game but had to pay some money up front in order to do so. How much would you pay? Would it be a good deal to pay \$5.00 for a chance to play this game? To address these questions we will start, as we did before, by considering a possible sequence of outcomes. Suppose in the first round you toss a tails and then a heads. Your payoff then is \$4. And this is also, at this point, your average payoff; you have played once and won \$4. This is illustrated in the first row of Table 19.2.

Suppose that in the next round you also toss tails and then heads. You will again get \$4, and the average winnings remains 4. If in the third round you tossed tails and then heads yet again, you would win another \$4. Your total winnings are now \$12, but the average is the same; $12/3 = 4$. This is shown in rows two and three in the Table 19.2. In this table I have listed the results from ten rounds of this game. Note that you get quite lucky in the sixth game, where you toss four tails before getting a heads. Thus, your payoff for this game is $2^5 = 32$. Accordingly, the average winning jumps sharply up at game number six.

What is the average winnings for the St. Petersburg game? What will happen if we play the game many, many times? Take a moment and make a conjecture. Looking at Table 19.2, it is not immediately clear what the average will be. So, as before, we turn to a plot of the average winnings as a function of the number of games played. Such a plot is shown in Fig. 19.4. This graph is somewhat perplexing. It appears as if the average winnings were may have been approaching 5. But then, around the 80$^\text{th}$ game, the average shoots way up. Apparently the player was very lucky here and tossed many tails before getting heads, thus winning a very large payoff. After this large payoff the average winnings again trend downward. So what is the average value? Perhaps the average winnings might be heading toward 9 or 10. What happens as we play the game more and more times?

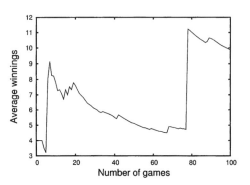

Fig. 19.4 Running average of the winnings for the St. Petersburg game.

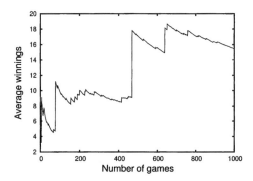

Fig. 19.5 Running average of the winnings for the St. Petersburg game.

Figure 19.5 shows the average winnings for up to 1000 games. Until around 500 rounds of the game it seems that the average might finally be leveling off near 10. The average is jumping around, but perhaps the fluctuations are getting smaller. But then at around the 500th game there is a huge upward spike. The player must have again been very lucky and received a huge payoff. But we still do not have a good sense of what the average winnings is. Evidently what is happening is that every now and again the player is extremely lucky. He or she gets a huge payoff, and this leads to a sudden spike in the average winnings. It is as if we were trying to figure out the average height of students on a college campus, and once in a rare while we come across a student who is 100 or 1000 feet tall.

So we need more data; we need to play the game many times to get a good estimate for the average winnings. If we play long enough that we observe a relatively large number of the extremely lucky outcomes, we will be able to account for them properly in our average. So, let us see what happens.

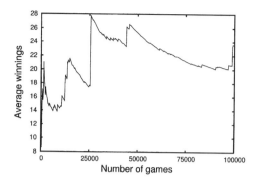

Fig. 19.6 Running average of the winnings for the St. Petersburg game.

In Fig. 19.6 I have again plotted average winnings as a function of the number of games played, this time up to 100,000 games. This plot is again somewhat surprising. We do not see the average leveling off. It continues to jump around. There are occasional large upward spikes associated with an enormous stroke of luck. The upward spikes are usually followed by long downward trends. But these downward trends are not perfectly smooth; they are interrupted by smaller upward spikes seemingly at random. In this sense, the graph of the average winnings as a function of the number of games played is fractal-like. The curve has upward jumps of many different sizes: a few huge ones, lots of medium-sized ones, many small ones, and so on.

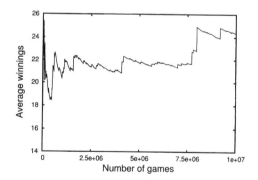

Fig. 19.7 Running average of the winnings for the St. Petersburg game. The numbers on the horizontal axis are written in scientific notation: "2.5e+06" means 2.5×10^6, or 2.5 million. "5e+06" is 5 million, and so on.

So what *is* the average? What happens if we play the game even more times? In Fig. 19.7 I have shown the results of playing the St. Petersburg game 10 million times. Amazingly, the average still has not settled down. Even after 7.5 million games, the average is still showing jumps. In sharp contrast, Fig. 19.3, a similar plot for the simpler coin-tossing game, shows that the average changes less and less as the number of tosses gets larger and larger. Moreover, in the St. Petersburg game not only does the average continue to fluctuate even after a great many tosses, it also seems to trend steadily upward. To see this, note that the vertical scales on the graphs shown in Figs. 19.4–19.7 are not the same.

19.3   Average Winnings for the St. Petersburg Game

Based on these computer experiments we suspect that something weird is going on. The average continues to grow and fluctuate no matter how many times we play the game. Let us see if we can gain additional insight by analyzing this situation theoretically—without using a computer to simulate the game.

Our goal is to come up with an equation similar to Eq. (19.1) that will give us a formula for the average winnings in the St. Petersburg game. To do so, we need to consider the probability of each outcome multiplied by the winnings for that that outcome. Doing so, I obtain:

$$
\begin{aligned}
\text{Average winnings} \quad = \quad & (\text{Probability of H} \times \$2) \\
+ \quad & (\text{Probability of TH} \times \$4) \\
+ \quad & (\text{Probability of TTH} \times \$8) \\
+ \quad & (\text{Probability of TTTH} \times \$16) \\
+ \quad & \ldots .
\end{aligned}
\tag{19.4}
$$

Note the "..." at the end of the equation. This indicates that the equation keeps on going. There are an infinite number of terms in the equation, one for each possible number of tails that could be tossed before the player finally gets a heads. It may seem odd to have an infinite number of terms like this. However, the probabilities of these longer and longer sequences of tails are getting smaller and smaller, so the terms farther and farther out in the equation matter less and less.[2]

In order to further analyze Eq. (19.4), we will need to figure out values for various probabilities. The probability of tossing a heads is 1/2. But what about the probability of tossing a tails and then a heads? The probability of getting heads on the first toss is 1/2. And the probability of getting tails on the second toss is also 1/2. To get the probability of getting tails and then heads we multiply these two probabilities together:

$$
\text{Probability of TH} = \left(\frac{1}{2}\right)\left(\frac{1}{2}\right) = \frac{1}{4}.
\tag{19.5}
$$

Similarly, the probability of getting tails, then tails, then heads, is:

$$
\text{Probability of TTH} = \left(\frac{1}{2}\right)\left(\frac{1}{2}\right)\left(\frac{1}{2}\right) = \frac{1}{8}.
\tag{19.6}
$$

And so on. Using these results in Eq. (19.4) we obtain:

$$
\begin{aligned}
\text{Average winnings} \quad = \quad & \left(\frac{1}{2} \times \$2\right) \\
+ \quad & \left(\frac{1}{4} \times \$4\right) \\
+ \quad & \left(\frac{1}{8} \times \$8\right)
\end{aligned}
$$

[2]This business of an infinite number of terms that matter less and less can be made mathematically precise. This is usually one of the main topics of the second term of a calculus sequence. We will not need any calculus to proceed here. However, it is worth nothing that handling equations with an infinite number of terms can be potentially subtle, and hence should be approached with some caution.

$$+ \quad \left(\frac{1}{16} \times \$16 \right)$$
$$+ \quad \dots . \tag{19.7}$$

Note that each term in the parenthesis in the above equation equals 1. What this means is that

$$\text{Average winnings} = 1 + 1 + 1 + 1 + \dots . \tag{19.8}$$

But what does *this* mean? On the right-hand side of the above equation we have an infinite number of 1's added together. Clearly this number keeps growing and growing; the right-hand side of Eq. (19.8) is infinite.

We have thus answered the question about the average winnings in the St. Petersburg game: the average winnings are infinite. Equivalently, one would say that the average does not exist. The averages, as plotted in Figs. 19.4–19.7 do not approach some finite value, no matter how many times we play the game. As a final illustration of this, the average winnings are shown plotted up to 300 million rounds of the game in Fig. 19.8.[3] Even after playing the St. Petersburg game 300 million times—approximately once for each person who lives in the United States—the average winnings are still fluctuating.

[3]Remarkably, this took only a little under 1 minute to simulate on my laptop computer.

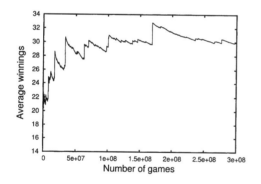

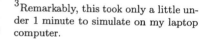

Fig. 19.8 Average winnings for the St. Petersburg game. Note that despite the very large scale on the horizontal axis, the average winnings still do not appear to be approaching a constant value. The numbers on the horizontal axis are written in scientific notation: "5e+07" means 5×10^7, or 50 million. "1e+08" is 100 million, and so on.

19.4 Implications

Let us step back for a moment and consider what this all means. The idea of an average size or an average weight or an average time is so basic that it is easy to take for granted. We expect, quite reasonably, that questions like "what is the average height of giraffes?", "how much does a typical bottle of craft beer cost?", or "how long do Armin van Buuren's DJ sets usually last?" are well posed and have a definite answer. We might need to measure quite a few giraffes, investigate more than a few bottles of beer, or listen to Armin van Buuren frequently. But the answer is out there; we just need to make some measurements and figure it out. Most crucially, if we want a more accurate average, all we have to do is make more measurements, and the average value will get closer and closer to the true average. This was the case in the first coin-tossing game we considered.

But for the St. Petersburg game, our intuition about averages leads us astray. Mathematically the average does not exist. In the limit that we play the game infinitely many times the average winnings will be infinite. We have seen the fact that an average does not exist illustrated graphically in Figs. 19.4–19.8.

However, in any experiment you will always have a finite amount of data. It is, of course, impossible to measure something infinitely many times. Suppose, for instance, that you played the game 100 times and recorded the payoff each time. You could then calculate the average of these 100 games, and this *would* be well defined mathematically. After all, a collection of 100 measurements most certainly does have an average. But this average is not that meaningful. It tells us something about the particular 100 measurements we made, but it is not a statement about the game in general.

For example, suppose we measure the height of 100 giraffes and then calculate the average height. We might publish this result in the *Journal of Ungulate Statistics*. In so doing, we would most likely not be making a claim about the 100 giraffes we measured. Rather, we would be making a statement about giraffes in general. For phenomenon like giraffe heights, this would almost surely be a safe generalization. However, we cannot do the same thing for the St. Petersburg game. If we observed 100 games, the average winnings would be easy to calculate. This would tell us something about the particular 100 games we happened to observe, but not so much about the St. Petersburg game in general.

Most phenomena are like the giraffes: there is a well-defined average that is not difficult to calculate. However, there are many phenomena that are like the St. Petersburg game: the average does not exist or is, at best, very misleading. Examples of phenomena that are generally believed to be like the St. Petersburg game include the frequency and severity of earthquakes, the popularity of websites, the sizes of corporations, the populations of cities, and the frequency of word usage. We will pursue these ideas further in the next chapter.

Exercises

(19.1) Suppose you toss a fair coin five times. The coin comes up heads with probability 0.5 and tails with probability 0.5. What is the probability of each of the following outcomes?

(a) HTHTH

(b) HTTTH

(c) HHHTT

(d) HHHHH

(19.2) Suppose you toss a fair coin five times.

(a) What is the probability that you get only one head out of the five tosses?

(b) What is the probability that you get only two heads out of the five tosses?

(19.3) Consider the simple coin-tossing game described in Sec. 19.1. Suppose that the coin is biased so that heads occurs with probability 0.75 instead of a probability of 0.5. What is the average winnings in this game?

(19.4) Ezra plays the simple coin-tossing game ten times. His outcomes are, in order: H, T, T, T, H, T, H, T, T, H. Construct a table of Ezra's average winnings after each game, as I did in Table 19.1.

(19.5) Lily plays the St. Petersburg game ten times. Her outcomes are, in order: TH, H, H, TTTH, H, TTTH, TH, H, TH, TH. Construct a table of Lily's average winnings after each game, as I did in Table 19.2.

(19.6) ♯ Consider a St. Petersburg game that is identical to the one described in the text, with one key difference. The coin is not fair; tails occurs with a probability of 0.75. Does the average winnings exist for this game? Why or why not?

(19.7) ♯ Consider a St. Petersburg game that is identical to the one described in the text, with one key difference. The coin is not fair; tails occurs with a probability of 0.25. Does the average winnings exist for this game? Why or why not?

Power Laws and Long Tails

20

In the last chapter we saw an example of a simple process, the St. Petersburg game, that does not possess an average. I concluded by suggesting that this state of affairs is not unusual for fractals; there are many phenomena for which the notion of an average or typical size is undefined or misleading. For example, the Sierpiński triangle (shown in Fig. 20.1) is made up of triangles of many sizes. There is one big triangle, three smaller ones, nine smaller still, 27 even smaller, and so on. It is thus not useful to describe a Sierpiński triangle by calculating an average triangle size.

Instead of thinking about the average size contained in a fractal, it is often more informative to look at the distribution of sizes. In this chapter we will look at distributions that are fractal in the sense that they are scale-free; they describe a phenomenon which does not have a characteristic size. Such distributions, often referred to as power laws, are the subject of this chapter.

In what follows I assume that you are familiar with using histograms to describe the frequency of different outcomes in a set of measurements and that you have had some experience thinking about distribution functions for continuous variables. An overview of these topics can be found in Appendix B.

20.1 The Central Limit Theorem and Normal Distributions

We begin by considering the normal, or Gaussian, distribution. This will provide a useful contrast to power-law, or fractal distributions. Moreover, normal distributions are important because the fact of their ubiquity has strongly shaped scientists' intuitions about statistics.

Suppose that one is interested in the distribution of the masses of cats. In this fictional example imagine I weigh 10,000 cats and form a histogram with the resultant data. The result might look something like Fig. 20.2. The histogram can be well approximated by the following function:

$$p(x) = \frac{1}{\sigma\sqrt{2\pi}} e^{\frac{-(x-a)^2}{2\sigma^2}} , \tag{20.1}$$

where $a = 5.0$ and $\sigma = 0.5$. We will not explicitly use this formula, but it plays such an important role in statistics and almost all branches of science that I think it is worth writing down.

Fig. 20.1 The Sierpiński triangle. Note that the triangle is made up of triangles of many different sizes.

The distribution of Eq. (20.1) is known as the **Gaussian** or **normal distribution**. It is also very commonly referred to as a **bell curve**, because it has a bell-like shape. I will use these three terms—the normal distribution, the Gaussian distribution, and the bell curve—synonymously. The quantity a in Eq. (20.1) is the average value for x. The quantity σ^2 is known as the **variance** and σ is known as the **standard deviation**. Both are related to the spread of the x values around the average. The larger the standard deviation (or variance), the more spread out the x values are. It turns out that around 68% of the quantity described by the normal distribution will be within one σ of the mean. For the cat example, this means that we expect 68% of the cats to have a mass between 4.5 and 5.5 kg.

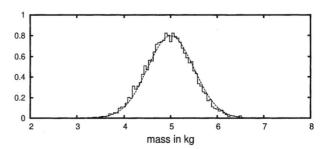

Fig. 20.2 A histogram of the masses of 10,000 cats. (This is the same plot shown on the bottom of Fig. B.6 in Appendix B.)

Gaussian distributions are remarkably common; they describe a vast number of different phenomena. Why is this? There is a remarkable theoretical explanation for the ubiquity of Gaussian distributions. The key result is as follows. Suppose we have a variable x that has some unknown distribution. Let us measure x several different times and add up the measurements. The resultant sum will approach a Gaussian distribution as we make more and more measurements. Amazingly, this is true regardless of the distribution of x.

Let us illustrate this with an example. Suppose a strange type of winter squash comes in three sizes. The smallest size weighs 1 pound, the middle size weighs 3 pounds, and the large size weighs 4 pounds. Suppose further that the large size occurs half the time and the other two sizes occur one quarter of the time. This situation is illustrated in Fig. 20.3. The idea here is that when we select a squash we do not know in advance how much it will weigh. With probability $\frac{1}{2}$ we get a 4-pound squash, with probability $\frac{1}{4}$ we get a 3-pound squash, and with probability $\frac{1}{4}$ we get a 1-pound squash.

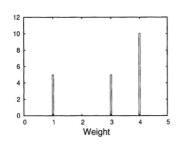

Fig. 20.3 The distribution for the winter squash example. There are three possible squash weights: 1, 3, and 4 pounds. The heavy squash are twice as common as the light or medium squash.

Note that the average squash weight is 3. To see this, observe that we would expect to have twice as many big squash as we will get medium and small squash. I.e., we expect to have squash with weights of 1, 3, 4, and 4. The average of this is 3, since

$$\text{average} = \frac{1+3+4+4}{4} = \frac{12}{4} = 3 \,. \tag{20.2}$$

Now let us imagine that we get five squash. How much would this bag of squash weigh? On average, we expect it to weigh 15 pounds, since the average squash weighs 3 pounds. But sometimes we will get a heavier bag, and sometimes a lighter bag. How will the weights be distributed? The answer is shown in Fig. 20.4, where I have plotted a normalized histogram of the weights of bags of five squash. To make this plot I wrote a short computer program that simulated grabbing five squash 100,000 times and plotted the histogram. It is possible, but tedious, to determine the histogram exactly in the limit that we sample an infinite number of bags of squash. It is much simpler to do it with a computer program. The histogram does not look much like a normal

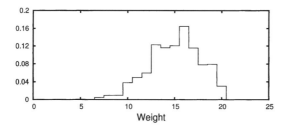

Fig. 20.4 A normalized histogram for the total weight of five squash, where the weight of each squash is distributed according to Fig. 20.3.

distribution—it does not have the characteristic bell shape. But note that it also clearly does not resemble the original distribution, shown in Fig. 20.3.

Let us repeat this experiment, but this time we will imagine filling our bag with twenty squash. If we do this many, many times, what will happen? The average weight should now be $3 \times 20 = 60$ pounds. But how will the bag weights be distributed? The answer is shown in Fig. 20.5. Note that now the histogram is certainly starting to look like

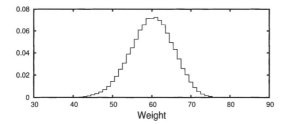

Fig. 20.5 A normalized histogram for the total weight of twenty squash, where the weight of each squash is distributed according to Fig. 20.3.

a Gaussian. It is getting smoother and appears symmetric about the average weight of 60.

Let us try this experiment once more. We now choose 100 squash at a time.[1] The average weight of our collection of 100 squash is now 300. The distribution of the squash weights is shown in Fig. 20.6. We see that the distribution closely resembles a Gaussian. Also in Fig. 20.6 I have plotted the Gaussian distribution, Eq. (20.1) with $a = 300$ and $\sigma = 12.2$. The histogram and the Gaussian curve do not exactly match, but the agreement is very good. I made this histogram by simulating 100,000 collections of 100 squash each, and then plotting the normalized histogram for them all. If I simulated more collections, the histogram and the curve would get closer and closer.

[1] The squash now do not fit in a bag. We will need a wheelbarrow or perhaps a car.

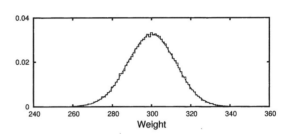

Fig. 20.6 A normalized histogram for the total weight of 100 squash, where the weight of each squash is distributed according to Fig. 20.3. The dashed line is a Gaussian distribution function, Eq. (20.1) with $a = 300$ and $\sigma = 12.2$.

To summarize, we started with a distribution of individual squash weights, Fig. 20.3, that is most definitely not Gaussian. Not only is it not a bell curve, it is not even a curve. The squash take on only three different values. I then imagined forming collections of successively larger number of squash: five, twenty, and then a hundred squash. For each I formed many such collections, measured the weights, and plotted a histogram. The result is that the distribution of weights of these squash collections gets closer and closer to a Gaussian curve.

The general version of this result is known as the **central limit theorem**:

> Let x be a random variable. Then the distribution of a sum of the random variables x approaches a Gaussian distribution, as the number of variables in the sum becomes large.

In practice, typically the number of variables does not have to be very large before the distribution is well approximated by a Gaussian. For example, in Fig. 20.5 the distribution was reasonably well approximated by a Gaussian, even though there were only twenty variables (squash) in the sum. It is remarkable that this result holds no matter what the original distribution of x is. The sum of a large number of *any* random variables will have a Gaussian distribution.

Moreover, the variables do not even have to have the same distribution. For example, we could imagine making a bag of ten vegetables by selecting one each of ten different types of vegetable: a squash, a potato, a rutabaga, and so on. Each of these vegetables would likely have a different distribution for its weight. Nevertheless, the sum of the weights of these ten different vegetables will nevertheless have a Gaussian distribution.[2] This is quite a strong result—it suggests that there is something truly universal about Gaussian distributions.[3]

The central limit theorem explains why Gaussian distributions are so common. Consider a quantity such as human height. In Fig. 20.7 I have plotted a histogram for a data set consisting of height measurements of $25,000$ people.[4] The data is very well approximated by a Gaussian. On the figure I have plotted a Gaussian distribution with $a = 67.99$ and $\sigma = 1.90$.

---

[2] This statement is only true if the random variables (each of the different type of vegetable) do not have weird distributions with infinite variance. But in practice, these limitations are not usually a concern. The details are rather technical; see a text on probability for details.

[3] It is important to note that the central limit theorem does not imply that *everything* is distributed according to a Gaussian. The central limit theorem only applies to a sum of random variables. It does not apply, e.g., to a product of random variables.

[4] The data that I used to make Fig. 20.7 were taken from the Statistics Online Computational Resource at the University of California Los Angeles. See `http://wiki.stat.ucla.edu/socr/index.php/SOCR_Data_Dinov_020108_HeightsWeights` for details and for access to the full data set.

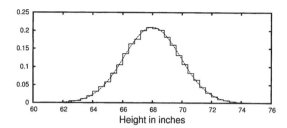

Fig. 20.7 A normalized histogram for the heights of 25,000 people. The solid line is a Gaussian distribution function, Eq. (20.1) with $a = 67.99$ and $\sigma = 1.90$.

What determines how tall an adult is? Nobody knows for sure, but it is certainly a combination of many factors: parental height, nutrition, overall physical and mental health, exposure to pollutants or toxins, and so on. If we assume that these effects are additive, then the central limit theorem assures us that height will be distributed according to a Gaussian. By additive, I mean that the effects add together to produce total height. I.e., the height contribution due to nutrition can be thought of as being added to the height contribution due to parental height. In general, when we have a quantity that is influenced by a number of different factors, it is very often reasonable to assume that they contribute in an additive way. Hence the near-ubiquity of normal distributions.

Even if a given phenomenon is not exactly described by a Gaussian, it is very often nevertheless Gaussian-like, in the sense that the distribution has a well-defined average and a relatively small variation about that average. For example, the average height of adult women in the U.S. is very close to 64 inches. There is relatively little variation about this mean. Ninety percent of all women are between 59 and 68 inches (McDowell *et al.*, 2008). So 90% of all women are within 5 inches of the average. Five is 7.8% of 64. So 90% of all U.S. women are within 7.8% of the average height. This is a fairly small range.

Another way of looking at the variation is to ask about the largest and smallest value in a set of measurements. The shortest human woman is about 23 inches and the tallest is around 97 (Glenday, 2010). Thus, the tallest woman is around 4.2 times taller than the shortest. This seems like quite a large range—imagining the tallest and shortest women in the world standing next to each other is an interesting image. However, compared to some other quantities this range is actually quite small. In the next section we will encounter a distribution whose largest member is over 14,000 times larger than the smallest.

20.2   Power Laws: An Initial Example

We now examine an example that is quite different from human heights or the weights of vegetables. Consider the frequency of words in the English language. Imagine a long text, such as the novel *Moby Dick* or a textbook on chaos and fractals. Choose a word from that text. How often does it appear? In this book the word "squash" appears 37 times, the word "hippopotamus" appears only once,[5] the word "rutabaga" appears

---

[5]This is its only appearance.

Table 20.1 A listing of word frequencies for the Poincaré quote.

6
5
3
3
3
3
2
2
2
2
2
2
2
2
2
2
1
1
1
1
1
1
1
1

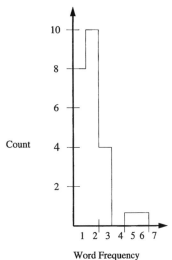

Fig. 20.8 A histogram for the data shown in Table 20.1.

twice, and the words "rabbit" or "rabbits" appear 232 times. Imagine doing this for all the words in a book. We could then determine the average number of times a word appears, and also the distribution of word frequencies.

Let us illustrate this process use two sentences written by Henri Poincaré, a mathematician who did groundbreaking work in the late 1800s and 1900s work on a number of questions in mathematics and physics, including dynamical systems. He writes:

> The scientist does not study nature because it is useful; he studies it because he delights in it, and he delights in it because it is beautiful. If nature were not beautiful, it would not be worth knowing, and if nature were not worth knowing, life would not be worth living. (Poincaré, 2001, p. 186)

There is one word (it) that appears six times in this passage and one word (not) that appears five times. There are four words that appear three times each (because, he, nature, worth). Ten words appear twice (and, be, beautiful, delights, if, in, is, knowing, were, would), and there are eight words that appear only once (does, life, living, scientist, study, studies, the, useful.) This information is collected in Table 20.1. We then use these data to form the histogram, shown in Fig. 20.8. I suggest taking a moment to make sure you see how to go from the Poincaré quote to the frequency data in Table 20.1 and subsequently to the histogram of Fig. 20.8.

What would happen if we tried a similar experiment but with a much larger text? What is the frequency of the average word? If I choose a word at random from a list of all the words in the text, how many times does that word appear overall? And what does the distribution of frequencies look like? Is it a Gaussian? Let us try it and see.

In a recent paper, Mark Newman (2005) measured the word frequencies for *Moby Dick*, the novel by Herman Melville. Using the same procedure that I did for the Poincaré passage, he determined the frequency of all words in the novel. The result is a list of numbers similar to Table 20.1, but much, much longer.[6] There are 18,855 different words in the novel. The total number of words is 209,994. The most common words are, in order, "the", "of", "and", "a", and "to." These words appear, respectively 14,086, 6414, 6260, 4573, and 4484 times. One can think of these words as being the "biggest" in the sense that that they occur the most frequently. The "smallest" words are those that occur only once. The average word frequency is around 11.1.

What about the frequency distribution? A histogram for the table of word frequencies is shown in Fig. 20.9. A few things are immediately apparent. First, there is no bump or central peak as there is for the other distributions we have looked at so far. By far the largest value

---

[6]This long table is available at http://www.santafe.edu/~aaronc/powerlaws/data/words.txt. It is part of a collection at http://www.santafe.edu/~aaronc/powerlaws/data.htm, where one can find quite a few other data sets which are distributed (or suspected of being distributed) according to a power law.

is for a frequency of 1. There are 9,161 words that occur only once in all of *Moby Dick*. This is 48.59% of the total words. The plot shown in Fig 20.9 is not the full histogram. I have plotted only the 100 most common frequencies.

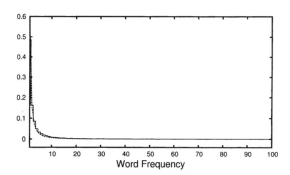

Fig. 20.9 A histogram for the frequency of occurrence of words in *Moby Dick*.

What to make of Fig. 20.9? Well, it clearly is not a Gaussian. Instead, it turns out that this distribution is well described by the following function:

$$p(x) = Ax^{-1.95} \tag{20.3}$$

where $A \approx 0.59$ (Clauset, Shalizi, and Newman 2009). The variable x is the word frequency. A plot of this distribution is shown superimposed on the data in Fig. 20.10. One can see that Eq. (20.3) agrees fairly well with the histogram generated from the data.

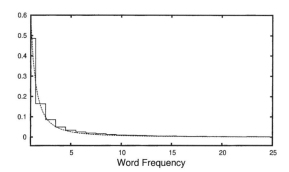

Fig. 20.10 A histogram for the frequency of occurrence of words in *Moby Dick*. The dashed line is the power law, Eq. (20.3).

20.3 Power Laws and the Long Tail

Equation 20.3 is an example of what is known as a **power law**. The general form is:

$$p(x) = Ax^{-\alpha} . \tag{20.4}$$

This is called a power law because the distribution of the variable x is given by x raised to some power. The quantity A is determined by the exponent α. We know the total area under the curve has to be one, so this determines A.[7] In other words, once we have figured out α, we can figure out A as well.

There are two important and noteworthy features of a power-law distribution. First, it decays rather slowly. This means that the frequency of observing a large x value, while small, is not tiny. In contrast, an exponential distribution decays much faster. An exponential distribution has the form

$$p(x) = Ae^{-ax} . \tag{20.5}$$

Note that in the exponential distribution the variable x is in the exponent. Exponential distributions are quite common—they frequently describe the waiting time between random events: the time between phone calls, or the time between accidents at a job site.[8] Exponential distributions arise when there is a fixed probability at every time interval that an event occurs, and so longer and longer waiting times are increasingly uncommon. Exponential and Gaussian distributions both have the property that it is extremely unlikely to observe events that are very different from the average.[9]

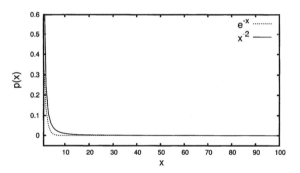

Fig. 20.11 A power-law and an exponential distribution.

An exponential and a power-law distribution are plotted in Fig. 20.11. The power-law exponent α is 2 and a in the exponential distribution is 1. One sees that the exponential decays faster than the power law. This faster decay is much more apparent, however, for larger x. To illustrate this, in Fig. 20.12 I have again plotted the two distributions, but this time x ranges from 50 to 100. Observe that the exponential distribution

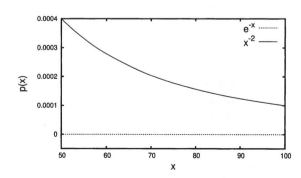

Fig. 20.12 A power-law and an exponential distribution.

is indistinguishable from zero, while the power-law distribution is small, but much larger than the exponential. One can see this numerically, as well. The probability of $x = 50$ for the power law is 0.0004. For the exponential distribution the probability[10] of $x = 50$ is around 2×10^{-22}.

This has important implications. If x is distributed exponentially, the probability of $x = 50$ is for all practical purposes zero. We would expect to never observe $x = 50$. On the other hand, if x is distributed according to a power law, then there is a 4 in 10,000 chance that $x = 50$. This is small, but certainly not microscopic. If there were some catastrophic event that has a 4 in 10,000 chance of occurring in the next decade, you might want to buy an insurance policy to protect you if such an event occurs. However, there is no need to worry about an event that happens with a probability of 2×10^{-22}, or 0.00000000000000000000002. This is why people buy fire insurance for their homes, but not insurance for getting hit by a meteorite.

Because power-law distributions decay slowly—and hence a plot of a power-law extends far to the right on a graph—they are said to have a **long tail**. Long-tailed distributions are generally taken to be those that decay more slowly than an exponential distribution.[11] For a situation described by a power law there is a relatively large probability of an extreme event—a large x value. In contrast, short-tailed distributions like the exponential predict vanishingly small probabilities for extreme events. The Gaussian distribution is, like the exponential, short-tailed. Much of our intuition—and much of the apparatus of statistics—is based on Gaussians. These work extremely well for many phenomena, but can be badly misleading when applied to long-tailed distributions.

[10] This assumes that the variable x is discrete and not continuous. If x is continuous the same general scenario holds, but one has to interpret $p(x)$ differently.

[11] There is not presently a standard technical criteria for long-tailed distributions. Such distributions are also known as fat-tailed distributions.

20.4 Power Laws and Fractals

A second noteworthy feature of power-law distributions is that, like fractals, they are self-similar. Another way to say this is that they are scale-free. They do not possess a meaningful average or typical size or scale. For fractals like the Sierpiński triangle, we observed self-similarity by zooming in on the shape and noticing that doing so left the shape unchanged. We can do a similar thing with a power-law distribution. In Fig. 20.13 I have plotted the same power-law distribution, $p(x) = Cx^{-2}$, from $x = 50$ to 100, and then from $x = 100$ to 200. The two plots look the same. The distribution is thus a fractal—it appears the same at different scales.

This scale-invariance does not occur for other distributions. To show this I have done the same thing with the exponential distribution $p(x) = Ae^{-x}$. In Fig. 20.14 I have plotted the exponential distribution for two different x ranges. Looking at the two plots we can clearly see that the distributions are not the same. Thus, exponential distributions are not scale-invariant in the way that power-law distributions are.

It turns out that power-law distributions are exactly those that are scale-free. Power laws are thus the signature of fractals, and vice-versa.

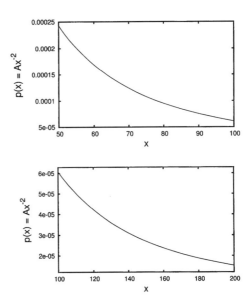

Fig. 20.13 A power-law distribution plotted on two different scales. The two plots look the same, illustrating that power-law distributions are scale-free. The numbers of the vertical axis are written in scientific notation. "2e-05" means $2 \times 10^{-5} = 0.00002$. "3e-05" means $3 \times 10^{-5} = 0.00003$, and so on.

[12]The general proof is rather technical and requires a knowledge of calculus. See Newman (2005, p. 11) or Frank (2009, p. 1566) for clear discussions of the relationship between scale-free distributions and power laws.

[13]I will assume that the branch lengths are always integers: 1, 2, 3, etc. The argument in the next several paragraphs works if the branches are non-integer values, but the mathematics becomes more subtle.

I have not mathematically proven this result, but Figs. 20.13 and 20.14 illustrate a particular instance of this phenomenon.[12] Power-law distributions are sometimes called **scaling distributions**. The reason is that they describe phenomena that scale—appear the same as one looks at the distribution at different ranges.

Here is another way to think about why power-law distributions are fractal. Suppose there is a tree with branches of many different lengths.[13] Let us suppose the distribution of branches is described by a power law with an exponent of 2. Specifically,

$$\text{Probability a branch has length x} = Ax^{-2} , \qquad (20.6)$$

where $A \approx 0.6079$. (One can calculate the value for A by requiring that the probability adds up to 1.) This distribution is plotted in Fig. 20.13. The distribution tells us that if we were to choose a branch at random, the probability that it has length 5 is 0.024, and the probability that it has length 10 is 0.006. I determined these numbers by plugging $x = 5$ and then $x = 10$ into Eq. (20.6). So branches of length 5 are four times more likely than branches of length 10. Equivalently, there are four times as many length-5 branches than length-10 branches. The distribution of Eq. (20.6) also tells us that the probability of a branch of length 20 is 0.00152, while the probability of a branch of length 40 is 0.00038. Thus, branches of length 20 are four times more common as branches of length 40. In both cases, there are four times as many branches that are half as large. This is true in general. For any size branch there will be four times as many branches that are half as long.

Now, imagine that some magic spell has been placed on you. You end up in the tree whose branch sizes are described by Eq. (20.6). You suspect that this magic spell has changed your size as well. You have no way of knowing for sure, however, since the tree contains no scale –

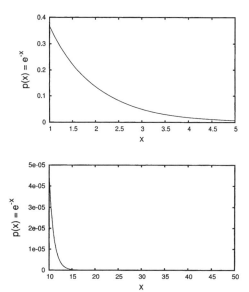

Fig. 20.14 An exponential distribution plotted on two different scales. Unlike Fig. 20.13, the two plots do not look the same. Thus, the exponential distribution is not scale-free.

no clues about size.[14] You might observe a number of branches that are about the same size as you. And you would observe that there are four times as many branches that are half your size. But this observation tells you nothing about *your* size. No matter what size you are, there are always four times as many branches that are half as large as you. It is in this sense that a power-law distribution is scale-free.

We have seen power-law relationships before. In Chapter 18 the equation used to determine the box-counting dimension D has the form of a power law. Equation (18.2) related D to the number of boxes $N(s)$ of size s needed to cover an object:

$$N(s) = k \left(\frac{1}{s} \right)^{D} . \tag{20.7}$$

This equation also indicates a scaling relationship; it says that no matter what the size of the box, if we make the box half as small, we will need 2^D more boxes to cover the shape. This relationship is independent of scale; it holds regardless of the initial box size s.[15]

In Chapter 18 we saw that if we took the logarithm of both sides of Eq. (20.7) and made a plot, the result was a straight line. The same is the case for a power-law distribution. To see this, start with the power-law equation

$$p(x) = Ax^{-\alpha} , \tag{20.8}$$

and take the logarithm of both sides. After doing so, and after simplifying the algebraic expressions, one obtains:

$$\log(p(x)) = \log(A) - \alpha \log(x) . \tag{20.9}$$

Thus, if one plots $\log(x)$ on the horizontal axis and $\log(p(x))$ on the vertical axis the result will be a straight line. This is illustrated in

[14]The tree is very large and dense. You cannot see out, nor can you tell how large the tree itself is. All you can see is lots and lots of branches, of many different sizes.

[15]This is a true mathematical statement about Eq. (20.7). In practice, however, this equation only applies if the box size is sufficiently smaller than the object.

Fig. 20.15 where I have plotted the histogram for the word frequencies from *Moby Dick* on a log-log scale. Note that the histogram now appears approximately linear. Log-log plots such as Fig. 20.15 are a standard way that power-law distributions are presented in the scientific literature.

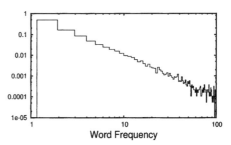

Fig. 20.15 A histogram for the frequency of occurrence of words in *Moby Dick* plotted on a log-log scale. Note that the histogram appears to be well approximated by a straight line. The same data was plotted on an ordinary scale in Fig. 20.9.

20.5   Where do Power Laws Come From?

A great many phenomena are believed to be distributed according to a power law or something close to a power law, including: the frequency of words in texts, the size of power blackouts, the populations of cities, books sales, the number of times scientific papers are cited, the size of forest fires, the size of earthquakes, the number of phone calls per fixed amount of time, the size of people's email address books, the number of links to a website, and the number of species per genus (Clauset, Shalizi, and Newman, 2009). What does all this mean?

As we have seen throughout the last several chapters, fractals are "easy" to make, in the sense that they can be generated with simple iterated rules. We have also seen in Chapter 19 that fractals can be produced by simple stochastic processes. There are many different ways that fractals can be generated, so I think we should not be surprised when we encounter them. And since power laws are just another manifestation of fractals, the same is true for them. There are many simple ways to generate power laws.

Somewhat surprisingly, this lesson is lost on some scientists, who posit that the existence of power laws indicates some overarching organizing principle. It seems to me that this is badly off the mark, since it is well known that there are many simple mechanisms that produce power laws. There is no reason to assume that all power laws are generated by the same mechanism, nor is there any reason to assume that power laws are evidence for a high degree of organization, complexity, or optimization. Power laws, and fractals, are interesting and noteworthy, but they should not be surprising. There is ample evidence that power laws and fractals are all around us.

Finally, a few important words of caution. Testing that a given data set really is power-law distributed is somewhat tricky. The natural thing to do is to take the histogram, plot it on a log-log scale, and see if it looks linear. If it does, it is often assumed that the data is a power law. One then estimates the slope using standard statistical techniques. The slope is the exponent α. Unfortunately, this method can be very misleading. It is a poor way to estimate α, and can lead one to incorrectly infer the presence of a power law that is not real.[16] Nevertheless, this method has been used by many scientists. As a result, some recent claims for the existence of power laws are almost surely wrong. See Clauset, Shalizi, and Newman (2009) for further discussion, including a detailed derivation of a more reliable method for estimating α and determining whether or not a power law is a good fit to a set of empirical observations.

What often is of interest is not whether or not a given data set is exactly a power law, but whether or not it has a long or heavy tail—i.e., if the probability of rare events decays significantly more slowly than an exponential distribution. Generically, long-tail distributions may not be fractal in the sense that they are scale-independent. But they still indicate a very different kind of statistics than Gaussian or exponential, and demand a different sort of intuition about the prevalence of rare events.

[16] A data set may be well-approximated by a power law, and yet another function might be an even better fit to the data. Ideally, one should test several alternative functions before asserting that data are power-law distributed.

Further Reading

The definitive treatment of how to test for power laws in empirical data is Clauset, Shalizi, and Newman (2009). An excellent review of different empirical power laws along with a discussion of power-law properties is Newman (2005). Chapter 17 of Mitchell (2009) has a good, largely non-technical discussion of power laws. Several papers critique the recent hype around power laws. Mitzenmacher's "A brief history of generative models for power law and lognormal distributions" (2004) and Keller's "Revisiting 'scale-free' networks" (2005), are both lucid and accessible. I particularly recommend the recent essay by Stumpf and Porter, "Critical truths about power laws" (2012). Frank (2009) is a clear and well written review of power laws and other distributions that occur commonly in biology.

Exercises

(20.1) Construct a word-frequency histogram similar to that of Fig. 20.8 using the first two sentences of this chapter.

(20.2) Below is a portion of the lyrics from *This Light Between Us*, by Armin van Buuren featuring Christian Burns (2010):

So we should dance like this forever
We're safer on the ground
When a million lights surround you

And you're moving to the sound

Don't waste another moment
It's waiting for you now
To dive in this new beginning
Let the colors show you how

Construct a word-frequency histogram similar to that of Fig. 20.8 using these lyrics.

(20.3) Fill in the algebraic steps between Eq. (20.8) and Eq. (20.9).

(20.4) Consider the probability distribution for the tree branches given in Eq. (20.6).

  (a) What is the probability of observing a tree branch of size 6?

  (b) What is the probability of observing a tree branch of size 12?

  (c) How much more common are size 6 branches than size 12 branches?

(20.5) In this exercise we will see that an exponential distribution is not scale-free in the way that a power law is. Suppose the number of days between accidents at a sawmill is well described by the following distribution $p(t) = \frac{1}{10}e^{-\frac{1}{10}t}$.

  (a) What is the probability that there are 3 days between accidents? What is the probability that there are 6 days between accidents? How many more times likely is it that there are 3 days between accidents than 6 days?

  (b) What is the probability that there are 10 days between accidents? What is the probability that there are 20 days between accidents? How many more times likely is it that there are 3 days between accidents than 20 days?

  (c) Use your answers to the above two questions to conclude that the exponential distribution is not scale-free.

Infinities, Big and Small

21

Fractals such as the Cantor set and the Koch curve were originally introduced not as geometrical objects, as we have viewed them thus far, but as a way to explore apparent paradoxes and difficulties in the areas of mathematics now known as set theory and analysis. A study of the Cantor set reveals some surprising and fun properties of infinite sets. This is the focus of this chapter. This chapter has a somewhat different character than most other parts of this book. It does not require any advanced background, but it is concerned with abstract mathematics and not dynamical systems. None of the subsequent chapters depend on this one; it is a bit of a tangent. But I have included it in this book because using the Cantor set to think about different sorts of infinities is fascinating and a lot of fun.

21.1 What is the Size of the Cantor Set?

The central question of this chapter is: How big is the Cantor set? Recall that the Cantor set is formed as follows: start with a line segment, remove the middle third of the line segment to produce two line segments, then remove the middle third of those two segments yielding four line segments, and so on. This process is illustrated in Fig. 21.1. The Cantor set is the collection of points that is left after this process has been repeated an infinite number of times.

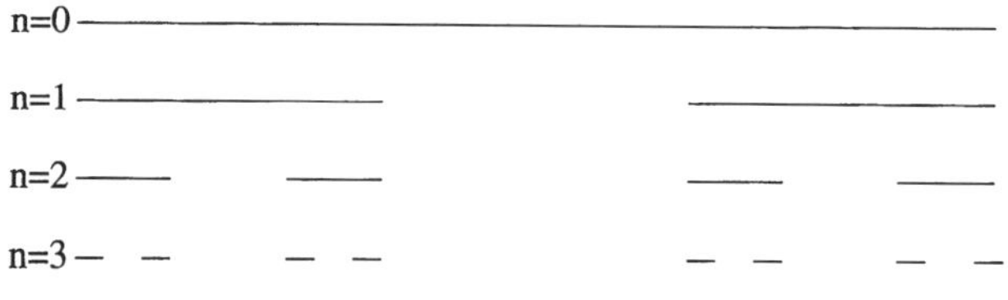

Fig. 21.1 The first several stages in the construction of the middle-thirds Cantor set.

At each step the number of line segments doubles while the length of each line segment decreases by a factor of three. This is shown in Table 21.1. At the nth step there are 2^n line segments, each of which has a length of $\frac{1}{3^n}$. To get the total length at each step in the construction of the set, one just multiplies the number of line segments by the length of each line segment. Thus, at step n the total length is $(\frac{2}{3})^n$.

As n goes to infinity the number of line segments also goes to infinity; 2^n gets larger and larger as n gets larger. However, the length of each segment goes to zero; $(\frac{1}{3})^n$ gets smaller and smaller as n gets larger. What about the total length? We have an infinite collection of tiny line

Step	# of Line Segments	Length of Each Segment	Total Length
0	1	1	1
1	2	$\frac{1}{3}$	$\frac{2}{3}$
2	4	$\left(\frac{1}{3}\right) \times \left(\frac{1}{3}\right) = \frac{1}{9}$	$\frac{4}{9}$
3	8	$\left(\frac{1}{3}\right) \times \left(\frac{1}{3}\right) \times \left(\frac{1}{3}\right) = \frac{1}{27}$	$\frac{8}{27}$
n	2^n	$\frac{1}{3^n}$	$\frac{2^n}{3^n} = \left(\frac{2}{3}\right)^n$

Table 21.1 The number of line segments, their length, and the total length, as the Cantor set is constructed.

segments. The length of the line segments goes to zero while the number of the segments goes to infinity. What is "$\infty \times 0$"? In this case, it turns out that the answer is zero. The total length is given by $\left(\frac{2}{3}\right)^n$. As n gets large, the total length gets smaller and smaller. We thus conclude that the length of the Cantor set—which is the collection of points left in the $n \to \infty$ limit in the construction illustrated in Fig. 21.1—is zero.

How many points are in the Cantor set? Thus far we have answered this question by simply saying that the number of points in the set is infinite. But there are also an infinite number of points in the line segment that we started out with. And in the process of forming the Cantor set we have removed a great many points. So even though there are an infinite number of points in the Cantor set, it seems that there might be a "smaller infinity" than there is in the initial line segment. But does this even make sense? Can we meaningfully talk about infinities of different sizes? It turns out that we can. In the following several sections we will explore this notion. We will then return to the Cantor set, and will be in for some surprises.

21.2 Cardinality, Counting, and the Size of Sets

Before we consider infinities, we will start by thinking about finite sets. A **set** is simply a collection of objects. The objects in a set are sometimes called elements. One way to specify a set is to simply list its elements. For example, the set $\mathcal{W}$ of days of the week is:

$$\mathcal{W} = \{\text{Sunday, Monday, Tuesday, Wednesday, Thursday,} \atop \text{Friday, Saturday}\} . \qquad (21.1)$$

It is conventional, as I have done above, to use a calligraphic letter to denote a set. There is no order to a set. That is, it does not matter in what order I list the elements. Thus, I could have written:

$$\mathcal{W} = \{\text{Wednesday, Sunday, Saturday, Monday, Friday,} \atop \text{Thursday, Tuesday}\} . \qquad (21.2)$$

Each element in a set appears only once. When working with sets we are just interested in a collection of things; we do not care about the frequency of occurrence of the things in the set.

What is the size of the set $\mathcal{W}$? Clearly, the answer is seven. But let us pause for a second and think about what this means. How do we know there are seven elements in $\mathcal{W}$? Well, we just count the elements. But what does it meant to count? This seems like a silly question, but thinking carefully about counting will be the key to understanding different types of infinity.

Let us take the counting numbers—1, 2, 3, and so on—as given. We will not try to derive these numbers; we will just accept that they exist. Our aim here is to establish that the set $\mathcal{W}$ has seven elements. To do this, let us consider the set $\mathcal{S}$, which we will define to consist of the first seven counting numbers:

$$\mathcal{S} = \{1, 2, 3, 4, 5, 6, 7\}. \tag{21.3}$$

There are clearly seven elements in this set. (Remember, we are taking the counting numbers as a given.)

Here comes the key step. When we say that the set $\mathcal{W}$ has seven elements, what we really mean is that for every element in $\mathcal{W}$ there is one and only one element in $\mathcal{S}$. We can illustrate this one-to-one relation as follows:

$$\text{Sunday} \longleftrightarrow 1 \tag{21.4}$$
$$\text{Monday} \longleftrightarrow 2 \tag{21.5}$$
$$\text{Tuesday} \longleftrightarrow 3 \tag{21.6}$$
$$\text{Wednesday} \longleftrightarrow 4 \tag{21.7}$$
$$\text{Thursday} \longleftrightarrow 5 \tag{21.8}$$
$$\text{Friday} \longleftrightarrow 6 \tag{21.9}$$
$$\text{Saturday} \longleftrightarrow 7 \tag{21.10}$$

This establishes that the two sets $\mathcal{W}$ and $\mathcal{S}$ have the same number of elements.

In set theory the size of a set is referred to as the set's **cardinality**. Two sets have the same cardinality if there is a one-to-one relationship between elements in the sets. In other words, to establish that two sets have the same size, or cardinality, we need to find a rule that assigns one element of one set to exactly one element of the other set. Equations (21.4)–(21.10) are an example of such a one-to-one relationship.

To summarize, the main idea is that we measure a set's size, or cardinality, by comparing it with another set. Two sets have the same cardinality if there is a one-to-one relationship between members of the sets. A set consisting of the first n counting numbers has cardinality n. This gives us a starting point. We then compare other sets to sets of counting numbers to determine their cardinality. I imagine that at this point the definition of cardinality seems pedantic and perhaps unnecessary. However, we shall see that this definition is quite useful when considering infinite sets.

21.3   Countable Infinities

For our first infinite set, let us consider the set of all counting numbers. We will call this set $\mathcal{N}$:

$$\mathcal{N} = \{1, 2, 3, 4, 5, \ldots\} . \tag{21.11}$$

This set is infinite; the counting numbers just keep on going and going and going. We will take the existence of this set as a given. We will not muse on exactly what the existence of this set means philosophically, but will accept that it exists. We shall refer to the cardinality of this set as **countably infinite**. Equivalently, one sometimes just calls such a set countable.

To show that a set is countable, all one needs to do is to come up with a way to count it. In other words, one needs to show that there is a one-to-one relation between elements in $\mathcal{N}$ and the set one is trying to count. As an example consider $\mathcal{E}$, the set of even numbers:

$$\mathcal{E} = \{2, 4, 6, 8, \ldots\} . \tag{21.12}$$

What is the cardinality of this set? Somewhat surprisingly, its cardinality is the same as that of $\mathcal{N}$. To establish this we need to find a rule that associates every element of $\mathcal{N}$ with one and only one element of $\mathcal{E}$. Such a rule is easy to find:

$$1 \longleftrightarrow 2 \tag{21.13}$$
$$2 \longleftrightarrow 4 \tag{21.14}$$
$$3 \longleftrightarrow 6 \tag{21.15}$$
$$4 \longleftrightarrow 8 \tag{21.16}$$

and so on. Since we have found a one-to-one relationship between elements of $\mathcal{N}$ and $\mathcal{E}$, the two sets are the same size; they have the same cardinality.

This seems counter-intuitive, as there are clearly elements of $\mathcal{S}$ that are not in $\mathcal{E}$. The numbers 3, 5, 7, and so on, are in the set of counting numbers yet are not in the set of even numbers. Given that the set of counting numbers contains objects that are not in the set of even numbers, how can it possibly be that the two sets are the same size?

At the root of these questions is the counter-intuitive nature of infinities. The cardinality of infinite sets is not like ordinary addition or subtraction. When we have a set with an infinite number of elements, we can remove a lot of those elements and still have an infinite number left. In fact, in the case we just considered we removed *half* of the elements of the set $\mathcal{N}$—i.e., we removed all odd numbers—and we ended up with the set $\mathcal{E}$ which has the same cardinality as $\mathcal{N}$. In effect, we have that $\infty - \frac{1}{2}\infty = \infty$. Strictly speaking, however, this is not a legitimate mathematical statement, since ∞ is not a number in the way that 4 or $100,000$ is. We cannot do arithmetic with ∞. We can, however, compare the cardinality of sets, as we have done above.

21.4　Rational and Irrational Numbers

As our next example of an infinite set, consider the set of *all* numbers between 0 and 1. By all numbers I mean exactly that—all real numbers. This set is known as the **unit interval**. I will denote this set by $\mathcal{L}$:

$$\mathcal{L} = [0, 1] \ . \tag{21.17}$$

In other words, $\mathcal{L}$ consists of all the points in a line segment of length one. This includes both rational and irrational numbers. **Rational numbers** are those that can be expressed as a fraction, or ratio, of two counting numbers. For example, $\frac{1}{3}$, $\frac{5}{8}$, and $\frac{44}{43}$ are all rational numbers. Rational numbers can be expressed as decimals that either repeat or terminate. For example,

$$\frac{1}{3} = 0.333333\ldots = 0.\bar{3} \ , \tag{21.18}$$

$$\frac{5}{8} = 0.625 \ , \tag{21.19}$$

$$\frac{44}{43} = 0.977272727\ldots = 0.97\overline{27} \ . \tag{21.20}$$

The bar above a number or numbers indicates that those numbers repeat indefinitely.

On the other hand, **irrational numbers** are those numbers that *cannot* be expressed as a fraction or a ratio of two counting numbers. Irrational numbers have decimal expansions that go on forever; they never repeat. For example the square root of 2 is an irrational number:

$$\sqrt{2} \approx 1.414213562\ldots \ . \tag{21.21}$$

I have used "$\approx$" instead of "$=$", because the decimal version of $\sqrt{2}$ goes on forever; it neither terminates nor repeats. Thus, 1.414213562 is just an approximation of $\sqrt{2}$.

There are clearly a lot of numbers between 0 and 1. But how many? What is the cardinality of $\mathcal{L}$? Is it countable, like the set of counting numbers $\mathcal{N}$? Or is it a different type of infinity? In order to answer this question we need to think about different bases in which we can represent numbers.

21.5　Binary

Let us start by thinking about base-10, the base system with which we are most familiar. As a concrete example, consider the number 457. What does this mean? It does not mean $4+5+7$. Rather, different digits have a different meanings. The "4" indicates 400, the "5" indicates 50, and the "7" simply indicates 7. This is written as:

$$457 = (4 \times 100) + (5 \times 10) + (7 \times 1) \ . \tag{21.22}$$

Another way of saying this is that the "4" is in the hundreds' place, the "5" is in the tens' place, and "7" is in the ones' place.

The number expressed in Eq. (21.22) is represented in **base-10**. To see why this is called base-10, we rewrite Eq. (21.22) as follows:

$$457 = (4 \times 10^2) + (5 \times 10^1) + (7 \times 1^0) . \tag{21.23}$$

This makes it clear that the number 10 is used as the base for the exponents that give meaning to the different digits in the number 457.

The number 10 is a particularly convenient number to use as a base. However, it is not the only choice. An alternative is the number 2. We can re-express the number 457 using base-2 as follows. It turns out that

$$
\begin{aligned}
457 &= (1 \times 2^8) + (1 \times 2^7) + (1 \times 2^6) + (0 \times 2^5) + (0 \times 2^4) \\
&\quad + (1 \times 2^3) + (0 \times 2^2) + (0 \times 2^1) + (1 \times 2^0) .
\end{aligned} \tag{21.24}
$$

You might wish to pause for a moment, grab a calculator or a pencil and some scrap paper, and convince yourself that the above equation really is true. To perform this verification is not difficult; all it takes is raising 2 to various powers and then doing some arithmetic.

The 0's and 1's in Eq. (21.24) constitute the **base-2**, or **binary**, representation of 457. That is:

$$457 = 111001001 , \tag{21.25}$$

where the left-hand side of the equation is understood to be in base-10, and the right-hand side in base-2.[1] The number 111001001 is shorthand for the right-hand side of Eq. (21.24):

$$
\begin{aligned}
111001001 &= (1 \times 2^8) + (1 \times 2^7) + (1 \times 2^6) \\
&\quad + (0 \times 2^5) + (0 \times 2^4) + (1 \times 2^3) \\
&\quad + (0 \times 2^2) + (0 \times 2^1) + (1 \times 2^0) .
\end{aligned} \tag{21.26}
$$

In binary, the first digit is the ones' place (since $2^0 = 1$), the second digit the twos' place (since $2^1 = 2$), the third digit is the fours' place (since $2^2 = 4$), the fourth digit is the eights' place (since $2^3 = 8$), and so on.

I figured out Eq. (21.24) via trial and error, experimenting until I found the right mix of 0's and 1's to make the equation true. There is a more systematic method for converting from base-10 to base-2, but this is not needed for what follows. The main point is that *any* integer can be expressed in base-10 or base-2, or any other base, for that matter. Moreover, this representation is unique. For any integer there is only one way to represent it in base-10, and there is also only one way to represent it in base-2.

Recall that the reason we are considering binary is that we want to understand the set $\mathcal{L} = [0, 1]$. Before we return to our investigation of the cardinality of $\mathcal{L}$, we need to figure out how to represent numbers between 0 and 1 in base-2. We start by thinking about decimals in familiar base-10. What does the number 0.593 mean? In this case the "5" is in the tenths' place, the "9" is in hundredths' place, and the "3" is in the thousandths' place. In other words,

$$0.593 = (5 \times \frac{1}{10}) + (9 \times \frac{1}{100}) + (3 \times \frac{1}{1000}) . \tag{21.27}$$

[1] Another way to write this is $457_{10} = 111001001_2$.

Using exponents, we can write this as:

$$0.593 = (5 \times 1^{-1}) + (9 \times 10^{-2}) + (3 \times 10^{-3}) \, . \tag{21.28}$$

If the negative exponents look strange, recall that $10^{-a} = \frac{1}{10^a}$. For a discussion of why negative exponents have this property, see Appendix A.1.

What about binary? The idea is the same as decimals, except that we use 2 as the base of our exponent instead of 10. For example, consider the binary number 0.1101. This means:

$$0.1101 = (1 \times 2^{-1}) + (1 \times 2^{-2}) + (0 \times 2^{-3}) + (1 \times 2^{-4}) \, . \tag{21.29}$$

The first "1" is in the halves' place (since $2^{-1} = \frac{1}{2}$), the second "1" is in the quarters' place (since $2^{-2} = \frac{1}{4}$), and so on. If one evaluates the right-hand side of Eq. (21.29), one finds that

$$0.1101 = 0.8125 \, , \tag{21.30}$$

where the left-hand side is understood to be in binary and the right-hand side in base-10. You might wish to take a moment and grab a calculator and confirm Eq. (21.29) for yourself.

Given a binary representation of a number between 0 and 1, one can convert to a base-10 representation, as we did in Eq. (21.30). One can also go the other way—from decimal to binary—although I will not go into the details here. The main point is that there is a well-defined rule that one can use to go back and forth between base-10 and binary numbers between 0 and 1.

21.6   The Cardinality of the Unit Interval

We are now in a position to return to the question of the size of the unit interval $\mathcal{L} = [0, 1]$. As noted previously, there are very, very many numbers between 0 and 1; the numbers are both rational and irrational. Some of the decimals repeat or terminate, but many do not.

It will be somewhat easier to picture $\mathcal{L}$ using binary. Converting the numbers in the set $\mathcal{L}$ from base-10 to base-2 will not change the size of the set, since every base-10 number has a unique binary representation, and vice-versa.[2] Similarly, if I took all of the students in a class of mine and replaced all vowels in their names with the letter "q", this would not change the number of students in my class.

Expressed in binary, then, the set of all the numbers between 0 and 1 is exactly the set of *all* possible infinite sequences of 0's and 1's. In this representation, those decimals that terminate just have an infinite number of 0's stuck on the end. For example, consider 0.75. The binary representation of this number is 0.11, since

$$0.75 = (1 \times 2^{-1}) + (1 \times 2^{-2}) \, . \tag{21.31}$$

To picture this as an infinite sequence of 0's and 1's we just write 0.75 = 0.110000000.... So now the question is: What is the cardinality of the

[2] Actually, this is not quite true. There are some subtleties associated with the representation of numbers between 0 and 1. The decimal expansion in base-10 is not unique. For example, 0.4 and $0.3\bar{9}$ represent the same number. The bar over the 9 indicates that the 9's repeat forever. With an infinite number of 9's after the 0.3, the number $0.3\bar{9}$ becomes identical to 0.4. In order to have a unique representation for each number between 0 and 1, we need to agree that when there is ambiguity— when there are two possible decimals for the same number—we will choose the decimal that does not have an infinite number of 9's at the end. Similar ambiguities arise in the binary representation of numbers between 0 and 1. So the representation of numbers between 0 and 1 is not unique in either base-10 or base-2. However, this non-uniqueness can be accounted for in such a way that it does not affect my assertion that there is a one-to-one relationship between base-10 and base-2 numbers between 0 and 1.

set of all possible infinite sequences of 0's and 1's? Is this set countable? To show that it is countable, all one has to do is come up with a scheme for counting, or listing, all elements of the set.

Let us imagine that someone claims to have such a list: a very long list of *all* possible numbers between 0 and 1. Such a list would, of course, be infinite. But is such a list possible? To investigate this question, let us imagine what this list might look like. Let x_1 denote the first number in the list, x_2 the second number, and so on. The list might look something like this:

$$
\begin{aligned}
x_1 &= 0.11100111\ldots \\
x_2 &= 0.10100100\ldots \\
x_3 &= 0.01111111\ldots \\
x_4 &= 0.01100111\ldots \\
x_5 &= 0.11100101\ldots \\
x_6 &= 0.01010110\ldots \\
x_7 &= 0.11100011\ldots \\
x_8 &= 0.10001101\ldots \\
\vdots\ & \qquad \vdots\ \vdots\ \vdots\ \vdots\ \vdots\ \vdots\ \ddots
\end{aligned}
\tag{21.32}
$$

Each number continues forever as an infinite sequence of 0's and 1's. And there are an infinite number of such numbers in the list. Whoever gave you this list claims that it is exhaustive—it contains all numbers between 0 and 1. We do not care about the order of the numbers; all we are interested in is the assertion that the list is complete.

How can you verify this claim? It certainly is not practical to check the entire list. And besides, what would you check it against—a list of all the numbers? If you had such a list, you would not be interested in the list shown in Eq. (21.32) in the first place.

It turns out that there is a clever way to show that such a listing of the numbers between 0 and 1 cannot possibly be complete. The first step is to consider the diagonal elements in the list of Eq. (21.32). These diagonal elements are shown in bold below:

$$
\begin{aligned}
x_1 &= 0.\mathbf{1}1100111\ldots \\
x_2 &= 0.1\mathbf{0}100100\ldots \\
x_3 &= 0.01\mathbf{1}11111\ldots \\
x_4 &= 0.011\mathbf{0}0111\ldots \\
x_5 &= 0.1110\mathbf{0}101\ldots \\
x_6 &= 0.01010\mathbf{1}10\ldots \\
x_7 &= 0.111000\mathbf{1}1\ldots \\
x_8 &= 0.1000110\mathbf{1}\ldots \\
\vdots\ & \qquad \vdots\ \vdots\ \vdots\ \vdots\ \vdots\ \vdots\ \ddots
\end{aligned}
\tag{21.33}
$$

We now form a number y by reading off the diagonal (bold in the above equation), and then choosing the opposite of each digit. Thus,

$$y = 0.01011000\ldots. \tag{21.34}$$

For example, the first digit along the bold diagonal of Eq. (21.33) is 1. Thus, the first digit of y is 0.

Is y a member of the list of numbers, Eq. (21.33)? Remember that the list is infinite, so this is not a trivial question. However, I claim that it is impossible for y to be in the list. Does y equal x_1? It cannot, because y and x_1 disagree in the first digit. Does $y = x_2$? Nope, because y and x_2 disagree in the second digit. In general, y and x_n can never be the same, no matter what n is. The reason for this is that we have explicitly constructed y to disagree in at least one digit with every x_n in our list.

Thus, there is no way that the list can be complete! Given any list we can always construct a number y that is not in the list. The inescapable conclusion is that such a list cannot exist: *it is impossible to list—and hence to count—the numbers between 0 and 1.* Thus, the set $\mathcal{L} = [0,1]$ is referred to as **uncountable**.[3] Uncountable infinities are a fundamentally different sort of infinity than countable infinities. The cardinality of the set $\mathcal{N}$ of counting numbers is sometimes denoted $\aleph_0$. This symbol is read "aleph naught". Aleph is the first letter of the Hebrew alphabet. The cardinality of the unit interval $\mathcal{L} = [0,1]$ is sometimes denoted $\aleph_1$.

To summarize, we have thus established that there are two different types of infinity: $\aleph_0$ or a countable infinity, which is the infinity associated with the integers; and $\aleph_1$ or an uncountable infinity, which is the infinity associated with the unit interval $[0,1]$.[4] More generally, any interval of real numbers, or, for that matter, the entire real number line, has an uncountable infinity of points. The cardinality $\aleph_1$ is thus also sometimes referred to as the cardinality of the continuum.

[3] The argument just used to show that the unit interval is uncountable is an example of a diagonalization argument. Such an argument was first published by Georg Cantor in 1891.

[4] The story does not stop here; there are even higher orders of infinity, denoted $\aleph_2$, $\aleph_3$, and so on. These infinities are associated with increasingly abstract and difficult-to-picture mathematical objects.

21.7 The Cardinality of the Cantor Set

In Section 21.1, the first section of this chapter, we saw that the length of the Cantor set was zero. But, viewed as a set, what is its size? In other words, what is its cardinality? Is it uncountable, like the unit interval $\mathcal{L}$? In constructing the Cantor set we started with $\mathcal{L}$ but then took away very many points—infinitely many, in fact. Did we take away so many points that its cardinality is no longer $\aleph_1$? To answer this question, we need to think geometrically about numbers between 0 and 1.

Let us start by thinking about familiar base-10 decimals and the number line. Two views of the number line are shown in Fig. 21.2. Numbers between 0 and 1 are represented in base-10 decimals. Each base-10 digit has the effect of dividing the number line into tenths, as illustrated in the figure. As one keeps zooming in on the number line, one would see this pattern repeating—each digit farther to the right of the decimal point partitions the line into finer and finer intervals. And each new interval is always one tenth of the size of the previous one.

A number such as 0.384 as specifies a location on the unit interval. Each digit determines the position of the point ten times more precisely

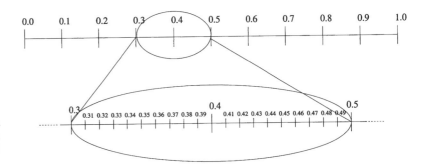

Fig. 21.2 The unit interval $[0,1]$ and a magnification of the unit interval from 0.3 to 0.5. Note that each base-10 digit divides the segment into tenths.

than the previous digits. The first digit tells us that the point is between 0.3 and 0.4, limiting the possible position of the point to one tenth of the unit interval. When the next digit is specified—in this case an 8—we now know that the point is between 0.38 and 0.39. Subsequent digits continue to specify the point's position more precisely. We can think of decimals as a type of address where each digit gives more detailed information. It is as if the first digit specifies the country, the next digit the state or province, the next digit the city, the next digit the neighborhood, and so on.

Next, we need to consider base-3 numbers.[5] This probably seems like a perverse thing to do, but it is necessary for our goal of figuring out the cardinality of the Cantor set. Just as base-10 numbers divide the unit interval into tenths, base-3 numbers divide the unit interval in thirds. This is illustrated in Fig. 21.3. For example, the first digit after the decimal point divides the unit interval into thirds. All base-3 numbers that start with 0.1 lie in the first third of the unit interval, all those that start with 0.1 lie in the second third, and all those that start with 0.2 lie in the third and final third. Take a moment to compare Figs. 21.2 and 21.3, and try to convince yourself that the two figures are essentially the same; we have chosen different conventions (or bases) for our digits, but the digits function in the same way.

[5]Base-3 is also known as **ternary**, just as base-2 is known as binary.

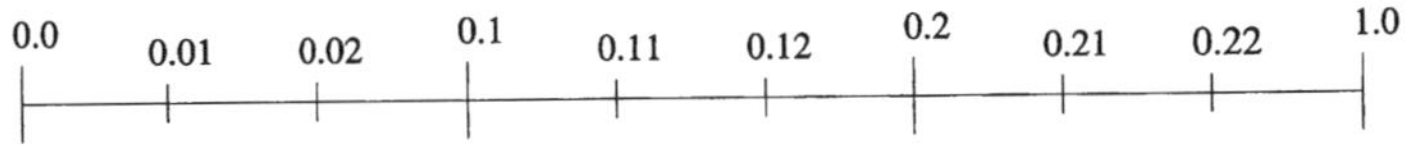

Fig. 21.3 The unit interval, with subdivisions in base-3, or ternary.

Let us now return to the Cantor set's construction, shown in Fig. 21.1. The first step in the construction of the Cantor set is to remove the middle third of the line segment. Looking at Fig. 21.3, we see that the middle third consists exactly of those points whose first ternary digit is 1. That is, the points 0.1001022020, 0.11, 0.102, and infinitely many others are all removed. What is remaining are the left and right thirds of the unit interval. The left third is made up of all those numbers whose first base-3 digit is zero. E.g., 0.012, 0.0220, 0.01212, and infinitely many other numbers. And the right third similarly consists of those numbers whose first base-3 digit is 2.

In the second step of the construction, we remove the middle thirds of the two remaining line segments. Looking again at Fig. 21.3, we see that the deleted segments correspond to those numbers whose second digit contains a 1. For example, when removing the middle third of the right line segment, we remove the numbers 0.211, 0.2102, 0.2102222, and infinitely many others. We remove all those numbers that start with 0.21.

The next step in the construction of the Cantor set entails removing all those numbers from the unit interval whose third base-3 digit is 1. And in general, the nth step of the process removes the numbers whose n^{th} digit is 1.

We can now give an alternative, but equivalent, description of the construction of the Cantor set. We start with all the numbers between zero and one. Express these numbers in base-3. This set consists of all possible infinite sequences of 0's, 1's, and 2's. In the first step in the construction of the set we remove all those numbers that have a 1 in the first digit. In the second step in the construction we remove all numbers that have a 1 in the second digit. And so on. This numerical process is equivalent to the geometric construction illustrated in Fig. 21.1.

The net result is that *the Cantor set consists of all numbers between 0 and 1 whose representation in base-3 contains no 1's.* Thus, the Cantor set consists of all infinite sequences of 0's and 2's. Recall that our goal in this section is to determine the cardinality, or size, of the Cantor set. There is one more step we need to perform before we can answer this question. Let us take the numbers in the Cantor set and erase every 2 and replace it by a 1. The result is a new set—one with the same cardinality as the previous set. This new set with 0's and 1's has the same number of elements as the Cantor set with 0's and 2's. Thus, the cardinality of the Cantor set is the same as the cardinality of this new set: the set of all infinite sequences consisting of 0's and 1's.

But wait a minute! We decided in Section 21.6 that the set of all infinite sequences of 0's and 1's was the same size as the set of *all* the numbers on the unit interval. We have thus arrived at the answer to the question posed at the beginning of this chapter: the cardinality of the Cantor set is the same as the cardinality of the unit interval itself. The unit interval is uncountably infinite. Thus the Cantor set is uncountable, as well.

This is an intriguing and perhaps perplexing result. We argued in the first section that the length of the Cantor set is zero. And hence the total lengths of the line segments that we removed when forming the Cantor set is one—equal to the length of the unit interval itself. But we have found that the Cantor set, despite having length zero, not only has the an infinite size, but it has the cardinality of the unit interval. The Cantor set is uncountable.

To recapitulate, we started with an infinite set, the unit interval [0, 1]. We subtracted from this set an infinite number of points, leaving us with a set—the Cantor set—that has length zero. Nevertheless, the cardinality of Cantor set is the same as the cardinality of the unit interval from which the Cantor set was derived.

21.8   Summary and a Look Ahead

In this part of the book I have introduced fractals: self-similar geometric or physical objects. A fractal's self-similarity can be exact or only approximate. We quantify this self-similarity via the self-similarity or box-counting dimension. Fractals consist of many copies of a shape or pattern, repeated at many different scales. As such, for many fractals it is not meaningful to speak of their average properties; in some cases, as was the case in the St. Petersburg game in Chapter 19, an average may not even exist. Finally, in this chapter we saw that fractals are not only useful objects for describing and understanding physical and natural phenomena. They are also useful constructions for probing our understanding of infinity. In the next part of the book we will consider Julia sets and the Mandelbrot set. We shall see that these are fractals of amazing complexity and beauty. Nevertheless, Julia sets and the Mandelbrot set are generated by very simple dynamical systems.

Further Reading

There is much more to be said about infinities in mathematics and elsewhere. Here are a few references I particularly recommend. Chapter 2 of Gary William Flake's *The Computational Beauty of Nature* (1999) is a clear and compelling discussion of number systems and infinities at roughly the level of this text. *Mathematics: A Very Short Introduction*, by Timothy Gowers (2002), contains an general discussion of the role of infinity in mathematics. Gowers does an impressive job of introducing non-mathematician readers to the abstract style of thought that is the essence of mathematics. John D. Barrow's, *Pi in the Sky: Counting, Thinking, and Being* (1992) is also a good introduction to number systems and mathematical thought. Chapter 2.1 of Peiten, Jürgens, and Saupe (1992) is a clear and thorough explication of the Cantor set.

Exercises

(21.1) Consider the set of all perfect squares: $\{1, 4, 9, 16, 25, \ldots\}$. What is the cardinality of this set?

(21.2) What is the cardinality of the following infinite set: $\{\frac{1}{2}, \frac{1}{3}, \frac{1}{4}, \ldots\}$?

(21.3) What is the cardinality of the following infinite set: $\{613, 614, 615, 616, 617, \ldots\}$?

(21.4) What is the cardinality of the following infinite set: $\{3^1, 3^2, 3^3, 3^4, 3^5, \ldots\}$?

(21.5) What is the cardinality of all numbers contained in the interval between 0 and $\frac{1}{2}$?

(21.6) What is the cardinality of all numbers contained in the interval between 0 and $\frac{1}{\sqrt{2}}$?

(21.7) In Eq. (21.24) I claimed that

$$457 = (1 \times 2^8) + (1 \times 2^7) + (1 \times 2^6)$$
$$+ (0 \times 2^5) + (0 \times 2^4) + (1 \times 2^3)$$
$$+ (0 \times 2^2) + (0 \times 2^1) + (1 \times 2^0).$$

Verify that this is the case. To do so, carry out the

arithmetic on the right-hand side of the equations and show that this yields 457.

(21.8) Convert the following numbers from base-10 to binary:

   (a) 8

   (b) 9

   (c) 48

   (d) 100

(21.9) Convert the following numbers from binary to base-10:

   (a) 100

   (b) 111

   (c) 1001

   (d) 10101

(21.10) Convert the following numbers from binary to base-10:

   (a) 0.1

   (b) 0.01

   (c) 0.001

   (d) 0.101

(21.11) Convert the following numbers from ternary (base-3) to base-10:

   (a) 2

   (b) 120

   (c) 0.21

   (d) 0.212

(21.12) Convert the following numbers from base-7 to base-10:

   (a) 0

   (b) 1

   (c) 24

   (d) 613

(21.13) Convert the following numbers from base-7 to base-10:

   (a) 0.4

   (b) 0.04

   (c) 0.22

Part IV

Julia Sets and the Mandelbrot Set

Introducing Julia Sets

22

In this part of the book we will encounter Julia sets and the Mandelbrot set—fractals of remarkable beauty and complexity. You are probably familiar with them; these are the psychedelic fractal images that have found their way onto posters, calendars, web pages, book covers, and so on. If you have not seen these images before, skip ahead and check out the Julia sets in Fig. 24.8. Chapter 25 is also festooned with images of the Mandelbrot set. You could also search the web and easily find a menagerie of colorful pictures of Julia sets and the Mandelbrot set. We shall see that these images arise quite naturally from simple iterated functions. Moreover, we will see some intriguing relationships between the geometry of the images and the dynamical systems that give rise to them.

In this brief chapter I introduce Julia sets via a few examples. In the subsequent chapter we will take a detour and cover complex numbers. We will then use complex numbers in Chapter 24 to generate a remarkable diversity of Julia sets. Attempting to classify this geometric diversity will lead us to the Mandelbrot set, which has been called the most beautiful object in all of mathematics.

22.1 The Squaring Function

We start with a familiar example, the squaring function, $f(x) = x^2$. As we have seen previously, any number larger than one or smaller than negative one will tend toward infinity. For example, the orbit of $x_0 = 2$ is:

$$2 \longrightarrow 4 \longrightarrow 16 \longrightarrow 256 \longrightarrow \cdots. \tag{22.1}$$

And the orbit of $x_0 = -3$ is:

$$-3 \longrightarrow 9 \longrightarrow 81 \longrightarrow 6561 \longrightarrow \cdots. \tag{22.2}$$

Any initial condition that is negative will become positive after one iteration, and will then remain positive for all future iterations.

So all initial conditions larger than 1 or smaller than −1 go to infinity. Conversely, initial conditions that are equal to or greater than −1, but not larger than 1, do not fly off to infinity. Another way to write this is that all initial conditions x_0 such that $-1 \leq x_0 \leq 1$ do not go to infinity. We can also write this interval as $[-1, 1]$. The square brackets mean that the interval contains its endpoints. For example, if we have an interval $I = [a, b]$, then a and b are taken to be part of the interval I. However, if $I = (a, b)$, then a and b are not part of I.

In any event, for the squaring function if the initial condition x_0 is in the set $S = [-1, 1]$, then it does not fly off to infinity when iterated. This set has a special name: the **filled Julia set**. The filled Julia set for a function f is the set of initial conditions which, when iterated with f, do not go to infinity. The initial conditions might get pulled to a fixed point or bounce around periodically or even chaotically; it does not matter. As long as the initial condition does not tend toward infinity, it is part of the filled Julia set.

Fig. 22.1 The filled Julia Set for the function $f(x) = x^2$. The Julia set is the line segment $[-1, 1]$.

$$-1 \qquad\qquad 1$$

The filled Julia set for the squaring function $f(x) = x^2$ is shown in Fig. 22.1. In this case, the filled Julia set is a simple geometrical object: a line segment. The Julia set, as opposed to the filled Julia set, is the boundary of the filled Julia set. In other words, the Julia set consists of the two points $+1$ and -1. In general, the Julia set is the boundary between two regions: initial conditions that fly off to infinity and those that do not. In this chapter and beyond we will be interested in filled Julia sets; all the points that do not go to infinity when iterated. Since "filled Julia set" is a somewhat cumbersome phrase, I will call them simply "Julia sets".[1]

[1] Julia sets are named for the French mathematician Gaston Maurice Julia. Julia introduced the idea behind the sets that now bear his name in an influential paper published in 1918. Pierre Fatou, another French mathematician, did similar work around the same time.

22.2   Other Examples

Before concluding this short chapter, let us do three examples. First, consider the function $g(x) = \frac{1}{2}x - 2$. This linear function has a fixed point at $x = -4$. And since the slope of this function is less than 1 and greater than -1, we know that the fixed point is attracting. All initial conditions are eventually pulled toward the fixed point, and thus no initial conditions tend toward infinity. Thus, the Julia set for this function is the entire number line; there are no initial conditions that tend toward infinity.

Our second example is the logistic equation with $r = 3.2$. We know from Section 9.1 that this function has an attracting cycle of period 2. All initial conditions between 0 and 1 are pulled to this cycle, so all of these points are in the Julia set. An initial condition of 0 will remain fixed at 0, so this point is also in the Julia set. If an initial condition is greater than 1 or less than 0 it will quickly tend toward negative infinity.[2] Thus, the Julia set for the logistic equation with $r = 3.2$ consists of the interval $[0, 1]$.

[2] To convince yourself of this, go back and do Exercise 9.7.

The third example is the function $f(x) = (2x)^2$. This function is plotted in Fig. 22.2 along with the $y = x$ line. We can see that there are two fixed points: one at $x = 0$ and another between 0.2 and 0.3.

Let us use algebra to find the exact value of the latter fixed point. The fixed-point equation is

$$f(x) = x . \tag{22.3}$$

Using $f(x) = (2x)^2$, we have:

$$(2x)^2 = x . \tag{22.4}$$

But $(2x)^2 = 4x^2$, so this becomes

$$4x^2 = x . \tag{22.5}$$

To solve for x we divide both sides of the equation by x to obtain:

$$4x = 1 . \tag{22.6}$$

So, the fixed point is at $x = \frac{1}{4}$.

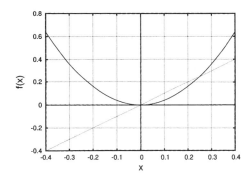

Fig. 22.2 The function $f(x) = (2x)^2$ and the $y = x$ line.

The fixed point at $x = 0.25$ is repelling. To see this, you can graphically iterate initial conditions on either side of $x = 0.25$. Or, one can note that the slope of the function at $x = 0.25$ is larger than 1. Since $x = 0.25$ is repelling, any initial condition larger than 0.25 will be pushed away and will tend toward infinity. The same conclusion holds for an initial condition that is smaller than $x = -0.25$. Such an initial condition will, after one iteration, be larger than 0.25. To see this, try graphically iterating an initial condition less than -0.25 on Fig. 22.2.

Thus, any initial condition that is equal to or greater than -0.25 but not larger than 0.25 does not tend toward infinity under iteration, and thus is in the filled Julia set. Hence, the Julia set for this function is the interval $[-0.25, 0.25]$.

22.3 Summary

I hope to have convinced you that the idea of a Julia set for a function is fairly straightforward: it is simply the collection of points that remain bounded—i.e., do not tend toward infinity—when iterated by that function. Determining the entire Julia set for a function may be a difficult task. However, it is not difficult to to test whether or not a particular point is in a Julia set: just iterate it and see if tends toward infinity. If it does, then it is not in the Julia set. In Chapter 24 we will see Julia sets that are complicated and interesting shapes. We will not be able to easily deduce the shape of these sets; instead we will build them one point at a time, checking each point to see whether or not it is in the set.

Exercises

(22.1) Consider the function $f(x) = (x-1)^2$. Calculate by hand or with a calculator the first few iterates of the following initial conditions and determine whether or not they are in the Julia set:

  (a) 0

  (b) 0.5

  (c) 2

  (d) 3

(22.2) Consider the function $f(x) = x^2 - 1$. Calculate by hand or with a calculator the first few iterates of the following initial conditions and determine whether or not they are in the Julia set:

  (a) −2

  (b) 0

  (c) 0.5

  (d) 2

(22.3) Consider the logistic equation with $r = 4.0$: $f(x) = 4x(1-x)$. For this parameter value we know that orbits are chaotic. What is the Julia set for this function?

(22.4) In Section 22.2 I stated that the Julia set for the function $f(x) = 3.2x(1-x)$ was $[0, 1]$. Verify that the endpoint of the interval, $x = 1$, is in the Julia set.

(22.5) Consider the function $f(x) = 5x(1-x)$. Calculate by hand or with a calculator the first few iterates

of the following initial conditions and determine whether or not they are in the Julia set:

  (a) 0.8

  (b) 0.5

  (c) 0.2

  (d) 0.1

(22.6) Determine the Julia set for the function $f(x) = \sqrt{x}$.

(22.7) Determine the Julia set for the function $f(x) = x^3$.

(22.8) Determine the Julia set for the function $f(x) = 2x^2$.

(22.9) Determine the Julia set for the function $f(x) = (3x)^2$.

(22.10) Determine the Julia set for the function $f(x) = (x+1)^2$.

(22.11) Determine the Julia set for the function $f(x) = 2x - 3$.

(22.12) Determine the Julia set for the function $f(x) = -\frac{1}{2}x - 3$.

(22.13) What are the possible shapes for the Julia sets of linear functions $f(x) = mx + b$? Is there a linear function that has a Julia set that is a line segment, such as $[3, 5]$? Why or why not? Is there a linear function whose Julia set is empty?

Complex Numbers

23

The fractal Julia sets of the next chapter and the Mandelbrot set of Chapter 25 arise from iterating functions using complex numbers instead of real numbers. Thus, in this chapter I introduce complex numbers, known also as imaginary numbers. If you have encountered complex numbers before and feel reasonably comfortable with them, you might want to skip ahead to the next chapter.

23.1 The Square Root of -1

What is the square root of -1? Is there any x such that

$$x^2 = -1 ? \tag{23.1}$$

The answer to this question cannot be $x = 1$, since $1^2 = 1$. And the answer also is not -1, since $(-1)^2 = (-1)(-1) = 1$. So we are stuck. The way to get out of this bind is to simply define a new number. We call this new number i, and it is defined by the fact that i squared is negative one:

$$i^2 = -1 . \tag{23.2}$$

Equivalently, this says that i is the square root of -1:

$$i = \sqrt{-1} . \tag{23.3}$$

The number i is often referred to as an imaginary number. However, the term **complex number** is more commonly used, at least in mathematics circles, for numbers of this sort. In contrast, *real numbers* are those numbers that are not complex. Thus far we have been working exclusively with real numbers. The number line, stretching from negative infinity to positive infinity, contains all the real numbers.

I think that the word "imaginary" carries some metaphysical or philosophical baggage that is not helpful. All numbers—complex and real— are imaginary. They are constructs and idealizations of ideas in people's heads. Numbers are not tangible things in the physical world like a tree or a rock or a table. In the world there might be seven rocks or seven trees or seven tables. But this is different from the pure number 7. You can find seven things, but this is not the same as finding 7. So I would say that, in some sense, the number 7 is "imaginary", in much the same way that the number i is imaginary.

But you might be wondering if there are any physical manifestations of i. We could find seven rabbits. Could we ever find i rabbits? The answer to this question is "no", at least as far as the rabbits are concerned. However, there are physical phenomena that are commonly described by complex numbers. One example is alternating current (AC) circuits. Determining the behavior of AC circuits with resistors, capacitors, and inductors is much more convenient if one uses complex instead of real numbers. And the theory of quantum mechanics also makes extensive use of complex numbers.

In any event, I will typically use z to refer to a complex number. This is standard, but not universal. Many authors use x or some other letter. A generic complex number is a combination of real numbers and imaginary numbers of the following form:

$$z = a + bi \,, \tag{23.4}$$

where a and b are real numbers. For example, we might have:

$$z = 3 + 4i \,. \tag{23.5}$$

The numbers a and b in Eq. (23.4) are referred to as, respectively, the **real part** and the **imaginary part** of z. The real part of z is sometimes denoted $\Re(z)$, and the imaginary part is denoted $\Im(z)$. I will not use this notation in this book, but it is possible that you will encounter them elsewhere. Note that the imaginary part of z is not imaginary. I.e., the imaginary part of $z = 3 + 4i$ is 4. We would not say that the imaginary part is $4i$.

23.2   The Algebra of Complex Numbers

We now consider performing basic mathematical operations with complex numbers: addition, subtraction, and multiplication. The key is to treat i as if it were a separate algebraic variable, like x, and make use of the fact that $i^2 = -1$.

Addition and Subtraction

To add two complex numbers, one adds their real and imaginary parts separately. For example, let

$$z_1 = a + bi \,, \tag{23.6}$$

and

$$z_2 = c + di \,. \tag{23.7}$$

Then

$$z_1 + z_2 = (a + bi) + (c + di) = (a + c) + (b + d)i \,. \tag{23.8}$$

So the real part of $z_1 + z_2$ is just the sum of the real parts of z_1 and z_2, and similarly, the imaginary part of $z_1 + z_2$ is just the sum of the imaginary parts. For example, let $z_1 = 4 - 3i$ and $z_2 = 3 + 2i$. Then

$$
\begin{aligned}
z_1 + z_2 \quad &= \quad (4 - 3i) + (3 + 2i) & (23.9) \\
&= \quad (4 + 3) + (-3 + 2)i & (23.10) \\
&= \quad 7 - i & (23.11)
\end{aligned}
$$

Multiplication

To multiply two complex numbers, multiply the two numbers together as you would binomials,[1] and then simplify using the fact that $i^2 = -1$. Suppose we want to multiply z_1 and z_2, given above in Eqs. (23.6) and (23.7):

$$
\begin{aligned}
z_1 z_2 \quad &= \quad (a + bi)(c + di) & (23.12) \\
&= \quad ac + adi + bci + bdi^2 & (23.13) \\
&= \quad ac + (ad + bc)i + bdi^2 & (23.14) \\
&= \quad ac + (ad + bc)i - bd & (23.15) \\
&= \quad (ac - bd) + (ad + bc)i . & (23.16)
\end{aligned}
$$

Note that to get from Eq. (23.14) to Eq. (23.15) I have used the fact that $i^2 = -1$.

Let us consider a numerical example. Suppose we want to multiply together the numbers $z_1 = 4 + 3i$ and $z_2 = 2 - 3i$:

$$
\begin{aligned}
z_1 z_2 = \quad &= \quad (4 + 3i) \times (2 - 3i) & (23.17) \\
&= \quad 8 + 4(-3)i + (3)(2)i + (3)(-3)i^2 & (23.18) \\
&= \quad 8 - 12i + 6i - 9^2 i^2 & (23.19) \\
&= \quad 8 - 6i - 9(-1) & (23.20) \\
&= \quad 17 - 6i . & (23.21)
\end{aligned}
$$

[1] You might have learned this process as FOILing. "FOIL" is a mnemonic device for accounting for the four terms that arise when multiplying binomials: Firsts, Outers, Inners, Lasts.

23.3   The Geometry of Complex Numbers

The real number line is a useful way of representing and visualizing the set of real numbers. We made use of this when drawing phase portraits in previous chapters. There is a similar way of representing and visualizing complex numbers: **the complex plane**. The complex plane is a natural generalization of the real number line. We plot the real part of a complex number on the horizontal axis and the imaginary part on the vertical axis.

This is illustrated in Fig. 23.1. There are four numbers plotted in this figure; each is shown with a different shape. The circle is $2 + 2i$, the triangle is $\frac{1}{2} - \frac{3}{2}i$, the square is $-i$, and the diamond is -2.5. Take a moment to examine Fig. 23.1 to ensure that you see the connection between the real and imaginary parts of a complex number and its representation on the complex plane.

There is another way of indicating the location of complex numbers on a plane. Rather than specifying a number's real and imaginary parts,

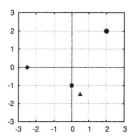

Fig. 23.1 The complex plane. Four numbers are plotted: $2 + 2i$ (circle); $\frac{1}{2} - \frac{3}{2}i$ (triangle): -2.5 (diamond); and $-i$ (square).

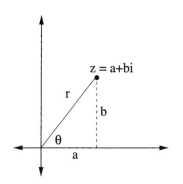

Fig. 23.2 Illustrating polar notation for points $z = a + bi$ on the complex plane.

a and b, we specify how far the point is from the origin, and the angle between the horizontal axis and a line drawn from the origin to the point z. This is illustrated in Fig. 23.2. In the figure the complex number $z = a+bi$ is plotted as a small circle. The distance from the origin to the point is denoted r. The angle θ is the angle between the horizontal axis and the line from the origin to the point. We can specify the complex number by giving the real and imaginary parts, a and b, or by giving its r and θ values. The a, b form is known as **Cartesian coordinates**; the r, θ form is known as **polar coordinates**. Both forms are equivalent; they are simply different representations of the same number. Which form we use will depend on the situation; in some cases Cartesian coordinates are easiest, while in other cases polar coordinates are more convenient.

Given the Cartesian coordinates for a complex number, one can determine the number's polar coordinates. For example, suppose we know a and b in Fig. 23.2 and we want to figure out r. Notice that a, b, and r form a right triangle, with r the hypotenuse. Thus, by the Pythagorean theorem:

$$r^2 = a^2 + b^2 . \tag{23.22}$$

Solving for r, we get:

$$r = \sqrt{a^2 + b^2} . \tag{23.23}$$

For example, consider the point shown as a circle on Fig. 23.1. This point is $2 + 2i$, so a and b are both 2. Thus, the r for this point is

$$r = \sqrt{2^2 + 2^2} = \sqrt{8} \approx 2.83 . \tag{23.24}$$

Geometrically, this means that the point indicated by the circle is approximately 2.83 units away from the origin. To determine θ given the Cartesian coordinates a and b requires trigonometry. The formula is:

$$\theta = \tan^{-1}(\frac{b}{a}) . \tag{23.25}$$

It is also possible to convert in the other direction—to go from polar coordinates (r,θ) to Cartesian coordinates (a,b). The formulas are:

$$a = r\cos\theta , \tag{23.26}$$
$$b = r\sin\theta . \tag{23.27}$$

[2]Equations (23.23) and (23.25)–(23.27) can be thought of as a "bilingual dictionary". With them one can translate a point from polar coordinates (r,θ) to Cartesian coordinates (a,b), or vice-versa.

We will not use these last three equations in this book, but I include them here for the sake of completeness.[2]

23.4   The Geometry of Multiplication

Multiplication and addition of complex numbers and the representation of complex numbers on the complex plane are sufficient background for understanding how to generate Julia sets and the Mandelbrot set. The next section aims to provide some additional geometric intuition to help you think about complex numbers. You can skip this section and go to Chapter 24 if you want.

Polar coordinates are particularly helpful when multiplying two complex numbers. We shall see that multiplication has a simple geometric interpretation in polar coordinates. To do so, we consider two examples. Our goal will be to use these examples to determine a rule that will let us perform multiplication using polar coordinates instead of Cartesian.

First, let

$$z_1 = 3, \quad \text{and} \quad z_2 = 2i. \tag{23.28}$$

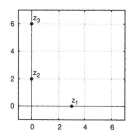

Fig. 23.3 Illustrating multiplication; $z_1 = 3$, $z_2 = 2i$, and $z_3 = z_1 z_2 = 6i$.

These two numbers are shown on the complex plane in Fig. 23.3. Note that for z_1, $r = 3$ and $\theta = 0$, while for z_2, $r = 2$ and $\theta = 90$ degrees. Let us multiply these two numbers together and call this new number z_3:

$$z_3 = z_1 z_2 = 3 \times 2i = 6i. \tag{23.29}$$

The number z_3 is also shown in Fig. 23.3. In polar coordinates, z_3 is given by $r = 6$ and $\theta = 90$. Note that the r for z_3 is just the r for z_1 times the r for z_2. And the θ for z_3 is the same as the θ for z_2.

For our next example, consider the following two numbers

$$w_1 = 2 + 2i, \quad \text{and} \quad w_2 = -1 + i. \tag{23.30}$$

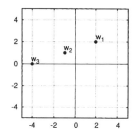

Fig. 23.4 Illustrating multiplication; $w_1 = 2 + 2i$, $w_2 = -1 + i$, and $w_3 = -4$.

(I am using w for complex numbers here to distinguish them from the z's of the previous example.) These numbers are shown in Fig. 23.4. Note that for w_1, $r_1 = \sqrt{8} \approx 2.83$ and $\theta_1 = 45$ degrees. To determine r, I used Eq. (23.23). And for w_2, $r_2 = \sqrt{2} \approx 1.41$ and $\theta_2 = 135$ degrees.

Let us multiply these two numbers together and see what happens:

$$\begin{aligned} w_3 &= w_1 w_2 = (2 + 2i)(-1 + i) \\ &= -2 + 2i - 2i + 2i^2 = -2 - 2 = -4. \end{aligned} \tag{23.31}$$

In polar coordinates, w_3 has $r_3 = 4$ and $\theta_3 = 180$.[3] As was the case in the previous example, multiplication has the effect of multiplying the two r's:

$$r_3 = r_1 r_2 = \sqrt{8}\sqrt{2} = \sqrt{16} = 4. \tag{23.32}$$

And the θ's get added:

$$\theta_3 = \theta_1 + \theta_2 = 45 + 135 = 180. \tag{23.33}$$

This is a general result. *To multiply two complex numbers, multiply their r's and add their θ's.* This gives us a way to multiply complex numbers that is often faster than using Cartesian coordinates. More important, it gives us a geometric insight into the effects of multiplication.

Finally, note that the above result for multiplying in polar coordinates leads to a simple view of squaring complex numbers. Squaring just means multiplying a number by itself:

$$z^2 = zz. \tag{23.34}$$

[3] Note that the radius r is, by definition, always positive. It may be tempting to say that the radius for w_3 is -4, but this would be incorrect.

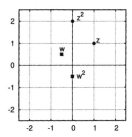

Fig. 23.5 Illustrating squaring complex numbers; $z = 1 + i$, $z^2 = 2i$, $w = \frac{1}{2} + \frac{1}{2}i$, $w = -0.5$.

Thus, to square a complex number, all one has to do is square the r value and double the θ value. For example, suppose $z = 1 + i$, so $r = \sqrt{2}$ and $\theta = 45$. Then, z^2 has an r of $(\sqrt{2})^2 = 2$ and a θ of 90. A θ of 90 indicates that the point is located "straight up". That is, the point is on the vertical (imaginary) axis. So it must be that $z^2 = 2i$.

We can verify this result by using Cartesian coordinates:

$$z^2 = (1+i)^2 \tag{23.35}$$

$$= (1+i)(1+i) \tag{23.36}$$

$$= 1 + i + i + i^2 \tag{23.37}$$

$$= 1 + 2i - 1 \tag{23.38}$$

$$= 2i , \tag{23.39}$$

which is the same as our result obtained via polar coordinates. Squaring z is illustrated in Fig. 23.5, where I have plotted z and z^2 on the complex plane. Note that we can see that the r for z^2 is equal to the square of the r of z, and that the θ for z^2 is twice that of z.

Also on Fig. 23.5 I have shown $w = -\frac{1}{2} + \frac{1}{2}i$ and w^2. The θ for w is 135, while r is given by:

$$r = \sqrt{\left(\frac{1}{2}\right)^2 + \left(\frac{1}{2}\right)^2} \tag{23.40}$$

$$= \sqrt{\frac{1}{4} + \frac{1}{4}} \tag{23.41}$$

$$= \sqrt{\frac{2}{4}} \tag{23.42}$$

$$= \frac{1}{\sqrt{2}} \tag{23.43}$$

$$\approx 0.707 . \tag{23.44}$$

Thus, w^2 has an r given by the square of this amount. Namely, $r = \frac{1}{\sqrt{2}^2} = \frac{1}{2}$. Note that the r of w^2 is less than the r of w. The θ for w^2 is twice that of w: namely, $\theta = 270$. So, w^2 has an r of $\frac{1}{2}$ and a θ of 270. This is shown in Fig. 23.5.

Exercises

For Exercises 23.1–23.4 let $z_1 = 3$, $z_2 = 2 - i$, $z_3 = 4i$, and $z_4 = -2 + 3i$.

(23.1) Calculate the following quantities:

    (a) $z_1 + z_2$

    (b) $z_1 + z_3$

    (c) $z_1 + z_4$

(23.2) Calculate the following quantities and plot the two numbers and their product on the complex plane.

    (a) $z_1 z_2$

    (b) $z_4 z_2$

    (c) $z_2 z_4$

(23.3) Calculate the following quantities:

(a) $5z_4$

(b) $z_4 + 5$

(c) $(z_3 - 1)(z_1 + 2i)$

(d) $(z_3 + z_4)z_2$

(23.4) Calculate the following quantities:

(a) z_4^2

(b) z_3^3

(c) z_1^2

(d) z_2^2

(23.5) Consider the function $f(z) = z^2$. Calculate the first three iterates of

(a) $z_0 = 0$

(b) $z_0 = i$

(c) $z_0 = 2$

(23.6) Four complex numbers are shown on the complex plane in Fig. 23.6. Determine the coordinates for each point, using Cartesian coordinates.

(23.7) Four complex numbers are shown on the complex plane in Fig. 23.6. Determine the coordinates for each point, using polar coordinates.

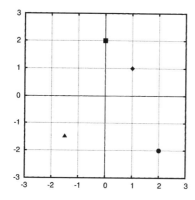

Fig. 23.6 Four complex numbers plotted on the complex plane. See Exercise 23.6 and 23.7.

(23.8) ⋆ Consider the squaring function $f(z) = z^2$. For each of the initial conditions listed below, do the following.

• Convert to polar representation.

• Calculate the first several iterates using the geometric method of Section 23.4.

• State what you think the long-term behavior of the orbit is.

(a) $z_0 = i$

(b) $z_0 = 1 - i$

(c) $z_0 = 2 + i$

(d) $z_0 = 2$

(e) $z_0 = -1$

(f) $z_0 = -1 + 2i$

(23.9) Figure 23.7 shows four different complex numbers. Use each as a seed for the squaring function $f(z) = z^2$. On the complex plane, sketch the first three iterates for each seed. Perform the iteration using the geometric view of squaring complex numbers.

(23.10) Based on the results of the previous two exercises, what do you think the Julia set is for the function $f(z) = z^2$, where z is a complex number? That is, what initial conditions z_0 have the property that, when iterated by f, they do not fly off to infinity?

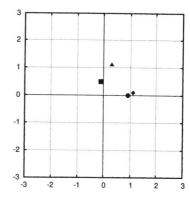

Fig. 23.7 Four different initial conditions plotted on the complex plane. See Exercise 23.9.

(23.11) ⋆ Consider the function $f(z) = z^2 - 1$. Calculate the first three iterates of

(a) $z_0 = 0$

(b) $z_0 = i$

(c) $z_0 = 2$

(23.12) Consider the function $f(z) = z^2 + i$. Calculate the first three iterates of

(a) $z_0 = 0$

(b) $z_0 = i$

(c) $z_0 = 2$

Julia Sets for the Quadratic Family

24

In Chapter 22 the concept of a Julia set was introduced: the Julia set for a function is simply the collection of all initial conditions that do not tend toward to infinity when iterated with that function. And in the previous chapter I introduced complex numbers. In this chapter we will look at the Julia sets that arise from iterating functions of complex numbers. We begin with a relatively simple example.

24.1   The Complex Squaring Function

As in many of the preceding chapters, we begin by considering the squaring function $f(z) = z^2$. This time, however, there is a twist: the numbers that we square can be complex. Recall that we saw in the previous chapter that squaring is particularly simple if we use the polar representation for the complex number. Namely, to square a complex number, square its r and double its θ. For example, consider the seed $z_0 = (r_0 = 2, \theta_0 = 45)$. The orbit of this point under the squaring function is:

$$(r = 2, \theta = 15) \quad \longrightarrow \quad (r = 4, \theta = 30) \longrightarrow (r = 16, \theta = 60)$$
$$\longrightarrow \quad (r = 256, \theta = 120) \longrightarrow \cdots . \qquad (24.1)$$

What is the long-term fate of the orbit?

Now that we are iterating complex numbers, we need to pause for a moment and consider what it means for a complex number to go to infinity. For real numbers, an orbit can tend toward positive infinity or negative infinity. But for complex numbers, which we visualize on the complex plane, there are many different directions in which a number could go to infinity. We shall say that an orbit tends towards infinity if its r value gets larger and larger. Geometrically, this means that the orbit is getting farther and farther away from the origin. The itinerary may spiral around while it gets farther away from the origin. Or it may get farther away from the origin along a straight line. In either case, we say that the orbit tends to infinity.

Returning to our example, the orbit of z_0 in Eq. (24.1) tends toward infinity, since its r value is getting larger and larger. In fact, any initial condition that has an r greater than 1 will fly off to infinity. The reason for this is that squaring complex number has the effect of squaring that number's r. And if we square a number larger than 1, the result is a larger number. Conversely, initial conditions that have an r equal to or less than 1 will *not* fly off to infinity. Hence, the Julia set for the complex squaring function consists of all points that have an r less than or equal to 1. This is illustrated in Fig. 24.1. The set of all points with an r less than or equal to 1 corresponds to a circle of radius 1. Thus, the Julia set for the squaring function is a circle of radius 1.[1]

[1]Strictly speaking the Julia set is the boundary of the circular region in Fig. 24.1, and the entire region is the filled Julia set. As in Chapter 22 I will, in a mild abuse of terminology, refer to filled Julia sets as simply *the* Julia set.

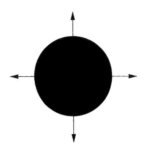

Fig. 24.1 The Julia set for the function $f(z) = z^2$. The radius of the circle is 1.

24.2 Another Example: $f(z) = z^2 - 1$

Having successfully determined the Julia set for the squaring rule, let us try a new function: $f(z) = z^2 - 1$. What might you guess the Julia set looks like for this function? This is similar to the squaring rule, for which the Julia set was a circle of radius 1. What effect does the -1 in $f(z) = z^2 - 1$ have on the Julia set? Does it shift the circle to the left or the right? Or does it do something more interesting?

To determine the Julia set for this function there is no quick calculation we can do. Instead, we need to choose some initial conditions, iterate, and see what happens. If the orbit flies off to infinity, the initial condition is not in the Julia set. And if it does not fly off to infinity, then it is in the Julia set.[2]

[2]Iterates of this function were explored in Exercise 23.11.

We start with $z_0 = 0$. Applying the function several times, we find:

$$0 \longrightarrow -1 \longrightarrow 0 \longrightarrow -1 \longrightarrow \cdots . \tag{24.2}$$

So the orbit of $z_0 = 0$ is periodic with period 2. And since the orbit does not tend toward infinity, the point $z_0 = 0$ is in the Julia set. We also know that $z_0 = -1$ is in the Julia set, since its orbit is also periodic.

Let us try some other initial conditions. How about $z_0 = i$? The orbit is:

$$i \longrightarrow -2 \longrightarrow 3 \longrightarrow 8 \longrightarrow 63 \longrightarrow \cdots . \tag{24.3}$$

So $z_0 = i$ flies off to infinity; it is not in the Julia set. This is perhaps somewhat surprising, as $z_0 = i$ was in the Julia set for the squaring function. Let us now try $z_0 = 1 + i$. This orbit is a little bit harder to calculate. It turns out that the orbit is:

$$1 + i \longrightarrow 2i \longrightarrow -5 \longrightarrow 24 \longrightarrow \cdots . \tag{24.4}$$

Thus, $z_0 = 1 + i$ is not in the Julia set, because its orbit tends to infinity. Finally, let us try the initial condition $0.8 + 0.2i$.

$$0.8 + 0.2i \quad \longrightarrow \quad 0.6 - 0.68i \longrightarrow 0.102 - 1.816i$$
$$\longrightarrow \quad -3.28 - 0.628i \longrightarrow 10.42 + 3.13i \cdots . \quad (24.5)$$

Although it took a few iterations be be certain, this orbit is indeed tending toward infinity. Thus, the initial condition $z_0 = 0.8 + 0.2i$ is not in the Julia set.

We could continue, choosing different initial conditions and then seeing if the orbit flies off to infinity. Every initial condition that does not fly off to infinity is in the Julia set. After checking a lot of different initial conditions, I can then make a rough plot of the Julia set by coloring in each initial condition that I have found that does not fly off to infinity.

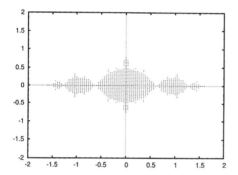

Fig. 24.2 The Julia set for the function $f(z) = z^2 - 1$.

The result of doing this is shown in Fig. 24.2. To make this plot I wrote a computer program to try around $10,000$ initial conditions. I then plotted only those points that are in the Julia set. For my experiment I found that around 880 of the $10,000$ initial conditions I tried turned out to be in the Julia set.

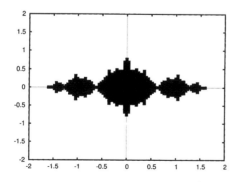

Fig. 24.3 The Julia set for the function $f(z) = z^2 - 1$.

Because the points in Fig. 24.2 are small it is hard to get a feel for the shape of the Julia set. So in Fig. 24.3 I have plotted the same data, but this time I have shown the points as large squares—large enough so that no white space shows between them. This helps us see the shape of the Julia set. It appears to be a fairly complex geometric object. It is certainly different from the circular Julia set for the squaring function, shown in Fig. 24.1.

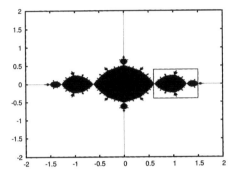

Fig. 24.4 The Julia set for the function $f(z) = z^2 - 1$. This is a higher resolution version of Fig. 24.3. A close-up of the boxed region is shown in Fig. 24.5.

To get a better view of the Julia set for $f(z) = z^2 - 1$ we need to test many more initial conditions. This will give us greater resolution in our image. In Fig. 24.4 I have shown the results of testing almost half a million different initial conditions. In this figure we see a fractal-like structure. The Julia set consists of a main oval, off of which hang other smaller ovals, off of which hang smaller ovals still, and so on. In Fig. 24.5 I have plotted a close-up of the Julia set shown in Fig. 24.4. We continue to see bulbs on top of bulbs on top of bulbs.

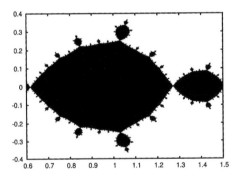

Fig. 24.5 A portion of the Julia set for the function $f(z) = z^2 - 1$. This is a close-up of boxed region of the Julia set shown in Fig. 24.4.

To summarize, we have plotted fairly detailed images of the Julia set for the function $f(z) = z^2 - 1$. I did so via a "brute force" method; I simply had a computer program test a vast number of initial conditions to see which do not tend toward infinity. The resultant shape, shown in Fig. 24.4 is a fractal.

24.3   Julia Sets for $f(z) = z^2 + c$

The Julia set of $f(z) = z^2 - 1$ yielded a surprising and, I hope, a somewhat pleasing result. This leads us to wonder about the Julia sets for other functions. It turns out that we can easily generate a fascinating menagerie of different Julia sets. To do so, we will consider functions of the form:

$$f(z) = z^2 + c. \tag{24.6}$$

We have already explored this function for two different values of c. If $c = 0$ we have the familiar squaring function whose Julia set is shown in Fig. 24.1. And if $c = -1$ we have the function we just considered, whose Julia set was illustrated in Fig. 24.4.

Fig. 24.6 The Julia set for the function $f(z) = z^2 - 0.84i$.

What about other values of c? Every c value we choose gives us a slightly different Julia set. In the exercises for this chapter you will investigate the Julia sets that result for different values of c. Here, we consider one additional example. Suppose that $c = 0 + 0.84i$. The Julia set for this function, namely $f(z) = z^2 + 0.84i$, is shown in Fig. 24.6. Note that this Julia set is not connected; it is not one continuous object. This appearance of non-connection is real; it is not simply that I have not tried enough initial conditions. A gallery of Julia sets for other values of c is shown in Fig. 24.8 at the end of this chapter.

24.4   Computing and Coloring Julia Sets

I conclude this chapter with some remarks about using computers to generate images of Julia sets. Often, images of Julia sets are plotted using a visually appealing set of colors. You may have encountered such images before, and you will see colored Julia sets when you use online programs to do the exercises at the end of the chapter. Here is one way to generate a colored Julia set. In colored images the Julia set itself is indicated by just one color, usually black. Points that are not in the Julia set are those which fly off to infinity when the function is iterated. However, different initial conditions fly off to infinity at different rates. It is this difference that is behind the colors on the images.

Even on a computer, we cannot iterate a point so many times that it literally flies off to infinity. Instead, we use some finite threshold. For example, we might take any point with $r = 10$ to essentially be at infinity.[3] Once an orbit has an r of 10, we would consider it to be at infinity, and we would thus classify that initial condition as not being in

[3]In practice, we can set our threshold values much lower. The reason for this is the following result. Let $f(z) = z^2 + c$. Let R be the larger of 2 or the r for c. Then any z with an r greater than R will grow without bound when iterated by $f(z)$. This can be proved analytically—i.e., by hand, without relying on computer experiments. See, for example, Peitgen, Jürgens, and Saupe (1992, p. 794).

the Julia set. However, as noted above, some initial conditions will reach this threshold value quickly, and others will take a long time. This is the basis for our color scheme. If an initial condition reaches the threshold in two or fewer iterations, we might color it blue. If it takes between three and ten iterations for the threshold to be reached we might color that point red. If it takes between eleven and twenty iterations, color it green, and so on. There are many different variations one can make on this basic idea.[4] The result is a rich array of different colored images. It is important to emphasize, however, that the Julia set itself is still just the collection of points, usually colored in black, that do not fly off to infinity.

[4] A clear and accessible introduction to a number of different coloring schemes for Julia and Mandelbrot sets is Rood (2004).

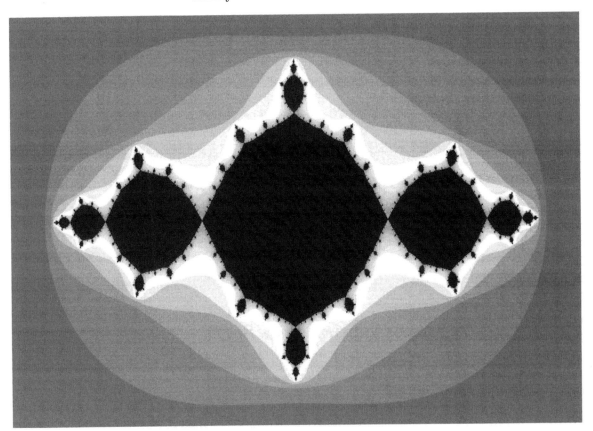

Fig. 24.7 A "colored" Julia set for $c = -1.0$. This image was made using the a program by David E. Joyce available at http://aleph0.clarku.edu/~djoyce/julia/juliagen.html. The same Julia set is shown in Fig. 24.4.

A "colored" Julia set image is shown in Fig. 24.7. Obviously, it is not really in color, since the picture is black and white. This is the same Julia set shown in Fig. 24.4. Note that the black central regions—the Julia set itself—are the same in both images. The "colored" Julia set differs in that points outside of the Julia set are different colors.

Exercises

For Exercises 24.4–24.6 you will need to use a program that will generate and display Julia sets. Links to such programs can be found on the webpage for the book `http://chaos.coa.edu`. You can also find many programs to generate Julia sets via a web search.

(24.1) Verify that the orbits of $z_0 = 0.8 + 0.2i$ are what I claimed they are in Eq. (24.5).

(24.2) Consider the function $f(z) = z^2 - i$. Calculate by hand the first few iterates for the following initial conditions. Which initial conditions are in the Julia set?

(a) $z_0 = i$

(b) $z_0 = 2$

(c) $z_0 = 0$

(d) $z_0 = 1$

(24.3) Consider the function $f(z) = z^2 + c$. Calculate by hand or using a calculator the first few iterates for the initial condition $z_0 = 0$ for the following c values. For which c values is z_0 in the Julia set?

(a) $c = -1$

(b) $c = -0.5 + 0.5i$

(c) $c = -1 + i$

(24.4) Eight different Julia sets are shown in Fig. 24.8 on the following page. These Julia sets were made with the following c values:

(a) $c = 0.0 - 0.72i$

(b) $c = -0.1 - 0.88i$

(c) $c = -0.6 + 0.2i$

(d) $c = -0.6 + 0.45i$

(e) $c = -1.35 + 0.02i$

(f) $c = 0.37 + 0.37i$

(g) $c = 0.2 + 0.57i$

(h) $c = -0.25 - 0.7i$

Determine which c value corresponds to which Julia set.

(24.5) Using a Julia set program, find a c value that yields a Julia set that you find particularly interesting. Choose a name for your Julia set, and briefly explain why you chose this name.

(24.6) ⋆ Use a program to plot Julia sets for the following c values:

(a) $-0.60 + 0.20i$

(b) $0.00 + 0.72i$

(c) $0.37 + 0.37i$

(d) $0.20 + 0.57i$

(e) $-0.10 + 0.88i$

(f) $-1.35 + 0.02i$

(g) $-0.60 + 0.45i$

Which Julia sets are connected—i.e., a single contiguous shape? Which are not connected, consisting of two or more separate parts?

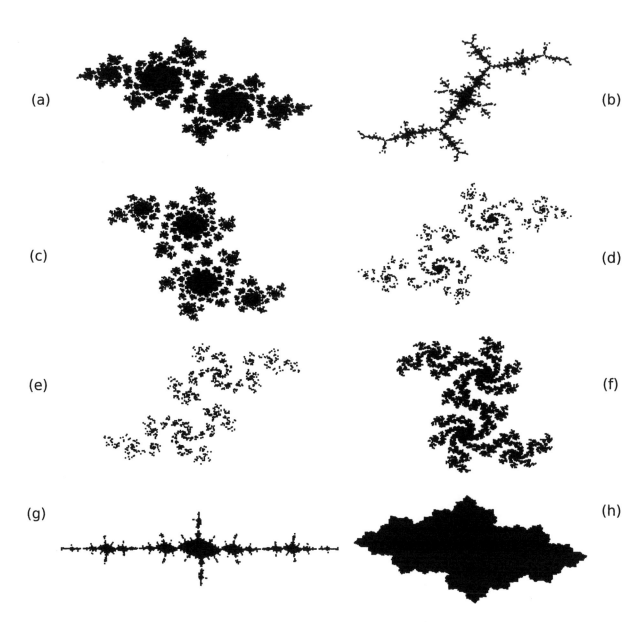

Fig. 24.8 Eight different Julia sets for $f(z) = z^2 + c$.

The Mandelbrot Set

25.1 Cataloging Julia Sets

In the previous chapter we considered Julia sets for the function

$$f(z) = z^2 + c \,, \tag{25.1}$$

where the variable z is a complex number and c is a parameter. Different values of c yield different Julia sets. Experimenting with different c values results in an impressive diversity of Julia sets. This is illustrated in Fig. 24.8, which shows eight different Julia sets.

A natural question at this point is whether or not we can classify or categorize Julia sets. At first blush this seems difficult, given the multiplicity of forms that the Julia sets can take. However, one simple dichotomy between different Julia sets is that some are connected, forming one contiguous shape, while others are disconnected. This basic observation will lead us to an amazingly complex and intricate mathematical structure.

We proceed with our categorization of Julia sets as follows. First, we choose a c value. As an example, let us choose $c = -0.6 + 0.2i$. The Julia set for this c is shown in Fig. 25.1. We see that this Julia set is

Fig. 25.1 The Julia set for $c = -0.6 + 0.2i$.

connected. It is a single, contiguous shape.

Next let us try another c value: $c = 0.0 - 0.72i$. This Julia set, shown

Fig. 25.2 The Julia set for $c = 0.0 - 0.72i$.

in Fig. 25.2, is disconnected. It is not one, single, connected shape. Such Julia sets are sometimes referred to as "dusts". If the resolution of a plot of such a Julia set is increased, one would observe the Julia set has a structure like a Cantor set, in that it consists of an infinite but totally disconnected set of points. What appears to be a connected region is actually disconnected upon closer examination. In any event, the Julia set for $c = 0.0 - 0.72i$ is most definitely not a single connected shape.

Let us try one more c value. Figure 25.3 shows the Julia set for $c = 0.37 + 0.37i$. We see that this Julia set is also connected. We can

Fig. 25.3 The Julia set for $c = 0.37 - 0.37i$.

continue this investigation by trying more c values. Each time we do so, we plot the Julia set and then note whether or not it is connected. In Table 25.1 I have kept track of the results of these experiments. The first three c values listed are the three whose Julia sets I plotted above. The other several are those c values you experimented with in Exercise 24.4 in the previous chapter. You can check my conclusions by plugging these c values into a program that generates Julia sets.

Table 25.1 Classifying Julia sets. Which are connected and which are not?

c Value	Connected?
$-0.60 + 0.20i$	yes
$0.00 + 0.72i$	no
$0.37 + 0.37i$	yes
$0.20 + 0.57i$	no
$-0.10 + 0.88i$	yes
$-1.35 + 0.02i$	yes
$-0.60 + 0.45i$	no

25.2   The Mandelbrot Set Defined

A table listing the connectedness for different c values will only get us so far. Is there any pattern or relationship between a parameter value and the appearance of its associated Julia set? To begin to answer this question, let us plot all the c values in Table 25.1 that give rise to connected Julia sets with one type of symbol. Those c values that yield a disconnected Julia set I will plot with a different symbol.

Fig. 25.4 A plot of the data from Table 25.1. Parameter values c are plotted on the complex plane; the real part of c is plotted on the x-axis and the imaginary part on the y-axis. Parameter values which give rise to a connected Julia set are shown as filled squares. Unfilled squares are parameter values for which the Julia set is unconnected.

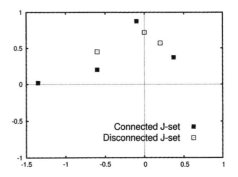

Such a plot is shown in Fig. 25.4. The four c values from Table 25.1 that yield connected Julia sets are plotted as solid squares. The three c values that yield disconnected Julia sets are shown as hollow squares. The dark points, those with connected Julia sets, are said to belong to the **Mandelbrot set**. In other words,

The Mandelbrot set consists of the set of all parameter values c for which the Julia set of $f(z) = z^2 + c$ is connected.

What does the Mandelbrot set look like? It is difficult to see a pattern in Fig. 25.4, since there are so few points. We would like to sample hundreds of different c values, plot them all, and see what we get. Before doing so, it will be helpful to formulate an alternate, but equivalent, specification of the Mandelbrot set.

25.3 The Mandelbrot Set and the Critical Orbit

The connectedness, or not, of the Julia set is the dichotomy that is used to define the Mandelbrot set. Determining the connectedness is potentially a lengthy process, as it requires making a reasonably detailed picture of the Julia set. Recall that to determine the Julia set, we need to test a large number of initial conditions to see which stay bounded and which tend toward infinity. Even when we have a computer do this work for us, it can still take a some time. Moreover, some Julia sets are just barely connected, so one needs to make a fairly high-resolution image of the Julia set in order to be certain about its connectivity.

It turns out that there is a much quicker method to figure out the connectivity of the Julia set. This method requires us to examine the fate of just one initial condition. As an example, let us suppose we want to figure out whether or not the Julia set for $c = -0.6 + 0.2i$ is connected. This is the c for the Julia set shown in Fig. 25.1. The method works as follows. We begin with the initial condition $z_0 = 0$. We then iterate, as usual, using the function $f(z) = z^2 + c$. The results of this iteration are shown in Table 25.2. We can see that the orbit does not tend toward infinity. In fact, it turns out that the orbit is pulled toward a fixed point at $z \approx -0.428 + 0.108i$. Since $z_0 = 0$ does not tend toward infinity, it is in the Julia set. Accordingly, in Fig. 25.1 we observe that $z_0 = 0$ is indeed inside the Julia set.

Let us try another c value: $c = 0 - 0.72i$. This c value gave us the Julia set shown in Fig. 25.2. The iterates for the initial condition $z_0 = 0.0$ are shown in Table 25.3. For this c value the orbit of $z_0 = 0$ tends toward infinity. Accordingly, the point $z = 0$ is not in the Julia set, as can be seen in Fig. 25.2.

It turns out that the fate of the orbit of $z_0 = 0$ tells us a great deal. For this reason, the orbit of this point is often called the **critical orbit**. In particular, the following turns out to be true:

> If, for a given c, the critical orbit (the orbit of $z_0 = 0$) for the function $f(z) = z^2 + c$ does not tend to infinity then the Julia set is connected. If the orbit does tend to infinity, then the Julia set is disconnected.

This was the case for the two examples considered in this section. For $c = -0.6 + 0.2i$ the critical orbit was bounded and the Julia set, Fig. 25.1,

Table 25.2 The orbit of $f(z) = z^2 + c$, for $c = -0.6 + 0.2i$ for the initial condition $z_0 = 0$.

Time	z_t
0	$0.0 + 0.0i$
1	$-0.6 + 0.2i$
2	$-0.28 - 0.04i$
3	$-0.5232 + 0.2224i$
4	$-0.375724 + -0.0327194$
⋮	⋮
50	$-0.42774 + 0.107785i$
51	$-0.428656 + 0.107792i$
52	$-0.427873 + 0.107588i$
53	$-0.428500\ 0.107932i$

Table 25.3 The orbit of $f(z) = z^2 + c$, for $c = 0.0 - 0.72i$ for the initial condition $z_0 = 0$. The orbit tends toward infinity.

Time	z_t
0	0.0
1	$0 + 0.72i$
2	$-0.5184 + 0.72i$
3	$-0.249661 + -0.026496i$
4	$0.0616288 + 0.73323i$
⋮	⋮
16	$9.79783 - 3.10886i$
17	$86.3324 - 60.2002i$
18	$3829.22 - 10393.7i$

[1]See, e.g., Peitgen, Jürgens, and Saupe (1992, Chapter 13).

was connected. And for $c = 0 - 0.72i$ the critical orbit tends toward infinity and the Julia set is disconnected. One can show that this must always be the case. However, doing so is somewhat beyond the scope of this book.[1] You can use a Julia set program to test this out for other c values. Look at many Julia sets, and you will notice that Julia sets that are connected always include the origin, the point $z = 0 + 0i$. Thus, the orbit of the origin must be bounded. Similarly, you will notice that disconnected Julia sets never contain the point $z = 0 + 0i$.

Let us now return to thinking about the Mandelbrot set. I initially defined the Mandelbrot set as the collection of c values for which the Julia sets are connected. We now know that a Julia set is connected if and only if the critical orbit is bounded. Thus, we can quickly check to see if c is in the Mandelbrot set by calculating the critical orbit and seeing if the orbit tends to infinity or not. If it does does, then that c value is not in the Mandelbrot set. In other words:

> If the orbit of $z_0 = 0$ for $f(z) = z^2 + c$ is bounded, then c is in the Mandelbrot set. If the orbit is not bounded, then c is not in the Mandelbrot set.

This gives us an efficient way to determine the Mandelbrot set. For a range of c values compute the fate of the critical orbit. If the orbit stays finite, then that c value is in the Mandelbrot set. We can do this for many different c values, record those c values that are in the Mandelbrot set, and then plot them.

The results of doing this are shown in Fig. 25.5. I tested around a quarter million c values by determining the fate of the critical orbit. Doing so took around four seconds on my desktop computer. I then plotted those c values to produce Fig. 25.5. The result is a surprising shape that has become one of the icons of chaos and fractals. It possesses an incredible intricacy and encodes a remarkable amount of geometrical information. In the next section we begin an exploration of this incredible mathematical structure.

25.4   Exploring the Mandelbrot Set

A first blush, the Mandelbrot set appears fractal-like, in the sense that it has repeating patterns at different scales. For example, in Fig. 25.5 we can see the repeating motif of bulbs with spokes emanating off of them. Let us zoom in and see what we see. Figures 25.6–25.9 show successive magnifications of a portion of the Mandelbrot set. Zooming in shows structure on finer and finer scales. Eventually, in Fig. 25.9 we see what looks like a small copy of the original Mandelbrot set. This is usually referred to as a baby Mandelbrot set. There are many such baby Mandelbrot sets hanging off the full Mandelbrot set. And there are baby baby Mandelbrot sets hanging off the baby sets, and so on.

The Mandelbrot set is indeed a fractal. In fact, the boundary of the set is so complex and intricate that it has a dimension of two.[2] Note,

[2]This was proved relatively recently, by Mitsuhiro Shishikura (1998).

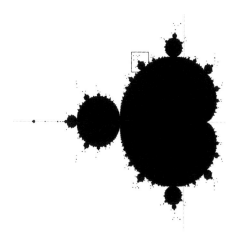

Fig. 25.5 A high-resolution plot of the Mandelbrot set. Over a quarter million c-values were sampled. The initial condition $z_0 = 0$ was iterated for each the c value. If the orbit remains bounded, then that c-value is in the Mandelbrot set. The zero axes are shown as dotted lines. The horizontal range on the plot is from -2 to 0.5. The boxed region is plotted in more detail in Fig. 25.6.

however, that unlike most of the fractals we have studied thus far, the smaller structures are not exact copies of the larger structures. As we zoom in, what we see looks familiar, but not identical to what we have seen before. Baby Mandelbrot sets are not exact replicas of the full Mandelbrot set. In contract, when we zoom in on Julia sets we see exactly the same shapes on different scales. An example of this can be found in Figs. 24.4 and 24.5. Zooming in on this Julia set we see the exact same fractal "cactus" structure repeating over and over.

The Mandelbrot set is an amazing mathematical object. It shows intricate order and structure, and zooming in continually reveals slightly new shapes and forms. Some have referred to it in almost hyperbolic terms:

> The Mandelbrot Set is the most complex object in mathematics, its admirers like to say. An eternity would not be enough time to see it all, its disks studded with prickly thorns, its spirals and filaments curling outward and around, bearing bulbous molecules that hang, infinitely variegated, like grapes on God's personal vine. ...[T]he Mandelbrot set seems more fractal than fractals, so rich is its complication across scales. (Gleick, 1987, p. 221)

This is perhaps over the top. But not by much. There is an amazing diversity of shapes and patterns in the Mandelbrot set. I encourage you to spend a while exploring it using one of the many programs on the web that will make Mandelbrot images for you.

As with the Julia set, it is possible to color the Mandelbrot set. There are a handful of ways of doing so, but usually it is done as follows. In colored Mandelbrot sets, it is typically the points *not* in the set that are colored. Recall that a point c is not in the Mandelbrot set if the orbit of $z_0 = 0$ tends toward infinity for the function $f(z) = z^2 + c$. Points that fly off to infinity are assigned different colors depending on how fast they do so. In practical terms, this is measured by how many iterations

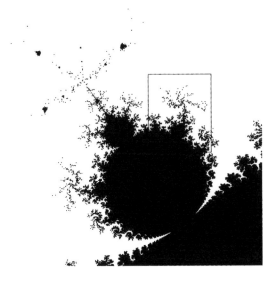

Fig. 25.6 A plot of a portion the Mandelbrot set. Over a quarter million *c*-values were sampled. The region plotted is the boxed region from Fig. 25.5. The rectangular region in this figure is plotted in Fig. 25.7.

Fig. 25.7 A plot of a portion the Mandelbrot set. Over a quarter million *c*-values were sampled. The region plotted here is the boxed region from Fig. 25.6. The rectangular region in this figure is plotted in Fig. 25.8.

it takes for the orbit to move to some predetermined distance from the origin. Usually this distance is taken to be 2.

For example, we might program the computer to assign the color pink to points that are a distance of 2 from the origin within five iterations. Blue points could be those that take up to twenty iterations to get 2 away from the origin. Points that take up to fifty iterations to get 2 away from the origin might be green, and so on. Which colors—and how many— one chooses is an art. Different choices yield different results. There are also a number of techniques for smoothing out the colors so the image appears as a continuous gradient of color and not as solid stripes.[3]

[3] A good, elementary introduction to Mandelbrot set graphics is (2004).

Fig. 25.8 A plot of a portion the Mandelbrot set. Over a quarter million c-values were sampled. The region plotted here is the boxed region from Fig. 25.7. The rectangular region in the middle of this plot is shown in Fig. 25.9.

Fig. 25.9 A plot of a portion the Mandelbrot set showing a baby Mandelbrot set. Almost a half million c-values were sampled. The region plotted here is the boxed region from Fig. 25.8. If the full Mandelbrot set of Fig. 25.5 was plotted at this scale, it would be as wide as a football field.

25.5   The Mandelbrot Set is a Julia Set Encyclopedia

Having enjoyed a journey through parts of Mandelbrot set, we now again consider the relationship between the Mandelbrot set and Julia sets. The Mandelbrot set is the set of c values for which the orbit of the critical point is bounded when iterated by $f(z) = z^2 + c$. Thus, each point on the Mandelbrot set is a particular c value, that we can then use to make a Julia set. We know that all the Julia sets so obtained will be connected. In fact, this is how we originally were led to the Mandelbrot set. Our initial definition of the Mandelbrot set was those c values for which $f(z) = z^2 + c$ had a connected Julia set. Thus, the Mandelbrot set is a listing of all the c values that have a connected Julia sets.

This is not surprising, since the Mandelbrot set was constructed to have exactly this property. What *is* surprising, however, is that there are deeper relationships between the Mandelbrot set and the Julia sets. The Mandelbrot set catalogs Julia sets in an unanticipated way. I will illustrate this with several examples. In Fig. 25.10 I have again plotted the full Mandelbrot set. I will choose regions of the Mandelbrot set and plot a Julia set for a c value chosen from that region.

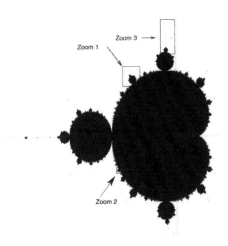

Fig. 25.10 Another plot of the full Mandelbrot set.

Fig. 25.11 The portion of the Mandelbrot set that is shown in the solid box labeled "Zoom 1" in Fig. 25.10.

Table 25.4 The orbit of $f(z) = z^2 + c$, for $c = -0.505 + 0.574i$ for the initial condition $z_0 = 0$. The orbit approaches a cycle of period 5.

Time	z_t
0	$0.0 + 0.0i$
1	$-0.505 + 0.574i$
2	$-0.579 - 0.00574i$
3	$-0.169 + 0.581$
4	$-0.814 + 0.377i$
5	$0.014 - 0.040i$
$\vdots$	$\vdots$
50	$0.015 - 0.035i$
51	$-0.506 + 0.573i$
52	$-0.172 - 0.006i$
53	$-0.172 + 0.581i$
54	$-0.813 - 0.035i$
55	$0.015 - 0.035i$

Consider first the region contained within the solid box labeled "Zoom 1" near the top of the Mandelbrot set in Fig. 25.10. A magnified view of this region is shown in Fig. 25.11. I then chose a c value from the middle of the bulb shown in this figure and made a Julia set. This Julia set is shown in Fig. 25.12. Note that the Julia set has five "arms". That is, the main pattern is five blobby arms joined together. This pattern then repeats at different scales. There are arms of many different sizes, and when they meet up, there are always five arms at the junction point. Now look at the Mandelbrot set in Fig. 25.11. One can see that the spokes or antennae that decorate the bulb also have five arms that meet at a central point. This is most evident in the long, dendritic arms on the top left of Fig. 25.11. When counting these arms, do not forget to include the arm that is attached to the bulb itself.

To recap, we have chosen a Julia set by picking a c value from the Mandelbrot set. And we have seen that the resultant Julia set, Fig. 25.12, resembles the region of the Mandelbrot set from which its c value was chosen. Both the Julia set and the Mandelbrot set are five-armed. The story does not end here. There is another feature of five-ness associated with this Julia set.

The c value for the Julia set of Fig. 25.12 is $c = -0.505 + 0.574i$. What happens if we plot the critical orbit for this c value? That is,

Fig. 25.12 The Julia set for a c value chosen from the center of the Mandelbrot bulb shown in Fig. 25.11. The exact c value is $-0.505 + 0.574i$.

what happens if we use the seed $z_0 = 0$ and iterate it with $f(z) = z^2 - 0.505 + 574i$? The result of doing this is shown in Table 25.4. We see that the orbit eventually becomes periodic with period 5. The Julia set for this c has five arms, the Mandelbrot set for c values near this value is a bulb from which five spokes emanate, and the critical orbit is periodic with period 5. The region of the Mandelbrot set contained in Fig. 25.11 is thus strongly associated with the number five.

Let us repeat this analysis for another region of the Mandelbrot set. In Fig. 25.13 I have plotted the small square region labeled "Zoom 2" in Fig. 25.10. We can see spiraling curls hanging off the bulb. In Fig. 25.14 I have plotted a Julia set for a c value of $-0.689 - 0.348i$, which is inside the bulb in Fig. 25.13. Again, we see that the Julia set bears a resemblance to the Mandelbrot set near where we found its c value. Although it is a little hard to see on the figure, both the Julia set and the decorations on the Mandelbrot set are nine-armed. And if one were to iterate $f(z) = z^2 + c$ using the seed $z_0 = 0$, one would find that the orbit is attracted to a cycle of period 9.[4]

[4]I will not list the itinerary, as it would take up too much space. But if you are skeptical you can try it out for yourself and you will see that the the critical orbit really does have a period of 9.

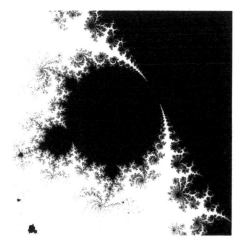

Fig. 25.13 The portion of the Mandelbrot set that is shown the solid box labeled "Zoom 2" in Fig. 25.10. Note the baby Mandelbrot set in the lower left corner.

Let us try this one more time. In Fig. 25.15 I have plotted a magnified view of the bulb on top the Mandelbrot set. The region I have plotted

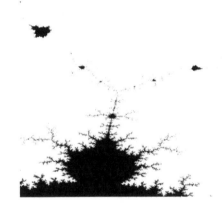

Fig. 25.14 The Julia set for a c value chosen from the center of the Mandelbrot bulb shown in Fig. 25.13. The exact c value is $-0.689 + -0.348i$.

is labeled "Zoom 3" in Fig. 25.10. Note that the spokes or antenna that branch off of this portion of the Mandelbrot set are all three-pronged. When the spokes join, it is always three spokes joining at the junction point. Note also that embedded in the antennae are baby Mandelbrot sets. In Fig. 25.16 I have plotted the Julia set for a c value chosen from the baby Mandelbrot set in the upper left of Fig. 25.15. The c value is $-0.157 + 1.031i$. Again, we see that the Julia set resembles the Mandelbrot set.

Fig. 25.15 The portion of the Mandelbrot set that is shown the solid box labeled "Zoom 3" in Fig. 25.10. Note the baby Mandelbrot set in the upper left.

Fig. 25.16 The Julia set for a c value chosen from the center of the baby Mandelbrot set in the upper left of Fig. 25.15. The c value is $-0.157 + 1.031i$.

If we iterate the seed $z_0 = 0$ with the usual quadratic function $f(z) = z^2 + c$ for this c value, we find that the orbit approaches a point of period

3. This can be seen in Table 25.5, where I have plotted the orbit for the critical seed $z_0 = 0$.

To summarize, then, we have seen that the Mandelbrot set does a lot more than just tell us whether or not a Julia set is connected. For every point in the Mandelbrot set there is a corresponding Julia set. We can look at the location in the Mandelbrot set that the Julia set came from and see that the Julia set resembles the structures found nearby in the Mandelbrot set. Moreover, the number of branches on the Mandelbrot antennae or the number of arms on the Julia set tell us the periodicity of the critical orbit. I.e., if the Julia set has five arms that branch from a junction point, as in Fig. 25.12, then the orbit of the seed $z_0 = 0$ approaches a cycle of period 5.

In a sense, then, the Mandelbrot set acts like a dictionary or an encyclopedia for the Julia sets. We can tell a lot about the structure of the Julia set by just knowing where on the Mandelbrot set it comes from. And this information can be read off the Mandelbrot set quite directly, by looking at the structure of the antennae or spokes that hang off the bulbs.

Table 25.5 The orbit of $f(z) = z^2 + c$, for $c = -0.505 + 0.574i$ for the initial condition $z_0 = 0$. The orbit approaches a cycle of period 3.

Time	z_t
0	$0.0 + 0.0i$
1	$-0.157 + 1.033i$
2	$-1.195 + 0.707i$
3	$0.772 - 0.660i$
$\vdots$	$\vdots$
50	$-1.195 + 0.707i$
51	$0.772 + -0.659i$
52	$-0.1571.031i$
53	$-1.195 + 0.707i$

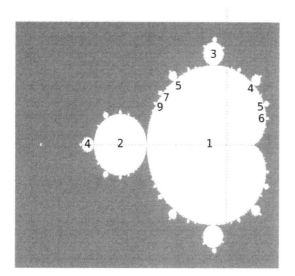

Fig. 25.17 The structure of the Mandelbrot set. The numbers indicate the period associated with each bulb. For example, the bulb on the top, labeled "3" is a period-3 bulb. Any Julia set with a c value chosen from this bulb will have a structure where three arms meet at a junction point, and the critical orbit will be drawn to a cycle of period 3.

Every bulb on the Mandelbrot set is associated with a particular period. This is shown in Fig. 25.17, where I have labeled several bulbs with their associated period. For example, the bulb on top labeled "3" is a period-3 bulb. Any Julia set drawn from this bulb will have a three-armed structure of some sort, and the critical orbit will be approach a cycle of period 3. To determine the period of a bulb, all one has to do is zoom in on the bulb and look at the structure of the spokes that emanate from the bulb. The spokes will branch, and the number of branches at a junction point is the period of the bulb.

25.6   Conclusion

The Mandelbrot set is a remarkable mathematical object. It combines regularity and novelty. There are structures that repeat on finer and finer scales, but the repetition is not exact. The baby Mandelbrot sets resemble the full set, but they are not perfect replicas. Spokes and antennae spiral off of the bulbs on the Mandelbrot set, but the numbers of spokes and the direction of their spirals vary from bulb to bulb. These variations are correlated with the structure of the Julia sets. Zooming in on the Mandelbrot set produces beautiful, other-worldly images.

Amazingly, all of this structure—the Julia sets and the Mandelbrot set—result from an extremely simple function:

$$f(z) = z^2 + c \, . \tag{25.2}$$

This quadratic equation, when iterated, gives rise to amazing complexity. We thus see yet again that simple, iterated systems can give rise to complex and complicated shapes and forms.

Further Reading

What I have covered in this chapter is just the beginning. There are many more fascinating, fun, and surprising properties of the Mandelbrot set. Many of the references listed in Appendix C have good discussions of the Mandelbrot set, as it is a central object of study in chaos and fractals. Of these, the discussion in Peitgen, Jürgens, and Saupe (1992) is particularly clear and thorough. The workbook by Devaney (2000) is a good, elementary introduction and contains numerous worksheets and suggestions for explorations and classroom activities. The accompanying video (Devaney, 1996) is also excellent. The nicely illustrated edited volume, *The Colours of Infinity* (Lesmoir-Gordon, 2004), has clear and accessible discussions of Julia sets and the Mandelbrot set, as well as fractals in general. This book is accompanied by an informative and entertaining documentary about fractals, narrated by Arthur C. Clarke. This documentary is also available on YouTube. The textbook by Falconer (2003) is a standard technical reference on fractal geometry and has good coverage of Mandelbrot and Julia sets. It is written at a level appropriate for junior mathematics majors. There are many websites that contain applets which allow you to explore and experiment with the Mandelbrot set. One of my favorites is `http://homepages.inf.ed.ac.uk/wadler/mandelbrot-maps/mmaps.html`.

Exercises

For Exercise 25.4 you will need to use a program that can make images of Julia sets and the Mandelbrot set. You can find links to such programs on the website for the book `http://chaos.coa.edu` or via a web search.

(25.1) Consider the function $f(z) = z^2 + c$ with $c = -0.6 + 0.2i$.

    (a) Show that $z = -0.428227 + 0.107732i$ is a fixed point for this function.

    (b) Verify that the first three orbits of z_0 are what I claimed they are in Table 25.2

(25.2) Verify that the first three iterates of $z_0 = 0 + 0i$ for $f(z) = z^2 + c$, where $c = 0.0 - 0.72i$, are what I claimed they are in Table 25.3.

(25.3) By hand or using a calculator, compute the first several iterates of $f(z) = z^2 + c$ for each of the following c values. Which c values are in the Mandelbrot set?

    (a) $c = 0$

    (b) $c = -1$

    (c) $c = -1.5 + 0.5i$

    (d) $c = -0.5 + 0.5i$

(25.4) Use the Mandelbrot set as an encyclopedia to find Julia sets with the following properties. For each, print out or sketch a picture of the Julia set, note the c value for the Julia set, and indicate where in the Mandelbrot set you found the c value:

    (a) The Julia set is a single connected blob.

    (b) The Julia set has three arms, i.e., three structures which join at a junction point.

    (c) The Julia set has eight arms.

    (d) The Julia set has eleven arms.

Part V

Higher-Dimensional Systems

Two-Dimensional Discrete Dynamical Systems

26

The dynamical systems we have studied in this book have almost all been iterated functions of one continuous variable. In other words, we have iterated a function whose value can be any among a continuous range of numbers. The main example of such a dynamical system is the iterated logistic equation, studied in considerable detail in Part II. This part of the book looks at other types of dynamical systems, including iteration in more than one dimension and iteration of a function that varies continually as opposed to changing at discrete time intervals. In so doing, we will see new examples in which simple iterated equations give rise to chaotic and surprisingly complex behavior. This chapter is about two-dimensional discrete dynamical systems. However, before making the jump to two dimensions, it will help to quickly review one-dimensional dynamical systems.

26.1 Review of One-Dimensional Discrete Dynamics

Our canonical example of a one-dimensional, discrete dynamical system is the logistic equation,

$$f(x) = rx(1-x) \,, \tag{26.1}$$

where x can take any value between 0 and 1. The variable r is a parameter; we vary r and see what dynamical behaviors result. Iterating this equation gives us a sequence of numbers. For example, we might consider the logistic equation with $r = 3.2$, choose the initial condition $x_0 = 0.7$, and obtain the itinerary shown in Table 26.1. Note that while x_n can take any value between 0 and 1, the time index is always an integer. That is, we can speak of the value of x at time $t = 1$ or $t = 2$ or $t = 613$. However, we cannot speak of the value of x at time $t = 1.5$ or $t = 6.13$.

Recall that we can graph the orbit in a time series plot. Such a plot for the iterates of Table 26.1 is shown in Fig. 26.1. We see that the long-term behavior of the orbit is periodic with a period of 2; the orbit oscillates between 0.513 and 0.799. As the last part of our review, recall that we can represent this long-term behavior with a final-state diagram in which the long-term behavior—in this case period 2—is shown on a

Table 26.1 The orbit of the initial condition $x_0 = 0.7$ when iterated by $f(x) = 3.2x(1-x)$.

Time	x_t
0	0.7
1	0.672
2	0.705331
3	0.665085
4	0.71279
5	0.655105

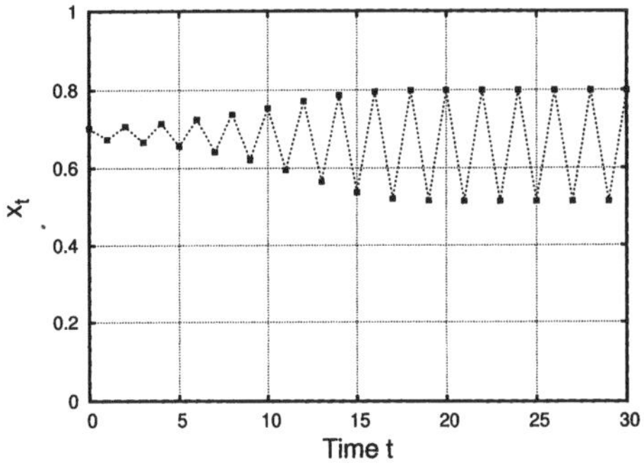

Fig. 26.1 A time series plot of the itinerary of 0.7 for the logistic equation with $r = 3.2$. The itinerary approaches the fixed point at $x = 1$. Numerical values for the first few iterates are given in Table 26.1.

number line. We say that this equation has an attractor of period 2, since nearby orbits are pulled toward the period-2 points at 0.513 and 0.799.

Fig. 26.2 The final-state diagram for the orbit graphed in Fig. 26.1. The behavior is periodic with period 2.

26.2   Two-Dimensional Discrete Dynamical Systems

We now turn our attention to two-dimensional discrete dynamics. In one-dimensional dynamics there was a single number x that varied over time. The dynamics are given by a function, such as that of Eq. (26.1). The function takes the current value of x as an input and returns the next value of x.

In two-dimensional discrete dynamics the scenario is the same, except that there are two numbers, x and y, that vary over time. There is a function that takes *two* numbers as input, and returns *two* numbers as output. Note that this is not the same as two separate functions. The process should be seen as one multi-input and multi-output function. The output for y depends not just on the input of y but also the input of x.

Fig. 26.3 A schematic representation of a function f that takes two numbers as input and returns two numbers as output.

As a simple example, consider the following function:

$$x_{n+1} = 1 + x_n + y_n , \tag{26.2}$$

$$y_{n+1} = 2 + \frac{1}{2}x_n - \frac{1}{2}y_n . \tag{26.3}$$

We can iterate this function just as we would a one-dimensional function. Since this is a two-dimensional function we need two initial conditions, x_0 and y_0. Suppose we choose $x_0 = 1$ and $y_0 = 2$. To determine x_1 we use Eq. (26.2):

$$x_1 = 1 + x_0 + y_0 = 1 + 1 + 2 = 4 \,. \tag{26.4}$$

Similarly, we determine y_1 by using Eq. (26.3):

$$y_1 = 2 + \frac{1}{2}x_n - \frac{1}{2}y_n = 2 + \frac{1}{2}1 - \frac{1}{2}2 = 2 + \frac{1}{2} - 1 = \frac{3}{2} \,. \tag{26.5}$$

Table 26.2 The orbit of the initial condition $x_0 = 1, y_0 = 2$ when iterated using Eqs. (26.2) and (26.3).

Time	x_t	y_t
0	1	2
1	4	1.5
2	6.5	3.25
2	10.75	3.625

We continue iterating as usual, using the output from one step as the input for the next step. The results of doing so are shown in Table 26.2. You might wish to check these numbers for yourself.

As in previous chapters, we will be concerned with the long-term behavior of orbits. Are there fixed points? Do initial conditions tend toward infinity? Are the orbits chaotic? Is there sensitive dependence on initial conditions? We will explore these questions using a dynamical system known as the Hénon map, introduced below.

26.3 The Hénon Map

The logistic equation was our standard example of a one-dimensional discrete dynamical system. We used it in previous chapters to explore chaos, the butterfly effect, and the universality of the period-doubling route to chaos. In a similar way, we will use the Hénon equation to explore and understand the basic phenomena exhibited by two-dimensional discrete dynamical systems. In Section 1.5 I mentioned that functions are also often referred to as maps. The logistic equation is often referred to as the logistic map, especially by mathematicians. I have been using the term equation throughout the book, since I think it is more familiar. However, the Hénon equations, given below, are almost always referred to as a map, and so I will use this terminology in this chapter.

The **Hénon map** is defined by the following:

$$\begin{aligned} x_{n+1} &= y_n + 1 - ax_n^2 \,, \\ y_{n+1} &= bx_n \,, \end{aligned} \tag{26.6}$$

where a and b are parameters. As we did for the logistic equation, we will explore what happens for different values of the parameters a and b.

You might be wondering where the Hénon map comes from. Who thought it up? You will not be shocked to learn that the answer to this question is Hénon—Michel Hénon, a French astronomer and mathematician. What led Hénon to this equation? The answer is somewhat complicated, and I think that explaining it in detail would be too much of a distraction. In brief, Hénon and other mathematicians and scientists in the mid-1970s were looking for simple examples of systems that display chaotic behavior. They began with equations that were

[1]Specifically, the Hénon map is constructed to be an approximation to the Lorenz equations, the topic of Chapter 31.

Table 26.3 The orbit of the seed $x_0 = 1, y_0 = 2$ for the Hénon map with $a = 0.155$ and $b = 0.6$.

Time	x_t	y_t
0	1	2
1	2.8450	0.6
2	0.345426	1.707
3	2.68851	0.207256
4	0.08690	1.6131
5	2.61193	0.0521436

relatively realistic models of the physical systems they were studying. However, these models proved to be somewhat unwieldy to work with, so they simplified their equations until they arrived at the form shown in Eq. (26.6).[1]

We begin by considering the dynamics for the parameter values $a = 0.155$ and $b = 0.6$. In this case, Eq. (26.6) becomes:

$$
\begin{aligned}
x_{n+1} &= y_n + 1 - 0.155x_n^2 \,, \\
y_{n+1} &= 0.6x_n \,.
\end{aligned}
\tag{26.7}
$$

Let us choose the initial condition $x_0 = 1, y_0 = 2$ and see what happens. Plugging into Eq. (26.7) and iterating, one obtains the itinerary shown in Table 26.3. To examine the long-term behavior, we need a lot more points, so we turn to a computer to do the iteration for us and then make a plot. The result of doing so is shown in Fig. 26.4. Looking at these plots we see that the orbit is pulled to a cycle of period 2. The x part of the orbit oscillates between 2.49 and 0.83, while the y part of the orbit oscillates between 1.50 and 0.05. We can represent the x and

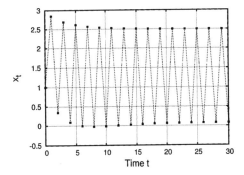
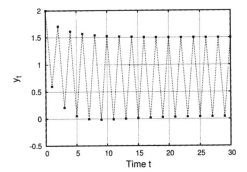

Fig. 26.4 A time series plot of the x- and y-itineraries for the Hénon map with $a = 0.155$ and $b = 0.6$. The initial conditions are $x_0 = 1, y_0 = 2$. The orbit is pulled toward a cycle of period 2. Numerical values for the first few iterates are given in Table 26.3.

y parts of the orbit together by plotting them on an x-y plane, as shown in Fig. 26.5. In this plot the x and y values are plotted together as a single point on the x-y plane. The time value is not plotted explicitly, but is indicated by the labels. The initial condition, labeled "0" on the figure, is $x = 1$, $y = 2$. The first iterate, $x = 2.8450$, $y = 0.6$ is labeled "1", and so on. Table 26.3, the time series of Fig. 26.4, and the plot in Fig. 26.5 all show the same information, but in different ways.

We can also plot just the final states of the orbit, much as we did for the one-dimensional system in Fig. 26.2. Here, however, the states are two-dimensional—each point on the itinerary is a pair of numbers, an x value and a y value. Thus, our final-state diagram is two-dimensional. The final-state diagram for the Hénon map considered in this section is shown in Fig. 26.6 The final-state diagram has two points, since the long-term behavior of the orbit is periodic with period 2.

In Fig. 26.5 I have drawn arrows connecting successive points on the itinerary. These arrows are designed to make evident the order in which

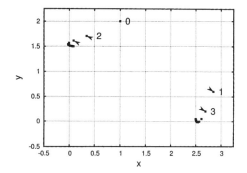

Fig. 26.5 A plot of the itinerary of the initial condition $x_0 = 1$, $y_0 = 2$ under the Hénon map with $a = 0.115$ and $b = 0.6$. The initial condition is labeled "0" on the figure. The first iterate is labeled "1", the second iterate is labeled "2", and so on. Numerical values for the first few iterates are given in Table 26.3. The orbit bounces around and gets pulled to the period-2 attractor.

the points occur. They do not, however, indicate the actual path taken from point to point. In fact, the idea of the path between points is meaningless, because the variable only has values at discrete times. E.g., the orbit is somewhere at $t = 2$ and then somewhere else at $t = 3$. At $t = 2.5$ it is not halfway between—it really is not anywhere. The orbit is only defined at discrete time steps. Nevertheless, I think that arrows like those on Fig. 26.5 are useful, and I will use them again in subsequent figures. Just please do not take them too literally.

26.4 Chaotic Behavior and the Hénon Map

We have now seen how to represent and think about periodic behavior for a two-dimensional iterated function. There is not much difference between one-dimensional and two-dimensional periodic behavior for discrete systems. In contrast, we will see that the chaotic behavior of the Hénon map holds some fun surprises.

We begin by returning for a moment to our standard example of a one-dimensional iterated function: the logistic equation, $f(x) = rx(1 - x)$. This equation is chaotic when the parameter $r = 4.0$. A typical time

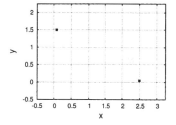

Fig. 26.6 The final-state diagram for the Hénon map with $a = 0.155$ and $b = 0.6$. The behavior is periodic with period 2.

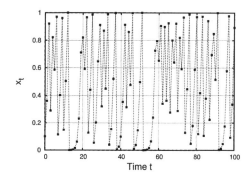

Fig. 26.7 A plot of the itinerary of initial condition $x_0 = 0.1$ under the logistic equation, $f(x) = 4x(1 - x)$. The orbit is aperiodic.

series for the chaotic logistic equation is shown in Fig. 26.7. Note that the orbit appears to be aperiodic—it does not repeat. If we looked at the itinerary of two nearby initial conditions, we would find that the difference between them grows rapidly as the function is iterated. I.e., the orbits display sensitive dependence on initial conditions, or the butterfly effect. The final state for this this system is shown in Fig. 26.8.

Fig. 26.8 A plot of the itinerary of the initial condition $x_0 = 0.1$. The orbit is aperiodic.

Since the orbit is aperiodic, it never settles down; it keeps bouncing around between 0 and 1. So the final states fill up the entire interval between 0 and 1.

Let us now return to the Hénon map. The rest of this chapter will focus on the behavior of the Hénon map for the parameter values $a = 1.4$ and $b = 0.3$. In Fig. 26.9 I have plotted the x and y time series for the initial condition $x_0 = 0.6$, $y_0 = 0.6$. Only the first thirty iterates are shown, but both orbits seem to be wandering around and not approaching a periodic attractor. To confirm this, let us look at similar plots, but for a longer time range.

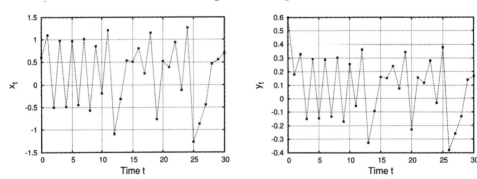

Fig. 26.9 Time series plots of the x- and y-itineraries for the initial condition $x_0 = 1, y_0 = 2$ under the Hénon map with $a = 1.4$ and $b = 0.3$. The orbit is aperiodic.

The results of doing so are shown in Fig. 26.10. Here I have plotted the first 150 iterates. Again, one sees that the two time series jump around and do not appear to be approaching a fixed point or a periodic attractor. If one plots more and more orbits, this remains unchanged. The orbits are aperiodic.

The chaotic orbits for the Hénon map look qualitatively similar to the chaotic orbit for the logistic map, as shown in Fig. 26.7. The time series is a jagged line that skips around and never repeats. However, what would happen if we plotted the x and y time series together as we did in Fig. 26.5? Take a moment and try to picture what this would look like for the chaotic times series of Fig. 26.10. One might expect that the plot on the x-y plane would look like a dense blob. As more and more iterates are plotted, the dots may scatter across the plot, eventually turning the picture into a dark cloud, much as the final states for the chaotic logistic equation fill up the interval.

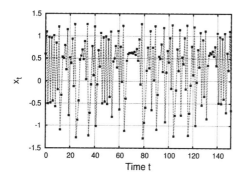 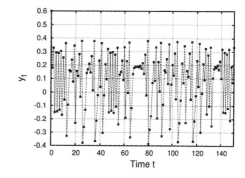

Fig. 26.10 Time series plots of the x- and y-itineraries for the initial condition $x_0 = 1, y_0 = 2$ for the Hénon map with $a = 1.4$ and $b = 0.3$. The orbit is aperiodic.

26.5   A Chaotic Attractor

However, if one plots x and y together for the Hénon map when $a = 1.4$ and $b = 0.3$, one finds a surprise. Such a plot is shown in Fig. 26.11. At first blush it does not look anything at all like what we saw when we plotted the x and y parts of the orbit separately in Fig. 26.10. This figure shows that even though the x and y time series are chaotic, there is a relationship that becomes apparent when x and y are plotted together. Remarkably, when the two aperiodic time series shown in Fig. 26.10 are plotted together, the result is the orderly shape seen in Fig. 26.11.

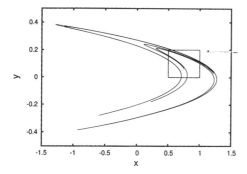

Fig. 26.11 The itinerary of the initial condition $(x_0, y_0) = (0.6, 0.6)$ for the Hénon map with $a = 1.4$ and $b = 0.3$. The plot shows $100,000$ iterates. A close-up of the boxed region is shown in Fig. 26.12.

Looking at Fig. 26.11 more closely, one sees that the structure is quite intricate. A magnified portion of this plot is shown in Fig. 26.12. Additional structure is apparent. Lines in Fig. 26.11 are, upon closer view, actually pairs of lines. Zooming in further, as in Fig. 26.13 we see even finer structure. Finally, in Fig. 26.14 we zoom in once more. Again, one sees finer and finer structure. By the time we get to the last zoom-in, Fig. 26.14, the plot appears to be grainy. The reason for this is that we have run out of points. The original plot, Fig. 26.11, contains 1 million points. This is certainly a lot. However, once we have zoomed in three times, the area plotted is small enough that it does not contain many points and so appears grainy. Had I plotted 1 billion points originally instead of 1 million, the lines in Fig. 26.14 would appear solid.

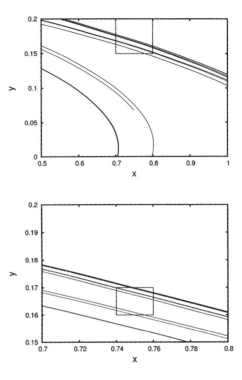

Fig. 26.12 A magnification of the boxed region in Fig. 26.11. Note that single lines on Fig. 26.11 resolve into pairs of lines when magnified. A close-up of the boxed region in this figure is shown in Fig. 26.13.

Fig. 26.13 A magnification of the boxed region in Fig. 26.12. Note that additional structure is present. The boxed region of this plot is shown in Fig. 26.14.

As you have probably noticed by now, this shape is a fractal. Its dimension is estimated to be 1.261 ± 0.003 (Russell, Hanson, and Ott, 1980). As expected, the dimension is between 1 and 2. It is two-dimensional, because it lives in the plane, but it is one-dimensional in the sense that the shape appears to consist solely of lines folded over on themselves many times.

What do these plots tell us about the dynamics of the Hénon map? And where did the chaos go? These plots arose from plotting simultaneously the x and y time series for a chaotic dynamical system. But the resultant plot, Fig. 26.11, does not appear to be chaotic. What is going on? The chaos is still there—it is just not readily apparent on the plot. To see the chaos, we need to follow the trajectory of a point over time,

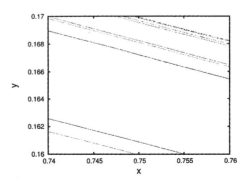

Fig. 26.14 A magnification of the boxed region in Fig. 26.12. Note that additional structure is present. If the full Hénon attractor was plotted at this scale it would be around 18 feet wide.

much as we did in Fig. 26.5 for the period-2 behavior. This will give us a more dynamical view. We will do so shortly. First, we consider the stability of the orbits.

The shape shown in Fig. 26.11 is an attractor, just like the set of two points on Fig. 26.6 is an attractor. In Fig. 26.6, nearby points are pulled to the period-2 attractor. Similarly, in the chaotic Hénon map, nearby points are pulled toward the fractal boomerang-like shape in Fig. 26.11.

This can be seen in Fig. 26.15 in which I have shown the first several iterates of three different initial conditions. These initial conditions are $(-1.0, 0.0)$, $(1.0, -0.4)$, and $(1.2, 0.5)$ and are labeled "A", "B", and "C", respectively.[2] Also shown are around $100,000$ iterates of the initial condition $(0.6, 0.6)$. These $100,000$ points form the boomerang-like shape seen in Fig. 26.11. Figure 26.15 shows that initial conditions are pulled to the boomerang. Each of the initial conditions lands on or very close to the boomerang after just two or three iterations.

[2]In this notation, $(-1.0, 0.0)$ means $x = -1.0$ and $y = 0.0$.

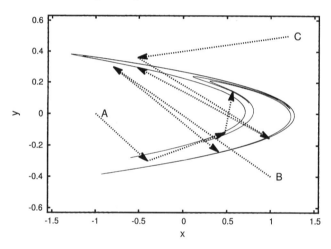

Fig. **26.15** A plot of the first several iterates of three different initial conditions: $(-1.0, 0.0)$, $(1.0, -0.4)$, and $(1.2, 0.5)$. These initial conditions are labeled "A", "B", and "C", respectively. Also shown are around $100,000$ iterates for the initial condition $(0.6, 0.6)$. Notice that the three different initial conditions are all quickly pulled into the attractor.

Thus, the boomerang shape is an attractor, referred to as the **Hénon attractor.** The Hénon attractor is the fractal shape to which almost all initial conditions are pulled when iterated using the Hénon map, Eq. (26.6), with $a = 1.4$ and $b = 0.3$. This phenomenon of a chaotic attractor is new. We have seen chaos before, and we have seen attractors. But previously it was periodic behavior that was attracting, not chaotic behavior.

Figure 26.15 shows us that orbits get pulled toward the attractor. What happens once they reach the attractor? How is it that a chaotic orbit traces out such a regular path? To investigate these questions, in Fig. 26.16 I have plotted a few iterates, but this time starting with the fiftieth iterate so that the orbit is already on the attractor. We can see in the figure that even though it is on the attractor, the orbit is still bouncing around. Notice, though, that the attractor is continuous, but motion on the attractor is not. That is, the attractor is a series of connected curves folded over themselves again and again, but the motion on the attractor is jumpy and discrete.

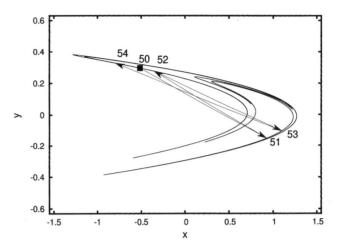

Fig. 26.16 The fiftieth through fifty-fourth iterates for the initial condition (0.6, 0.6). The fiftieth iterate is indicated by the solid square. Once the orbit is on the attractor it continues to jump around. The motion on the attractor is aperiodic, and the orbit does not trace out the lines on the attractor in order. The fiftieth iterate is shown as a square.

Recall that a dynamical system is defined to be chaotic if it is a deterministic function whose orbits are bounded, aperiodic, and possess sensitive dependence on initial conditions (SDIC). Does the Hénon map have SDIC? The answer to this question is "yes". To demonstrate this, in Fig. 26.17 I have plotted two different orbits. One of the orbits starts with $(-0.514651, 0.297419)$, which is the fiftieth orbit of the initial condition $(0.6, 0.6)$. I chose this as my initial condition because I know that for these x and y values the orbit is already very close to the attractor. The x and y orbits for this initial condition are plotted in Fig. 26.17 using square points connected with a dashed line. Figure 26.17 also shows the orbit for a slightly different initial condition. I just rounded a little and used $(-0.515, 0.297)$. The orbit for this initial condition is plotted using circles connected with a solid line.

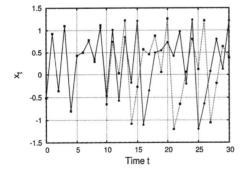 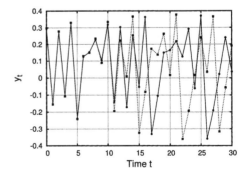

Fig. 26.17 The first thirty x and y iterates for two sets of almost identical initial conditions. The orbit for the initial condition $(-0.514651, -.297419)$ is shown as squares connected with a dashed line. The initial condition $(-0.515, 0.297)$ is plotted with circles connected with a solid line. The two orbits begin very close to each other, but they pull apart around the fifteenth iterate.

In both time series the two orbits begin very close to each other. For the first ten iterates the orbits are so close that they appear identical on the plot. But around the tenth iterate the two orbits become distin-

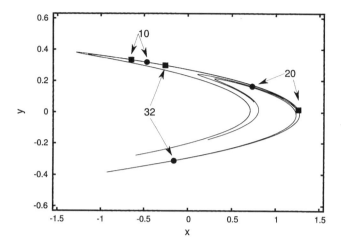

Fig. 26.18 The tenth, twentieth, and 32nd iterates for the two initial conditions $(-0.514651, -.297419)$ and $(-0.515, 0.297)$. The iterates for the two initial conditions are plotted as squares and circles, respectively. Note that the distance between the two orbits increases, but nevertheless the orbits remain on the attractor.

guishable, and by the fifteenth orbit the two orbits have pulled apart. These plots show sensitive dependence on initial conditions. Two initial conditions that differ by a very small amount—around one part in one thousand—eventually get pulled apart under iteration by the Hénon map. Yet, despite the fact that the difference between the two orbits grows, both orbits stay on the attractor. This is hard to see on the time series plots of Figs. 26.17 and 26.17. However, if we look at the attractor in the x-y plane in Fig. 26.18 we can see that the two points get pushed apart from each other although they both remain on the attractor.

26.6 Strange Attractors Defined

There are three interesting and noteworthy properties of the Hénon attractor. First, it is an attractor. Almost all initial conditions get pulled toward it. Two initial conditions that start away from the attractor will both move closer to it. Second, the behavior on the attractor itself is chaotic. The orbits are aperiodic and show sensitive dependence on initial conditions. Third, the attractor is a fractal. By zooming in successively we have seen that the boomerang shape is actually a self-similar structure of folds.

Objects such as the Hénon attractor are known as **strange attractors.** There does not appear to be a completely standard definition for this term. But strange attractors are generally taken to be attractors that are fractal in structure and on which the dynamics are chaotic. Strange attractors are an interesting mix of order and disorder. The dynamics of a system with a strange attractor is fully chaotic. It possesses the butterfly effect, and so long-term prediction is impossible. However, there is considerable order to strange attractors, too. The existence of the attractor, to which almost all initial conditions are drawn, means that the system is constrained despite being chaotic. In the long term the orbit never repeats, but it also never strays from the attractor. The structure of the attractor often displays an intricately ordered fractal shape. And the fact that a strange attractor attracts means that this shape is robust or stable; almost all initial conditions will, in the long run, result in the same attractor.

The exact origin of the term "strange attractor" is somewhat hazy. It is agreed that the term first appeared in a 1971 paper by David Ruelle and Floris Takens (Ruelle and Takens, 1971). However, it is not clear which of the two originally coined the phrase. Ruelle, writing in 1980 states that "I asked Floris Takens if he had created this remarkably successful expression [strange attractor]. Here is his answer: 'Did you ever ask God whether he created this damned universe? ... I don't remember anything...I often create without remembering it...' The creation of strange attractors thus seems to be surrounded by clouds and thunder." (Ruelle, 1980). In 1993 Ruelle writes that "the term [strange attractor] was new, and nobody now remembers if Floris Takens invented it, or I, or someone else" (Ruelle, 1993).

Regardless of exactly who coined the term, it is worth reflecting on the use of the word strange. One reason mathematical objects like the Hénon attractor might be called strange is that their shapes are unusual. The image in Fig. 26.11 perhaps simply looks odd. But I think the term strange was invoked also to indicate surprise that chaotic behavior could be associated with a stable structure such as an attractor. As noted throughout this book, one of the fundamental new realizations emerging from the study of dynamical systems is that order and disorder are frequently mixed together. They are not mutually exclusive opposites. Strange attractors are almost the apotheosis of this idea—they are intricately ordered structures arising from a deterministic system, on which a dynamical system behaves chaotically.

Strange attractors such as the Hénon attractor studied in this chapter occur quite commonly in two-dimensional discrete dynamical systems. For example, the paper by Sprott (1993) shows dozens of different strange attractors arising from simple two-dimensional iterated quadratic functions.

We will encounter strange attractors again in Chapter 31. In the next chapter we will study a different type of dynamical system, known as cellular automata. We will see that these systems produce periodic and chaotic behavior and also some complex phenomena that are not easy to classify.

Exercises

(26.1) Verify that the numerical values contained in Table 26.2 are correct.

(26.2) Consider the two-dimensional discrete dynamical system given by:

$$x_{n+1} = y_n^2 + 1 - x_n^2 , \qquad (26.8)$$
$$y_{n+1} = \frac{1}{2}x_n + \frac{1}{2}y_n . \qquad (26.9)$$

(a) Determine the first four iterates of $x_0 = 0, y_0 = 0$.

(b) Determine the first four iterates of $x_0 = 1, y_0 = -1$.

(26.3) Consider the following two-dimensional dynamical system:

$$x_{n+1} = y_n , \qquad (26.10)$$
$$y_{n+1} = -x_n . \qquad (26.11)$$

(a) Determine the orbit of $x_0 = 1, y_0 = 1$.

(b) Determine the orbit of $x_0 = 1, y_0 = 0$.

(c) Determine the orbit of $x_0 = 2, y_0 = 2$.

(d) Does this dynamical system have any fixed points? If so, what are they? Are these fixed point(s) attracting, repelling, or neutral?

(26.4) Consider the following two-dimensional dynamical system:

$$x_{n+1} = \frac{1}{2}y_n , \qquad (26.12)$$
$$y_{n+1} = \frac{-1}{2}x_n . \qquad (26.13)$$

(a) Determine the fate of the orbit of $x_0 = 4, y_0 = 4$.

(b) Determine the fate of the orbit of $x_0 = 4, y_0 = 0$.

(c) Determine the orbit of $x_0 = -4, y_0 = 6$.

(d) Does this dynamical system have any fixed points? If so, what are they? Are these fixed point(s) attracting, repelling, or neutral?

(26.5) Consider the Hénon map with $a = 0.155$ and $b = 0.6$, as in Eq. (26.7). Compute the first five iterates of the seed $x_0 = 1$, $y_0 = 1$. Plot the orbit on the x-y plane, as was done in Fig. 26.5. What do you think is the long-term behavior of this orbit?

Cellular Automata

27

In the previous chapter we considered two-dimensional, discrete dynamical systems. These dynamical systems have two continuous variables, x and y, that get updated at discrete time intervals via a deterministic function. The variables x and y are continuous in the sense that they can assume any value, not just integers. In this chapter we will look at another type of dynamical system—one which has a large number of discrete variables arranged in an array or grid. These variables are then updated at discrete time steps via a local, deterministic rule. Such a system is known as a cellular automaton (CA). We begin with a simple example from this class of models.

27.1 One-Dimensional Cellular Automata: An Initial Example

For our initial example, we will consider a cellular automaton in which the variables can take on only two different values, visualized as white or black boxes. CAs whose variables assume only two values are known as binary CAs. The variables are aligned in a sequence of length N. The state of the CA is given by specifying all the values of all the variables. For example, Fig. 27.1 shows a sample state for a binary CA with twenty variables. These individual variables are often called cells.

Fig. 27.1 A sample state for a binary, one-dimensional CA with $N = 20$. There are twenty variables, or cells. Each cell can take on one of two values, white or black.

Having defined the variables for cellular automata, we now consider how these variables change. The dynamics for a CA are specified by a local update rule. Such a rule is shown in Fig. 27.2. The idea is that the subsequent value of a variable is determined by its current value and the value of the neighboring sites one to the left and one to the right.

The rule is applied as follows. Let us use the state shown in Fig. 27.1 as the initial condition and then apply the rule shown in Fig. 27.2. The first step in this situation is shown in Fig. 27.3. The initial condition is labeled $t = 0$. The next step, $t = 1$, is drawn directly below the initial condition. (In the figure, the next step has not been filled in yet.) It is the rule that tells us how to fill in the next step. For example,

Fig. 27.2 A rule for a cellular automaton. There are eight possible three-site neighborhoods, indicated by the eight blocks of three variables in the figure. Below each block is the output associated with that neighborhood.

suppose we want to figure out the value of the site marked A at $t = 1$. At $t = 0$ site A was black, while its left and right neighbors were white. This corresponds to the □■□ neighborhood that is third from the left in Fig. 27.2. Looking at that figure, we see that the next value of site A should be black.

Fig. 27.3 The first step in applying the rule shown in Fig. 27.2 to the initial condition of Fig. 27.1.

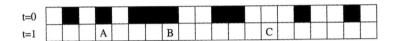

What about site B? Its neighborhood is ■■□. Looking up the output for this neighborhood in Fig. 27.2, we see that output value is white. Site C has a neighborhood of □□□, which, according to the rule, gives an output of zero. We can repeat this procedure for all the sites, and in this way fill out all the variables for the CA for $t = 1$. The result of doing so is shown in Fig. 27.4. I suggest taking a moment to make sure you understand how the values of the variables were determined for $t = 1$.

Fig. 27.4 The result of applying the rule, Fig. 27.2, to the initial condition.

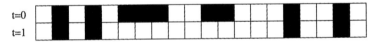

There is one subtlety associated with determining the value of the cells at $t = 1$. What happens to the cells on the edges? They do not have the full complement of neighbors, since they are on the boundary. What is usually done is to use periodic boundary conditions. This means that the row of cells is viewed as wrapping around itself, so that the neighbor of the right-most cell is the left-most cell. This is illustrated in Fig. 27.5. For example, the left-most cell has a neighborhood of □□■ and thus becomes □ at $t = 1$. And the right-most cell has a neighborhood of ■□□, and so it also maps to □ at $t = 1$.

Fig. 27.5 A schematic illustration of periodic boundary conditions.

Once we have the full configuration at $t = 1$, we iterate the process. We apply the same rule to the configuration at $t = 1$ to get a new configuration. We then apply the rule again and get another configuration, and so on. The result of this process is shown in Fig. 27.6. One can see that the CA quickly reaches a fixed point. After $t = 1$ the configuration no longer changes.

Let me summarize this example and, at the risk of being pedantic, clarify some terminology. Cellular automata are a type of dynamical system in which a grid of discrete variables which are updated at discrete time intervals according to a local, deterministic rule. Automata is

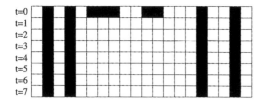

Fig. 27.6 The result of iteratively applying the rule, Fig. 27.2, to the initial condition. The CA quickly reaches a fixed point. After $t = 1$ the configurations do not change.

the plural form of automaton. So the phrase "cellular automata" refers to the collection of such models, just as "iterated functions" refers collectively to the set of functions.[1] A particular iterated function is specified by stating what that function is—e.g., $f(x) = x^2$. Similarly, a particular CA is specified by stating what the rule is, such as the rule given in Fig. 27.2.

In order to iterate a function, one needs an initial condition. The initial condition plus the function determines the orbit. Similarly, to determine the dynamics of the CA, one needs an initial condition. For CAs, the initial condition is the value of an entire row of cells, as in Fig. 27.1. The CA rule and the initial condition determine the orbit of the dynamical system. In this case, the orbit is a sequence of rows of discrete variables. This is almost always drawn with time moving down the page, as in Fig. 27.6. When iterating functions, one often arbitrarily chooses a few numbers, iterates, and sees what happens. For CAs one can follow a similar approach by choosing a long initial condition that is random. I.e., choose a long sequence of black and white squares by tossing a coin repeatedly to determine the color of each square. Then iterate, and see what happens.

The result of doing this is shown in Fig. 27.7. The initial condition is a randomly generated sequence of 200 black and white boxes. The rule in Fig. 27.2 is then applied to this sequence 200 times. The result is plotted in a manner similar to that of Fig. 27.6. Figure 27.7 is just like a time series plot for an iterated function, except that time goes down the page instead of a cross, and instead of plotting a single value one displays an entire row of cells. It is somewhat difficult to resolve individual cells in this figure, but the overall behavior is clear: the system reaches a fixed point after one iteration. The fixed point manifests itself as vertical stripes on Fig. 27.7. An unchanging configuration is simply copied verbatim down the page, yielding vertical stripes.

Figures like Fig. 27.7 are known as space-time diagrams for CAs. The term space-time diagram was originally used in Einstein's special theory of relativity, where diagrams are needed to visualize the trajectory of a particle as it moves through both space and time. Much of the early work on CAs was done by physicists who were familiar with space-time diagrams from relativity, so presumably the term occurred to them when they were looking for a name for figures such as Fig. 27.7. In any event, the term "space-time" diagram is now the standard term for such figures.

Before considering other CAs, a few remarks on the terms "cellular" and "automaton". Why was this mathematical model given this name?

[1] The acronym CA is usually taken to be singular: CA = cellular automaton. The acronym is pluralized by adding an s: CAs = cellular automata.

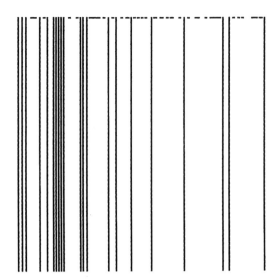

Fig. 27.7 The result of iteratively applying the rule, Fig. 27.2, to a randomly generated initial condition. The CA quickly reaches a fixed point. After $t = 1$ the configurations do not change.

The cellular part is perhaps straightforward—the variables are spread out across space, but consist of discrete boxes or cells. "Automaton" is a general term for a self-governing or independent machine or computing device. An automaton is autonomous in that it just does its thing without needing commands or coordination from a central controller. Thus, CAs can be viewed as a model of distributed computation. Each cell is updated depending only on the values of the cells in its local neighborhood. There is no other communication between the cells other than this. We shall see that despite this locality, CAs are capable of producing large-scale patterns and structures.

27.2 Surveying One-Dimensional Cellular Automata

Having analyzed one CA, Fig. 27.2, let us see what sort of behavior we find for other rules. As usual, we will be mainly interested in global behavior. Are there any fixed points? Any attractors? And is there chaos? We shall see that there is a range of different CA behaviors and that these are more difficult to categorize than the dynamical behaviors we have encountered in our study of iterated functions.

For the first example in this section we consider the rule shown at the top of Fig. 27.8. There is a standard numbering scheme for CAs, and this rule happens to be known as rule 32. The details of the numbering convention are not important, but if you are curious I explain it in Section 27.5, below. The space-time diagram for rule 32 is shown in Fig. 27.8. One can see that quite quickly the pattern "dies off". After three time steps all cells are white. Note that this outcome is not immediately obvious. Looking at the rule, Fig. 27.8, we see that there is one output that is non-white, as was the case for the rule for our initial example, Fig. 27.2. Nevertheless, the dynamics of rule 32 are such that the configuration very rapidly converts to all white. This happens regardless of the initial condition used.

Fig. 27.8 Space-time diagram for cellular automata rule 32. A random initial condition of 100 sites was used. After three time-steps all cells are white.

For our next example, consider the rule shown at the top of Fig. 27.9. In the figure I also show the result of iterating a random initial condition. The CA configuration quickly settles into a pattern, but unlike Fig. 27.8 the pattern is not static. The configuration at a given time is the same as the previous time, except all cells are shifted to the left by one unit. Note that in Fig. 27.9 it is easy to see the effect of the periodic boundary conditions—the stripes reappear on the right of the space-time diagram immediately after disappearing from the left.

Fig. 27.9 Space-time diagram for cellular automata rule 46. The rule is shown at the top of the figure. A random initial condition of 200 sites was iterated for 200 time steps.

The space-time behavior of Fig. 27.9 is stable, in the sense that different randomly generated initial conditions produce statistically similar results. The space-time diagram always shows tilted stripes, although the exact distribution of those stripes depends on the particular initial condition. This space-time behavior is simple and periodic. The stripes repeatedly wrap around the lattice. The exact periodicity depends on the size of the system. But the main point is that this sort of behavior is generally analogous to the stable, periodic behavior that we have seen for iterated functions.

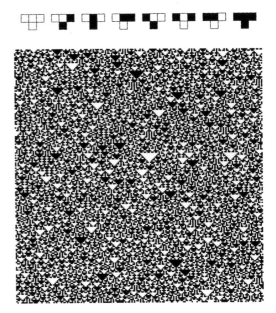

Fig. 27.10 Space-time diagram for cellular automata rule 150. The rule is shown at the top of the figure. A random initial condition of 200 sites was iterated for 200 time steps.

For our next example, consider the rule shown on the top of Fig. 27.10. Applying this rule to a random initial condition yields the space-time diagram shown in Fig. 27.10. It is immediately apparent that this behavior is very different from the previous two space-time diagrams we have looked at. How can we describe or characterize this behavior? The CA most certainly does not appear to be heading toward any sort of fixed point. Nor do the configurations appear to be periodic. It thus seems reasonable to call this CA chaotic. But what does it mean for a system such as a CA to be chaotic? It turns out that a rigorous definition for the chaotic behavior of a CA is a subtle matter. I will return to this issue briefly in section 27.3. For now, however, let us take Fig. 27.10 as being intuitively chaotic and continue with our survey of CAs.

The last example of this section is the rule shown in Fig. 27.11. The space-time diagram for this CA, seeded with a randomly generated initial condition, is shown in Fig. 27.11. How can we describe this behavior? It is certainly not periodic, so it is not like the space-time behavior seen in Fig. 27.9. It appears to be aperiodic, but not in the same way as the previous example, Fig. 27.10, which looks like a uniform assortment of black and white triangles. If one squints a little, Fig. 27.10 resembles foam or perhaps something that would be produced by a printer gone haywire.

In contrast, in Fig. 27.11 there is clearly some structure amid the chaos; there are periodic regions, mostly chaotic regions, and a complicated structure of boundaries and edges between these different regions. It appears that the CA does not settle in to either a chaotic state or a periodic state. Further evidence for this can be seen in Fig. 27.12, the space-time diagram for the same rule, but for a system with 500 sites iterated for 500 time steps. One can see that the complicated structures

Fig. 27.11 Space-time diagram for cellular automata rule 110. The rule is shown at the top of the figure. A random initial condition of 200 sites was iterated for 200 time steps. A larger space-time diagram is shown in Fig. 27.12.

persist. This CA, known as rule 110, has been much studied. There have been hundreds of scientific papers written about various properties of this rule. The space-time pattern shown in Fig. 27.11 is often described as being complex, in contrast to being chaotic. The pattern clearly has elements of chaos in it, in that it is aperiodic and in a sense unpredictable. However, it is also patterned or structured in a way that suggests something beyond simple chaos.

27.3   Classifying and Characterizing CA Behavior

What types of dynamical behaviors are CAs capable of? A review of the examples we have looked at thus far suggests a possible categorization of CAs into four classes.

(1) The first class consists of those CAs whose space-time diagrams quickly turn all white or all black. Rule 32, Fig. 27.8, is an example of this sort of behavior. For this class of CAs the randomness of the initial condition is quickly forgotten and the configuration ends up at a very simple fixed point.

(2) The second class consists of CAs whose configurations freeze into some sort of regular, periodic pattern. The space-time diagram for rule 46, Fig. 27.9, is an example of behavior in this class. Note that while the pattern is quite regular, some vestiges of the random initial condition remain, as reflected in the distribution of the stripes. If a different random initial condition is used the resultant space-time diagram will exhibit similar but not identical stripes.

Fig. 27.12 The space-time diagram for cellular automaton rule 110. A complex pattern can be seen. The figure is 500 × 500 sites.

The exact number and location of the stripes depends on the initial condition.

(3) The third class consists of rules like that shown in Fig. 27.10. These rules can be said to be chaotic. They do not repeat and their long-term behavior is homogeneous. For example, in Fig. 27.10 there are many different black and white triangles, but there are no boundaries dividing distinct regions. The space-time diagram looks like one large sea of boiling triangles. These CAs can be thought of re-membering their random initial condition. The process of iteration does not induce much, if any, order in the configuration.

(4) The fourth class consists of rules which are aperiodic but are more structured or organized than the chaotic rules. An example of such a CA is Fig. 27.11. These rules are often said to be complex and not necessarily chaotic.

This classification scheme, introduced by Stephen Wolfram in a highly influential 1984 paper (Wolfram, 1984), captures some qualitative, general distinctions among different types of CA behavior. However, in the subsequent decades it has generally be recognized that this classification is rather coarse, and that much more can be said, both quantitatively and qualitatively, about the behavior seen in CA's space-time diagrams. The Wolfram classification scheme is subjective, in the sense that it relies mainly on visual inspection of space-time diagrams. In particular, it is not clear how to objectively distinguish between class 3 and class 4 behavior.

A full discussion of characterizing, quantifying, and categorizing CA behavior beyond the scope of this text. Indeed, some aspects of this are topics of current research. Nevertheless, here are a few general comments on these issues that will not take us too far afield and which perhaps will shed light on some of the themes of the text.

First, let us think about chaos. In Section 9.3 I defined a dynamical system as being chaotic if it possesses all of the following properties:

(1) The dynamical rule is deterministic.
(2) The orbits are aperiodic.
(3) The orbits are bounded.
(4) The dynamical system has sensitive dependence on initial conditions.

This definition was developed in the context of iterated functions. How might it apply to CAs? Cellular automata are certainly deterministic. The rule that is used to iterate the configurations are a deterministic function of the cell's local neighborhood. The orbits of a CA are bounded; since there are only two possible values for each site there is no infinity for them to move toward. The notion of aperiodicity applies to CAs as expected; we say an orbit is aperiodic if the configuration does not repeat.[2] But what about SDIC? How might this phenomenon manifest itself in CAs? Recall that an iterated function had SDIC if two initial conditions that started off very close got far apart. We can do the same thing with CAs. Choose two different initial conditions that vary only by a single cell, and iterate them forward. Compare the two space-time diagrams. If they are significantly different, then we would say that the rule is chaotic. There are some subtleties associated with actually carrying out this procedure, but hopefully the general idea is clear.

But what about behavior like that seen in Fig. 27.12? There is more than just chaos going on here. Chaos and order seem to be combined in an interesting way. Such behavior is often called complex. But how could complexity be defined? It seems an inherently subjective quality: different observers likely will not agree on which patterns are the most complex or interesting or structured. And for that matter, what is a pattern? How can we be certain we are seeing a pattern? And how can we discover or come to know patterns that we have not seen before? These are rich and vexing questions that are topics of current research.[3]

[2] Strictly speaking it is impossible for a CA to be truly aperiodic. Since there are a finite number of possible configurations the CA will eventually return to a configuration it has been in previously. To get around this, one studies larger and larger systems as a way to approximate a system with an infinite number of sites.

[3] An accessible and engaging discussion of these questions is the essay "Is Anything Ever New? Considering Emergence", by James Crutchfield (1994).

Let it suffice to say that CAs produce space-time diagrams that appear structured in a way that combines elements of both chaos and regularity. Many refer to this sort of behavior as *complex*.

Regardless of how one defines complexity, I hope it is clear that CAs provide another example of a simple, deterministic dynamical system that produces apparently random behavior. Chaos for CAs looks a little different than it did for the iterated functions we studied earlier in the book. But the basic message is the same. In addition, CAs show us yet again that simple, deterministic systems can exhibit behavior that is intricate or complicated or complex. Finally, note that it is difficult to predict the behavior of a CA simply by looking at the rule. If you are handed a CA rule, it is usually not obvious what the resultant space-time diagram will look like. Will it be chaotic? Will all the cells turn white? The best way to figure this out is to start iterating and see what happens. This is similar to the situation with the logistic equation. It is not obvious at first blush that $f(x) = rx(1 - x)$ is chaotic if $r = 3.7$ but is periodic if $r = 3.835$. But it is not too difficult to figure this out if one iterates a few initial conditions several hundred times on a computer. This is a common feature of dynamical systems. Their long-term dynamical behavior is difficult to deduce from the equation or rule, and so often the best strategy is to start iterating with a computer and see what happens.

27.4 Behavior of CAs Using a Single-Cell Seed

Thus far we have considered what happens when we use a random sequence as the initial condition for a CA. It is also interesting and fun to look at the effects of applying a CA rule to a configuration that begins with a single black cell. Not surprisingly, we will tend to see simpler and more ordered patterns in the space-time diagram, reflecting the fact that the initial condition is just about as simple as can be.

Let us first investigate rule 46, shown in Fig. 27.9. We have already seen that a random initial configuration yields a space-time diagram that is a series of stripes. Not surprisingly, starting with a single black cell yields a single diagonal stripe. This can be seen in Fig. 27.13. The initial condition of a single black cell leads to a stripe that moves right to left. When the stripe reaches the end of the grid on the left, it wraps around to the right due to the periodic boundary conditions.

Let us try a few other rules with a single-cell seed and see what happens. First, consider rule 50, shown in Fig. 27.14. Applying this rule to a single black cell yields the pattern shown in Fig. 27.14. The result is a checkerboard pattern that expands to both the right and the left. Looking at the rule it is not too hard to see where this pattern comes from. If you do not see this, you might want to take a moment and sketch out the first few iterations of the CA rule, starting with a configuration consisting of a single black square. One way to think of rule

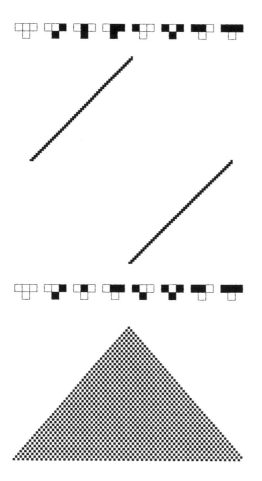

Fig. 27.13 CA rule 46 and the result of iteratively applying that rule to an initial condition consisting of a single black site. The stripe moves to the left, reaches the boundary of the CA, and then wraps around to the right due to the periodic boundary conditions. This CA was applied to a random initial condition in Fig. 27.9.

Fig. 27.14 CA rule 50 and the result of iteratively applying that rule to an initial condition consisting of a single black site.

50, then, is that it is a concise set of instructions for drawing a checkerboard triangle, starting at the point of the triangle and moving down the page. Drawing such a triangle is not a difficult task—one certainly does not need a CA to do it for you. But it is nevertheless perhaps of some interest that Rule 50 has formalized this checkerboard-triangle-making process.

As our final example for this section will consider rule 22, shown in Fig. 27.15. If we use this rule to iterate a configuration that initially has only one black cell, what do you think will happen? The result of doing this is shown in Fig. 27.15. Like magic, a Sierpiński triangle has appeared. This result is, I think, rather surprising. I certainly do not look at rule 22 and immediately suspect a Sierpiński triangle. This way of making a Sierpiński triangle suggests some interesting things about both CAs and the Sierpiński triangle. This is yet another example of a simple dynamical system producing a complicated or surprising result. The CA makes the Sierpiński triangle via a rather different procedure than the one we first used to construct the triangle. Originally, we started with a full triangle and successively removed inner triangles from each triangle. This procedure is illustrated in Fig. 16.4.

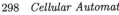

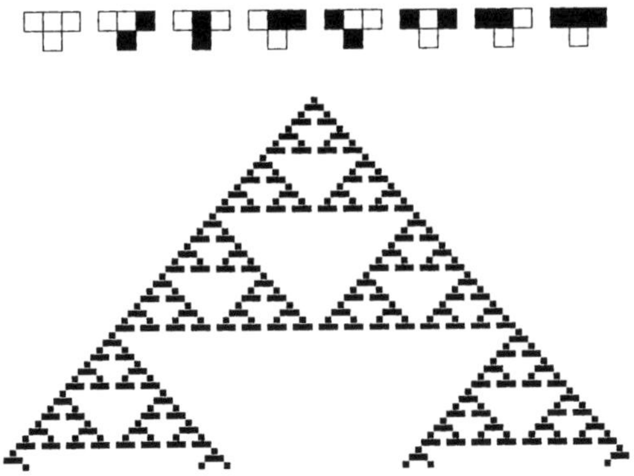

Fig. 27.15 CA rule 22 and the result of iteratively applying this rule to an initial condition consisting of a single black site.

The CA, however, produces the Sierpiński triangle top to bottom, one line at a time. At first blush, this seems like a complicated feat. Imagine that you were tasked with producing a Sierpiński triangle on a piece of graph paper and you could only write one horizontal line at a time. This would not be too difficult, but you would likely find yourself scanning up and down and to the left and the right to make sure that the triangles were the correct size. However, the CA does not have the benefit of hindsight. All the CA can do is look at the current row and use this information, and this information only, to determine the next line. Moreover, CAs are local. The next value of a cell is determined solely by that cell's current value and the value of its two nearest neighbors. Nevertheless, CAs are capable of producing "global" patterns—patterns such as the Sierpiński triangle, where there are correlations or structure at all scales.

This method of producing a Sierpiński triangle also gives us another way of thinking about fractals. Most of the classic fractals in Chapter 16 were produced by an iterative process that begins with a complete shape and then alters it by removing or bending portions of that shape. Then in Chapter 17 we saw that the chaos game—a random dynamical system—generates the Sierpiński triangle, and I suggested that this was a general result. The collage theorem (see Section 17.6) says that almost any fractal can be generated by a random process of some sort. Our study of CAs indicates that there is a very different sort of simple procedure—one which involves generating the pattern line-by-line and not working with a full shape—that can make fractals. Throughout this text we have seen several quite different mechanisms that are capable of producing fractals. This might explain fractals' ubiquity: although they may appear complicated, they are a generic form in the sense that they result from many different types of simple processes.

27.5   CA Naming Conventions

You may have been wondering where I was getting the rule numbers for the CAs in the previous sections. Why is the rule depicted in Fig. 27.11

known as rule 110? The answer is fairly simple, but I have delayed discussing it until now, since it has nothing to do with the main thrust of this chapter, which is looking at CAs as an example of a dynamical system. But the naming scheme is an interesting application of binary arithmetic, so I think it is worth going over. If you want, you can skip this section; what I am about discuss has absolutely no bearing on what follows in subsequent chapters.

The standard naming scheme for CAs works as follows. Specifying a particular CA rule requires specifying the output values for the eight possible neighborhoods. These outputs have been depicted as white or black, but we can just as well use the numbers 0 and 1, respectively. Since the eight neighborhoods are listed in standard order, there is not necessarily a need to show the neighborhoods when specifying the rule. For example, we could write rule 110, shown on the top of Fig. 27.11, as

$$\square\blacksquare\blacksquare\blacksquare\square\blacksquare\blacksquare\square \ . \tag{27.1}$$

Or, using 0 and 1 instead of $\square$ and $\blacksquare$, rule 110 is:

$$01110110 \ . \tag{27.2}$$

The last step is to convert the string of 0's and 1's into a number. To do this, one views the sequence of Eq. 27.2 as a binary number and converts it to base-10.

The binary, or base-2, number system was introduced in Section 21.5. In binary, digits correspond to 1, 2, 4, 8, and so on. In the more familiar base-10 system, digits correspond to 1, 10, 100, etc. For example, the number 613 in base-10 is:

$$613 = (6 \times 10^2) + (1 \times 10^1) + (3 \times 10^0) = 600 + 10 + 3 \ . \tag{27.3}$$

Similarly, the binary number 101 is:

$$101 = (1 \times 2^2) + (0 \times 2^1) + (3 \times 2^0) = 4 + 0 + 1 = 5 \ . \tag{27.4}$$

In other words, 101 in binary is equal to 5 in base-10.

We are now in a position to convert the rule, Eq. (27.2), into a base-10 number. The idea is to read the 0's and 1's in Eq. (27.2) left to right, interpret them as binary digits, and then convert to base-10.

$$
\begin{aligned}
01110110 &= \ (0 \times 2^0) + (1 \times 2^1) + (1 \times 2^2) \\
&\quad + (1 \times 2^3) + (0 \times 2^4) + (1 \times 2^5) \\
&\quad + (1 \times 2^6) + (0 \times 2^7) \tag{27.5} \\
&= \ 2 + 4 + 8 + 32 + 64 \tag{27.6} \\
&= \ 110 \ . \tag{27.7}
\end{aligned}
$$

Thus, the rule of Fig. 27.11 or Eq. (27.1) is known as rule 110.

Let us do one more example. The rule in Fig. 27.10 is:

$$\square\blacksquare\blacksquare\square\blacksquare\square\square\blacksquare \ . \tag{27.8}$$

Converting, this yields

$$01101001 = (0 \times 2^0) + (1 \times 2^1) + (1 \times 2^2)$$
$$+ (0 \times 2^3) + (1 \times 2^4) + (0 \times 2^5)$$
$$+ (0 \times 2^6) + (1 \times 2^7) \tag{27.9}$$
$$= 2 + 4 + 16 + 128 \tag{27.10}$$
$$= 150 . \tag{27.11}$$

So this rule is known as rule 150.

There is nothing meaningful about the number of the rule. Rule 150 just means that it is the 150th rule in this naming scheme. It does not mean that the rule has anything to do with "150-ness." Similarly, I do not expect the sixth student on my class list to have six toes or be born on June 6. As noted above, this naming scheme is entirely standard. Books, articles, and websites often refer to CAs by number without explicitly listing the outputs of the rule.

27.6   Other Types of CAs

There are many different sorts of CAs. The type we have been considering in this chapter are known as **elementary cellular automata**, usually abbreviated as ECAs. ECAs have the following properties: they have only two states, black or white; the configurations are one dimensional; and the neighborhood used when determining the next value of a cell is three cells wide. There are thus eight different neighborhoods. As we have seen, specifying a rule requires specifying the output for each of these eight particular neighborhoods, equivalent to specifying a sequence of eight 0's and 1's, as in Eq. (27.1). There are $2^8 = 256$ such sequences.[4] This is a relatively small number. Although we will not do it here, it is possible to use a computer to investigate all 256 ECAs and compare and contrast their properties.

ECAs are not the end of the CA story. It is possible to have CAs with larger neighborhoods. The neighborhood size is specified by stating how far to the left and right the neighborhood extends. This quantity is known as the CA's **radius**, commonly denoted by r. For ECAs the radius is 1, since the neighborhood for a cell consists of the cell itself and the sites one to the left and one to the right for a total neighborhood size of three sites. A radius-2 neighborhood is illustrated in Fig. 27.16.

The neighborhood for an $r = 2$ CA has five sites, so there are thus $2^5 = 32$ different particular neighborhoods. So to specify a rule, we need to specify the outputs for all thirty-two of these possibilities, just as we had to specify the output for all eight possibilities for the ECAs. For example, one possible $r = 2$ rule is:

$$01101110010000100110010111001110 0 . \tag{27.12}$$

This is, of course, just one among a vast many of possible $r = 2$ rules. How many such rules are there? The answer is 2^{32}. This is a very large

[4]To see this, note the flowing. There are two ways to have one symbol (0 or 1), four ways to have two symbols (00, 01, 10, 11), eight ways to have three symbols, and so on. In general, there are 2^n possible sequences of zeros and ones of length n.

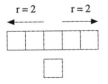

Fig. 27.16 An illustration of the neighborhood for a radius-2 CA.

number:

$$2^{32} = 4294967296 \approx 4.3 \times 10^9. \tag{27.13}$$

There are thus almost 4.3 billion different CAs with radius 2. If you took 1 second to study each such CA it would take over 136 years to get through them all. Thus, it is not feasible to exhaustively study radius-2 CAs. Instead, one can only study a subset of them. Radius-2 CAs are a relatively simple type of mathematical model, yet it would take several lifetimes to explore them all.

There are additional variations possible on the types of CAs we have been discussing. So far we have considered CAs where each cell can take on one of two states: black and white, or, equivalently, 0 and 1. Instead, one could allow the states to have more than two different values. The number of states is usually denoted by K. For example, if $K = 4$, one could think of the states as having four different colors or numerical values. For larger K, the number of different CAs becomes astronomically large. For example, consider a $r = 2$, $K = 3$ CA. There are five cells in the neighborhood, as indicated in Fig. 27.16. Now, however, each of these cells can take on three different values. So the number of distinct neighborhoods is $3^5 = 243$. So, specifying a rule requires specifying the outputs for these 243 different neighborhoods. And each output can take on three different values. Thus, instead of a sequence of thirty-two 0's and 1's, as was the case for the $k = 2$ radius-2 CA in Eq. (27.12), a rule consists of a sequence of 243 0's, 1's, and 2's. There is a mind-bogglingly large number of such sequences:

$$3^{243} \approx 8.7 \times 10^{115} . \tag{27.14}$$

It is hard to know how to even begin to think about this number. There have been around 10^{17} seconds since the big bang. So 10^{115} is unthinkably large. In general, for radius-r, K-state CAs the number of possible CAs is:

$$\text{Number of rules} = K^{K^{2r+1}} . \tag{27.15}$$

This number grows extremely quickly as r and K are increased.

Finally, another variation on the CAs we have considered in this chapter is to increase their dimension. The CAs we have looked at have been one-dimensional, in the sense that their configurations are a one-dimensional sequence of cells. The resultant space-time diagram is two-dimensional; the horizontal direction corresponds to space while the vertical direction shows the time evolution of the CA. However, there are also two-dimensional CAs. Here, a configuration is a two-dimensional grid or lattice of cells. The CA rule determines the next value of a cell based on the value of cells in the neighborhood. The result is a sequence of two-dimensional grids of cells. For the most part, one-dimensional CAs are used as quite abstract models, in the sense that they are not taken to correspond, even loosely, to phenomena or processes in the physical or biological realm. Two-dimensional CAs, however, are often used to model particular phenomena in biology and physics.

Further Reading

This chapter is just the tip of the iceberg. A great deal has been written about the mathematical properties of cellular automata and their applications in physics, biology, and other areas of science. Daniel Kaplan and Leon Glass's *Understanding Nonlinear Dynamics* (1995) and Heinz-Otto Peitgen, Hartmut Jürgens, and Dietmar Saupe's *Chaos and Fractals* (1992) each have chapters on CAs. Although the level of mathematics in these books is more advanced than in this one, they nevertheless should be accessible to most readers. Melanie Mitchell's *Complexity: A Guided Tour* (2009), has an excellent, mostly non-technical discussion of CAs. I also recommend the chapter on CAs in Nino Boccara's *Modeling Complex Systems* (2004).

Stephan Wolfram's *New Kind of Science* (2002) is largely focused on CAs. This book garnered considerable attention and some accolades from the popular press. However, in the science and mathematics communities, reviews have been mixed, at best. I recommend not relying on this book as your main reference for CAs, as I think it does not provide a balanced overview of current research in CAs and related fields.

Exercises

(27.1) Consider rule 250, shown in Fig. 27.17. Starting with a single black cell, iterate using rule 250 for twenty or so time-steps.

Fig. 27.17 CA rule 250.

(27.2) Consider rule 250, shown in Fig. 27.17. Starting with a random initial configuration of sixteen cells, iterate using rule 250 for ten time-steps.

(27.3) Consider rule 182, shown in Fig. 27.18. Starting with a single black cell, iterate using rule 192 for twenty or so time-steps.

(27.4) Find an initial configuration of six sites that does not lead to an all-white state when iterated with rule 32, Fig. 27.8

(27.5) Is it possible to find a seven-site configuration that does not turn all white when iterated by rule 32, Fig. 27.8? Why or why not?

Fig. 27.18 CA rule 182.

(27.6) Consider rule 182, shown in Fig. 27.18. Starting with a random initial configuration of sixteen cells, iterate using rule 182 for ten time-steps.

(27.7) Write down the rule (I.e., the series of outputs as in Fig. 27.2) for:

  (a) Rule 0

  (b) Rule 3

  (c) Rule 133

  (d) Rule 200

(27.8) How many CAs are there if the number of states $K = 3$ and the radius $r = 3$?

(27.9) How many CAs are there if the number of states $K = 2$ and the radius $r = 4$?

Introduction to Differential Equations

28

Thus far in this book we have looked exclusively at systems which change discretely, in jumps from one time interval to the next. In this chapter I introduce dynamical systems that change continuously, where the variable of interest—perhaps the temperature of a cup of coffee or the position of a ball that is thrown through the air—has a value at every instant. Dynamical systems that describe this sort of change are known as differential equations. After introducing differential equations in this chapter, in the subsequent three chapters we will explore their dynamical properties in one, two, and three dimensions.

28.1 Continuous Change

Before beginning our foray into continuous systems, it will be helpful to quickly review discrete dynamical systems. Consider yet again the logistic equation,

$$f(x) = rx(1 - x). \tag{28.1}$$

Introduced in Chapter 7, the logistic equation is a very simple model of the dynamics of a population that has some limit to its growth. Iterating the logistic equation yields an orbit, or itinerary, which is a listing of the successive values of the function. For example, choosing $r = 3.9$ and a seed of $x_0 = 0.8$, one obtains the itinerary shown in Table 28.1. Recall that the population value x in the logistic equation is expressed as a fraction of the maximum possible population. So a population of 0.8 indicates that the population is 80% of its maximum value. The first fifteen iterates are shown in the time series plot of Fig. 28.1. We are by now quite familiar with such plots. Recall, however, that the iterates only have a value at discrete time steps. That is, iterating the equation gives the population at the first generation, the second generation, the third generation, and so on. It does not give the population at every instant of time. Thus, in Fig. 28.1, the values on the graph only make sense at integer values of t. That is, the population is 0.915 at $t = 2$ and 0.303 at $t = 3$. There is a dotted line connecting these two points on the figure. But this line is just a visual aid, designed to help make the plot easier to read. The straight line does not mean that the population passes through all intermediate values between 0.915 and 0.303.

Table **28.1** The orbit of the initial condition $x_0 = 0.8$ when iterated by $f(x) = 3.9x(1 - x)$.

x_0	0.8
x_1	0.624
x_2	0.915
x_3	0.303
x_4	0.824
x_5	0.566
x_6	0.958

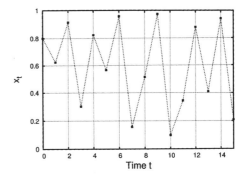

Fig. 28.1 A time series plot of the orbit of 0.8 for the logistic equation with $r = 3.9$. The value of the function is only defined at discrete time steps: $t = 0, 1, 2, \ldots$.

In contrast, consider a warm cup of coffee that slowly cools, eventually reaching room temperature. Here there is no natural notion of a generation or a time step. The temperature of the coffee is changing continuously. If, for example, the coffee started at 80 degrees Celsius and a little while later was at 40 degrees Celsius, it must have been the case that the coffee was, for at least an instant, at any and all temperatures between 80 and 40 degrees. The coffee cannot skip any temperatures as it cools.

This is illustrated in Fig. 28.2, which is what a plot of temperature T versus time t might look like for a cooling cup of coffee. Note that the plot is a smooth, continuous curve. There are no gaps or missing values as the coffee cools. The temperature is a function of time. This

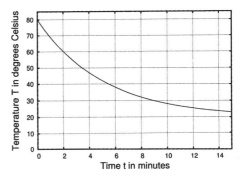

Fig. 28.2 A plot of the temperature of a cup of coffee as it cools. The temperature changes continuously. At every instant of time it has a well-defined temperature, and the temperature cannot skip any values as it cools.

function is written $T(t)$. Capital T is the temperature, and lowercase t is the time. The idea is that $T(t)$ is a function which takes as an input the time t and outputs the temperature T of the coffee at that time.

28.2   Instantaneous Rates of Change

How might we model this situation? A continuously changing temperature is a quite different sort of thing than a population that changes in discrete time-steps. To describe this situation mathematically we need to think carefully about rates of change. A rate tells us how fast something is changing. For example, if a tree is 10 meters tall today and is

40 meters two years from now, the average growth rate over those two years is 15 meters/year. We write this as:

$$\text{Average growth rate} \quad = \quad \frac{40 \text{ meters} - 10 \text{ meters}}{2 \text{ years}}$$

$$= \quad 15 \text{ meters/year} . \tag{28.2}$$

In general,

$$\text{Average growth rate} = \frac{\Delta h}{\Delta t} , \tag{28.3}$$

where h represents the height of the tree and Δ stands for "change in". Read aloud, the above equation is: The average growth rate equals the change in height h divided by the change in time t. I.e., the change in height divided by the time interval during which that change occurs.

The above equation gives the average growth during a time interval Δt. However, the growth rate need not be constant over this time interval. For example, the tree could grow very fast the first year and slower the second year. Suppose we want to know the *instantaneous* growth rate—how fast the tree is growing at a particular moment in time. This immediately poses a quandary. Equation (28.2) makes it clear that the growth rate is a statement about the difference in heights at two different times. But an instantaneous growth rate should be well defined at a particular instant, without needing to make reference to some secondary instant. We cannot use the same two instants in Eq. (28.3), because doing so would yield a Δt of zero, and dividing by zero is undefined. On the other hand, the essence of a growth rate—or any other rate for that matter—is that some change is taking place. Hence, it appears that the notion of a rate requires two separate instants. Is it possible to speak of an instantaneous rate at all?

It turns out that the answer to this last question is "yes": one can indeed give meaning to an instantaneous rate of change. Suppose we want to know the growth rate at the exact instant where $t = 1.0$. The trick is to use Eq. (28.3) to find the growth rate using the heights at two instants, perhaps $t = 1.0$ and 1.1. Then find the growth rate using two instants that are closer together: $t = 1.0$ and $t = 1.01$. We make the two instants closer and closer together, and in this way we can sneak up on an instantaneous growth rate.

This might seem a little bit like magic, or perhaps like cheating. We are essentially dividing by zero without actually dividing by zero. It may seem like lots could go wrong in this process, but it turns out that this can be done in a consistent and well-defined way. The branch of mathematics that makes all this possible is calculus—specifically differential calculus, which is the central topic of the first part of most multi-term calculus sequences. A full discussion of calculus is well beyond the scope of this text.[1] However, we will be able to introduce and understand a new class of dynamical systems without having to dig deeply into the mechanics of calculus.

In any event, the central thing to take from this discussion is that in order to describe and understand quantities that change continuously

[1] Chapter 11.2 of Flake (1999) is an accessible, short overview of calculus. By far the best book-length primer on calculus that I know of is *Calculus Made Easy* (1998). Written by Silvanus Thompson, this book was first published in 1910. (Its full title is *Calculus Made Easy: Being a Very-Simplest Introduction to those Beautiful Methods of Reckoning which are Generally Called by the Terrifying Names of the Differential Calculus and the Integral Calculus.*) In 1998 a version was published that includes some additional introductory material by Martin Gardner. But Thompson's prose, now a century old, remains a remarkably readable and engaging introduction to calculus.

we will need to make use of the instantaneous rate of change. Although the mathematical procedure for defining this quantity may be subtle, I hope that its interpretation is fairly intuitive. Before proceeding, however, there is some technical terminology and notation for instantaneous growth rates that will be essential to what follows.

To introduce this notation, let us continue with the example of the growing tree. Let $h(t)$ represent the height of the tree at a time t. Then the instantaneous growth rate is denoted

$$\text{instantaneous growth rate of } h(t) = \frac{dh}{dt} \,. \tag{28.4}$$

This equation is almost identical to Eq. (28.3). The only difference is that there are lowercase d's instead of capital deltas (Δ). The little d's are an indication that we are interested in tiny little changes in h and t. (Strictly speaking, the little changes are infinitesimal.) That said, it is best to think of $\frac{dh}{dt}$ not as a fraction but as a single symbol which stands for the instantaneous rate of change of the function $h(t)$. The instantaneous rate of change $\frac{dh}{dt}$ is known as the **derivative** of $h(t)$. The first term of a calculus sequence is concerned primarily with derivatives: their definition, techniques for calculating them, and their applications.

Finally, it is useful to know that there are alternative notations for the derivative. The derivative is also indicated with a symbol that looks like an apostrophe:

$$h'(t) = \frac{dh}{dt} = \text{instantaneous growth rate of } h(t) \,. \tag{28.5}$$

The symbol on the left-hand side of the above equation would be read "h prime of t". I will not use this notation, but it is quite common, and so it is good to know about it, as you might encounter it elsewhere. In physics and engineering, rates of change are sometimes denoted with a dot:

$$\dot{h}(t) = \frac{dh}{dt} = \text{instantaneous growth rate of } h(t) \,. \tag{28.6}$$

28.3 Approximately Solving a Differential Equation

We now use the derivative, i.e. the instantaneous rate of change, to model a quantity that changes continuously. To do so, we return to the example of a cooling cup of coffee. Let $T(t)$ represent the coffee's temperature, in degrees Celsius. We will assume that room temperature is 20 degrees C. The rate at which the coffee cools is proportional to how much warmer it is than room temperature. That is, if the coffee is 40 warmer than room temperature it will cool twice as fast as if it is 20 degrees warmer than room temperature. This seems reasonable; a hot beverage cools quite quickly at first, but then cools less quickly the closer it gets to room temperature. This result—that the rate of cooling is proportional to the temperature difference—is known as Newton's law of cooling.[2]

[2]The same Issac Newton from Chapter 8 who came up with the laws of motion and the universal law of gravitation.

This law can be written in terms of an equation involving the temperature and its derivative. If the coffee is cooling in a 20-degree room, this equation is:

$$\frac{dT}{dt} = -0.2(T(t) - 20) . \tag{28.7}$$

This an example of a **differential equation**, an equation that relates a function to its derivative. In this case, the function is $T(t)$. The solution to a differential equation is a function—here the temperature T as a function of time.

Let us analyze this equation piece by piece. The left-hand side of the equation, dT/dt is the rate at which the coffee cools. The term on the right-hand side, $(T(t) - 20)$, is the difference between the current temperature of the coffee $T(t)$ and room temperature, 20. The minus sign in the equation makes the rate of change negative; since the coffee is cooling, its temperature is decreasing. The number 0.2 is a factor that depends on the manner in which the coffee exchanges heat with the surroundings. The better the insulation, the more slowly the coffee will lose heat, and the smaller this number will be.[3]

What can we do with Eq. (28.7)? How can we use it to figure out the temperature of the coffee as a function of time so as to make a plot of $T(t)$ as in Fig. 28.2? In order to start, we need to know the initial temperature of the coffee. For this example, let us assume that the coffee is initially at 80 degrees. Thus, $T(0) = 80$. Now what? Well, Eq. (28.7) tells us the instantaneous rate of temperature change at $t = 0$. We plug in $T(0) = 80$ on the right-hand side and obtain

$$(\text{Rate of temp change at } t = 0) = -12 \, \frac{\text{C}}{\text{min}} . \tag{28.8}$$

Given this information, can we figure out $T(1)$, the temperature 1 minute later?

This may seem to be a straightforward task, since we know the current temperature and how fast the temperature is changing. But the catch is that -12 degrees C/min is the *instantaneous* rate of change. It is how fast the temperature is changing at *exactly* $t = 0$. Even a few seconds later this rate will be less. How can we cope with a rate of change that is always changing?

One coping mechanism is to momentarily ignore this changing rate of change and pretend that cup of coffee is cooling at -12 degrees C/min for the entire first minute. Given this, it then follows that at $t = 1$ the temperature of the coffee is $80 - 12 = 68$ degrees. Thus, $T(1)$ is approximately equal to 68 degrees. Next we seek the temperature at $t = 2$. We could just pretend that the temperature is still decreasing at 12 C/min. But we can do better, by going back to Eq. (28.7) and determining the cooling rate at $t = 1$. To do so, we plug in $T(1) = 68$ and obtain:

$$(\text{Rate of temp change at } t = 1) = -9.6 \, \frac{\text{C}}{\text{min}} . \tag{28.9}$$

[3]In this case I just made up this number to give a reasonable cooling rate. In a particular application one would need to infer this quantity by measuring the cooling rate. Also, for some simple physical situations these numbers have been tabulated, and so one might be able to use such a table to estimate the number. A clear introduction to mathematical models of heating and cooling can be found in Chapter 9 of Barnes and Fulford (2002).

Table 28.2 The temperature of a cup of coffee, approximated using Eq. (28.7) and a time interval of 1 minute.

Time t	Cooling Rate	Temp. $T(t)$
0	−12.00	80.00
1	−9.60	68.00
2	−7.68	58.40
3	−6.11	50.72
4	−4.91	44.58
5	−3.93	39.66

As expected, the coffee is cooling less quickly now. We now use this new rate to estimate the temperature at $t = 2$. As before, this is only an approximation. The coffee is not cooling at 9.6 C/min for the entire minute, as its rate of cooling is continually decreasing. However, we can pretend that this rate is constant and estimate that the temperature of the coffee after two minutes is $68 - 9.6 = 58.4$ C. We continue in this fashion, approximating the temperature of the coffee minute by minute. The result of doing this is shown in Table 28.2. You might want to take a moment and verify the temperature values for $t = 3, 4$ and 5.

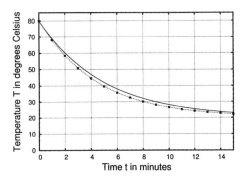

Fig. 28.3 A plot of the temperature $T(t)$ of a cup of coffee as it cools. The solid curve is the exact result, obtained using calculus. The squares are the approximate results obtained using Euler's method. The first several approximate values are listed in Table 28.2.

In Fig. 28.3 I have plotted the approximate results for the temperature of the coffee $T(t)$ listed in Table 28.2. (I continued determining approximate value out to $t = 15$ for the plot, but I did not include them in the table.) Also in the figure I have plotted as a solid curve the exact result for the temperature of the coffee as a function of time. I will say a little bit at the end of this chapter about how one can obtain an exact result such as this.

Figure 28.3 shows that our approximate result is pretty good—perhaps better than could be expected given the pretending. The rate of cooling is always decreasing, but we pretended that this rate was constant over each 1-second time interval. The result is that we overestimate how much the coffee cools off in each time interval. Accordingly, the approximate solution, the squares connected by dashed lines, is below the exact solution.

Can we improve our approximation? Yes. To do so, we will consider smaller time intervals. Let us use a time interval of 0.5 seconds. We will pretend that the rate of cooling is constant over this interval instead of pretending it is constant for a full second. This is clearly a better approximation, as the cooling rate changes less in half a second than it does in a full second. Thus, our approximation will be better. The price we have to pay, however, is that it takes more work to get the approximate answer.

Let us give it a try. Initially the temperature of the coffee is 80 degrees. We plug this in to Eq. (28.7) to determine the cooling rate. We have done this already in Eq. (28.8), where we found that the rate is -12 degrees C/min. The cooling rate is constantly changing, but we pretend that it is constant for 0.5 min, during which the change in the coffee temperature is

$$-12\frac{C}{\min} \times 0.5\min = -6 \min . \tag{28.10}$$

Thus, after 0.5 minutes, the temperature of the coffee is 74 degrees.

Next, we recalculate the cooling rate using Eq. (28.7) with $t = 0.5$ and $T(0.5) = 74$. Doing so, we get

$$(\text{Rate of temp change at } t = 0.5) = -10.8 \frac{C}{\min} . \tag{28.11}$$

We pretend that the cooling rate is constant during the next 0.5 minutes. This gives a change in temperature of -5.4, so the temperature of the coffee is now $74 - 5.4 = 68.6$. And in general,

$$\text{New temperature} = \text{old temp} + \left(\frac{dT(t)}{dt} \times (0.5)\right) . \tag{28.12}$$

Continuing, we find the temperature values shown in Table 28.3.

In Fig. 28.4 I have plotted the exact result for $T(t)$, the approximation for $T(t)$ using a time step of 1 minute, and the approximation using a time step of 0.5 minutes. As anticipated, the smaller time step leads to a more accurate result. We can get better approximations to the exact

Table 28.3 The temperature of a cup of coffee, approximated using Eq. (28.7) and a time interval of 0.5 minutes.

Time t	Cooling Rate	Temp. $T(t)$
0.0	−12.00	80.00
0.5	−10.80	74.00
1.0	−9.72	68.60
1.5	−8.75	63.74
2.0	−7.87	59.37
2.5	−7.09	55.43
3.0	−6.34	51.89

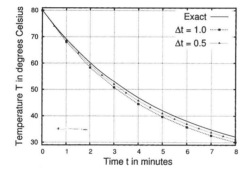

Fig. 28.4 A plot of the temperature $T(t)$ of a cup of coffee as it cools. The solid curve is the exact result, obtained using calculus. The squares are the approximate results obtained using Euler's method with a time step of one minute and the triangles are the results obtained using Euler's method with a time step of 0.5 minutes. A smaller time step leads to a more accurate approximation.

function by choosing smaller and smaller values for our time step. The approximation gets better, because the fiction that the rate is constant over the time interval becomes less and less of a lie as the time step is smaller. The smaller the time step, the less time the rate has to change, and so treating the rate as approximately constant is more justified. For example, it is definitely not true that you have grown at a constant rate your entire life, or even over one year. However, your growth rate is reasonably constant if measured over a month, and is essentially indistinguishable from constant if measured over a week or a day.

How small a time step is needed so that the approximation is reasonable? The answer depends on how fast the rate of growth is changing: one wants a time step to be small enough so that we can treat the growth as essentially constant over the time interval. In practice, one might choose smaller and smaller values for Δt until the solution does not change significantly. The main point is that by choosing smaller and smaller time steps, one can get a solution that is a better and better approximation to the exact solution.

28.4   Euler's Method

[4]Euler's method is named after Leonard Euler, a prolific Swiss mathematician and physicist in the 1700s. His last name is pronounced "oiler", not "youler".

The method just described is known as **Euler's method**.[4] Having seen an example in which we used Euler's method to approximate the solution to a differential equation, I will now describe Euler's method more generally. We will consider differential equations of the following form:

$$\frac{df}{dt} = F(f(t)) . \tag{28.13}$$

The picture here is that $f(t)$ is an unknown function—perhaps the temperature of a cooling cup of coffee. And $F(f(t))$ is some expression involving the unknown function $f(t)$. For example,

$$\frac{df}{dt} = f(t) - 613 + (f(t))^2 , \tag{28.14}$$

has the form of Eq. (28.13). When faced with a differential equation, the task is to find $f(t)$. That is, it is the entire function $f(t)$ that is unknown.

In order to solve for $f(t)$, one needs the initial condition—the value of $f(t)$ when $t = 0$. The differential equation then lets us figure out the rate of change of $f(t)$ by plugging in to the right-hand side of Eq. (28.13). So we determine how the function is changing and then use this to figure out the value of the function. But we need an initial value to start this process. This is similar to how we need an initial condition or seed when iterating a function.

In order to employ Euler's method to find $f(t)$, we also need to choose the step size Δt. Once we have an initial condition $f(0)$ and the step size, we can begin using Euler's method. Here is an outline of the procedure:

(1) Choose an initial condition $f(0)$ and a step size Δt. Set $t = 0$.

(2) Evaluate the rate of change of the function by plugging in to the right-hand side of the differential equation, Eq. (28.13). Call this rate of change $\frac{f(t)}{dt}$.

[5]Compare this with Eq. (28.12), which we repeated used to approximate the temperature of the cooling cup of coffee.

(3) Use this rate of change to figure out the next value for the function:[5]

$$f(t + \Delta t) = f(t) + \left(\frac{df(t)}{dt} \times \Delta t \right) . \tag{28.15}$$

(4) Increase t by Δt.

(5) Go to step 2.

Euler's method for approximating solutions to differential equations is, I think, quite logical. It follows from the idea that a differential equation is a relationship between a function and its rate of change, and thus we can utilize this relationship step by step to infer the value of a function.

Loosely speaking, a differential equation such as Eq. (28.7) or (28.13) is a set of directions. The equation is giving you directions to somewhere, but it does so indirectly, by telling you the rate of change of the function at every instant in time. Euler's method turns this indirect information—the rate of change—into a sequence of approximate values of the function.

To summarize, we began with a new mathematical entity—a differential equation describing an unknown, continuous function $f(t)$. A differential equation is a formula for $f(t)$'s rate of change. We then need to determine the function $f(t)$. We used Euler's method to convert the differential equation into a discrete, iterated function. The result is a time series of numerical values at discrete time intervals. An example is Table 28.3 or Fig. 28.4. However, the results from Euler's method must be interpreted differently from the time series we have studied in previous chapters. Time series from Euler's method are an approximation to a continuously varying function. For example, in Table 28.3 we see that the temperature is 80 at $t = 0$ and 74 at $t = 0.5$. This means that in between these two times the temperature must have passed through all values between 80 and 74. The temperature does not instantaneously jump from 80 to 74 but does so continuously, sliding through all intermediate temperature values.

28.5 Other Solution Methods

In Figs. 28.3 and 28.4 I plotted the exact solution in addition to the approximate solutions using Euler's method. Where does this exact solution come from? The short answer is that one needs the branch of mathematics known as calculus to solve differential equations exactly. Indeed, calculus was invented in part for just this purpose. Calculus is an incredibly powerful mathematical framework that relates functions to their rates of change. Calculus is one of the pillars of mathematics and it has widespread application in almost any scientific field.

Unfortunately, explaining where the exact solution to Eq. (28.7) comes from is beyond the scope of this book. I just do not see any way to explain it without calculus. However, knowing calculus is most definitely not needed in order to understand what a differential equation is and what its solution means. What *is* needed is an understanding of what differential equations are and how to interpret their solutions. I think that knowing how Euler's method works is helpful for understanding what solutions to differential equations mean, but to understand the dynamical systems in the next several chapters it is not necessary to become an expert at using Euler's method.

Finally, I should mention that Euler's method is not a particularly efficient technique for coming up with an approximate solution to a differential equation. Its main benefit is that it is simple to implement and fairly clear conceptually. There are a number of other approximation schemes that are more efficient, in the sense that they take less computing power to get a solution to a specified accuracy. The most common of these are a family of techniques known as Runge–Kutta methods. It turns out that most differential equations cannot be solved exactly even with calculus, so approximate methods are often a necessity, even for those who do know calculus. These approximate methods are almost always carried out with a computer.

<hr>

Exercises

(28.1) Verify the entries in Table 28.2 for $t = 3, 4,$ and 5.

(28.2) Verify the entries in Table 28.3 for $t = 1.5, 2.0, 2.5,$ and 3.0.

(28.3) Consider the following differential equation:

$$\frac{df}{dt} = 3 . \tag{28.16}$$

Let the initial condition be $f(0) = 2$.

 (a) Use Euler's method with $\Delta t = 2.0$ to determine an approximate solution to the differential equation.

 (b) Use Euler's method with $\Delta t = 1.0$ to determine an approximate solution to the differential equation.

 (c) Does your solution depend on Δt? Why or why not?

 (d) Determine a formula for the solution $f(t)$ to this differential equation.

(28.4) Repeat Exercise 28.3, but use the differential equation:

$$\frac{df}{dt} = 0 . \tag{28.17}$$

(28.5) Repeat Exercise 28.3, but use the differential equation:

$$\frac{df}{dt} = -2 . \tag{28.18}$$

(28.6) Consider the following differential equation:

$$\frac{df}{dt} = 2f(t) . \tag{28.19}$$

Let the initial condition be $f(0) = 2$.

 (a) Use Euler's method with $\Delta t = 2.0$ to determine an approximate solution to the differential equation.

 (b) Use Euler's method with $\Delta t = 1.0$ to determine an approximate solution to the differential equation.

 (c) What is the long-term behavior of solutions to this differential equation? Explain.

 (d) What is the solution to the differential equation if $f(0) = 0$?

(28.7) Consider the following differential equation:

$$\frac{df}{dt} = 2f(x)(1 - f(x)) . \tag{28.20}$$

This can be thought of as the differential equation version the logistic equation. The population in instance varies continuously instead of in discrete steps. For each of the following initial conditions $f(0)$ and time steps Δt, use Euler's method to determine an approximate solution to the differential equation.

 (a) $f(0) = 0.5$ and $\Delta t = 1.0$
 (b) $f(0) = 0.5$ and $\Delta t = 0.5$
 (c) $f(0) = 2.0$ and $\Delta t = 1.0$
 (d) $f(0) = 2.0$ and $\Delta t = 0.0$

How would you describe the global behavior of this equation. Are there any attractors? (This exercise can be somewhat tedious to do by hand. You might want to use a spreadsheet to automate the calculations.)

(28.8) Repeat Exercise 28.7, but use the differential equation

$$\frac{df}{dt} = 4f(x)(1 - f(x)) . \tag{28.21}$$

Is the behavior qualitatively different from that which you found in Exercise 28.7? The logistic equation with $r = 4$ is chaotic. Do you see chaotic behavior here? Why or why not?

One-Dimensional Differential Equations

29

In the previous chapter I introduced differential equations. These dynamical systems are used to model phenomena that change continuously. In this short chapter we will take a look at some one-dimensional differential equations. In so doing we will gain a sense of the types of dynamical behavior that are (and are not) possible for these types of dynamical systems. In the subsequent two chapters we will investigate two-dimensional and then three-dimensional differential equations.

29.1   The Continuous Logistic Equation

Throughout this book I have repeatedly used the logistic equation as one of the central examples of a chaotic system:

$$f(x) = rx(1 - x) \,. \tag{29.1}$$

Iterating this equation gives us times series that show a range of behaviors depending on the value of the parameter r. We now analyze the **logistic differential equation**:

$$\frac{dP}{dt} = rP(t)\left(1 - \frac{P(t)}{K}\right) \,, \tag{29.2}$$

where $P(t)$ is the population, r is a parameter related to the growth rate of the population, and K is, for reasons which we shall see below, known as the carrying capacity. What are the solutions to this differential equation and how do they depend on the parameter r? Equations. (29.1) and (29.2) look similar and they have similar origins; both were constructed to be simple models of a population that has some limit to its growth. Does the differential equation Eq. (29.2) exhibit chaos like the discrete logistic function?

We begin our analysis by noting that Eq. (29.2) is of the following form:

$$\frac{dP}{dt} = F(P) \,. \tag{29.3}$$

In other words, dP/dt, the rate of change of the population, depends entirely on P, the current value of the population. We can make this concrete by choosing values for the parameters. I will choose $r = 4$ and $K = 100$. Thus, Eq. (29.2) has the form:

$$\frac{dP}{dt} = 4P\left(1 - \frac{P}{100}\right). \tag{29.4}$$

Again, note that this equation says that if we know P at an instant we can determine dP/dt at that instant.

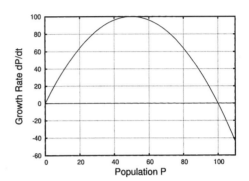

Fig. 29.1 A plot of $F(P)$, the right-hand side of Eq. (29.4). This shows how the growth rate $\frac{dP}{dt}$ of the population depends on $P(t)$, the value of the population.

The key to determining the behavior of $P(t)$ is to plot the right-hand side of Eq. (29.4). This is done in Fig. 29.1. We can learn a great deal about $P(t)$ by looking at this plot. First, notice that the growth rate dP/dt is positive if P is greater than zero and less than 100. This means that the population grows for any P value in the range 0 to 100. The growth rate dP/dt has a maximum when $P = 50$, so the population increases the most rapidly at $P = 50$. As P approaches 100 the growth rate decreases and eventually reaches zero. Thus the population will level off at 100.

When P is greater than 100, dP/dt is less than zero, and so the population is decreasing. A population that starts off larger than 100 will decrease, at first rapidly, and then less rapidly, until it levels off at 100. This analysis suggests that 100 is an attractor.

The above analysis gives us a qualitative, global picture of the differential equation. We expect an attracting fixed point at $P = 100$. Population values above 100 decrease, while population values below 100 increase. We can now use Euler's method to solve for $P(t)$. The $P(t)$ curves for four different initial conditions are shown in Fig. 29.2. As anticipated, all curves approach the fixed point at $P = 100$. The population increases for $P < 100$ and decreases for $P > 100$. Note that the lower-most curve, with $P(0) = 10$, increases the most rapidly when $P = 50$, as expected.

The dynamics of this differential equation are summarized with a phase line, shown in Fig. 29.3. This is similar to the phase lines that were introduced in Chapter 3. However in this case, since $P(t)$ is a continuous function and not a discrete time series, $P(t)$ does not move in jumps. Rather, it can only slide up and down the phase line; it cannot skip any values.

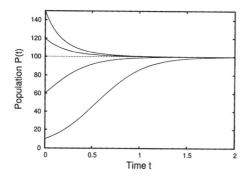

Fig. 29.2 A plot of $P(t)$, the population as a function of time, determined using Eq. (29.2) with $r = 4$ and $K = 100$. Shown are the solutions $P(t)$ for four different initial conditions, $P(0) = 150, 120, 60$, and 10. Note that all curves approach the attracting fixed point at $P = 100$.

Fig. 29.3 The phase line for the differential equation of Eq. (29.1). There is an attracting fixed point at $P = 100$.

It is possible to use algebra to solve for the fixed point at $P = 100$. A fixed point occurs when $P(t)$ does not change. This will be the case when

$$\frac{dP}{dt} = 0 .$$ (29.5)

Thus, we need to find the value(s) of P that make the above equation true. In contrast, for discrete iterated systems like the logistic equation, the equation for finding fixed points was

$$f(x) = x ,$$ (29.6)

first introduced in Section 3.4. The two fixed-point equations, Eq. (29.5) and Eq. (29.6), are different because they apply to different types of dynamical systems. Equation (29.5), the fixed-point equation for a differential equation, says that we need to find the P value such that the rate of change of P is zero. Equation (29.6), the fixed-point equation for discrete iterated systems, says that we need to find the x value such that x is unchanged after f acts on it.

In any event, let us use Eq. (29.5) to find the fixed points of Eq. (29.4). We need to solve the following equation for P:

$$4P\left(1 - \frac{P}{100}\right) = 0 .$$ (29.7)

We can see that $P = 0$ is one solution. The other solution occurs when the term in parentheses is zero:

$$\left(1 - \frac{P}{100}\right) = 0 .$$ (29.8)

A few steps of algebra shows that $P = 100$ is a solution to the above equation. Thus, $P = 100$ is a fixed point.

29.2   Another Example

Let us do one more example. Consider the following differential equation:

$$\frac{dX}{dt} = \frac{1}{5}X(3 - X)(8 - X) . \tag{29.9}$$

Note that in this example the unknown function is $X(t)$ instead of $P(t)$ as it was in the previous section. The right-hand side of this equation is a function of X. That is,

$$\frac{dX}{dt} = F(X) , \tag{29.10}$$

where

$$F(X) = \frac{1}{5}X(3 - X)(8 - X) . \tag{29.11}$$

As in the previous section, we can gain a qualitative, global understanding of the solutions to this equation by plotting $F(X)$.

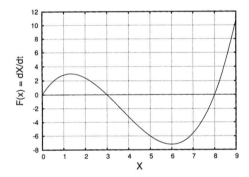

Fig. 29.4 A plot of $F(X)$, Eq. (29.11). The differential equation has three fixed points. There are unstable fixed points at $X = 0$ and $X = 8$, and a stable fixed point at $X = 3$.

Such a plot is shown in Fig. 29.4. We see that $F(X)$ is positive if X is between 0 and 3, negative if X is greater than 3 but less than 8, and positive if X is greater than 8. Since X increases if $F(X)$ is positive and decreases if $F(X)$ is negative, we can immediately draw the phase line for the solutions to Eq. (29.9), shown in Fig. 29.5.

Fig. 29.5 The phase line for the differential equation of Eq. (29.9).

One can now use either the phase line or the plot in Fig. 29.4 to sketch approximate solutions to the differential equation. We know that any X that starts between 0 and 3 will grow and approach the attracting fixed point at $X = 3$. And any X that starts between 3 and 8 will decrease and approach 3. Finally, any X that starts above 8 will grow without bound and will tend toward infinity. This is illustrated in Fig. 29.6. If we needed greater accuracy or if we wanted to check our work, we could use Euler's method to obtain an approximate solution to the differential equation. However, if all we are interested in is classifying the equation's global behavior—determining the fixed points and their stability—Euler's method is not necessary, at least for simple differential equations of the sort we have studied in this chapter and the previous one.

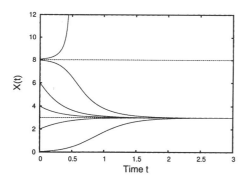

Fig. 29.6 Solutions to the differential equation, Eq. (29.9). Shown are solutions $X(t)$ for six different initial conditions: $8.05, 7.95, 6, 4, 2, 0.05$. The phase line is shown in Fig. 29.5. There is an attracting fixed point at $X = 3$ and repelling fixed points at $X = 0$ and $X = 8$.

29.3 Overview of One-Dimensional Differential Equations

In this chapter we have looked at differential equations of the form:

$$\frac{dX}{dt} = F(X) . \tag{29.12}$$

What type of dynamical behaviors are such systems capable of? One-dimensional iterated functions are capable of periodic behavior and chaos. Can one-dimensional differential equations do the same? The answer to this question is "no". A differential equation that has the form of Eq. (29.12) is deterministic and the solutions are continuous. As a result, solutions to Eq. (29.12) can never change direction, and so chaos and periodic solutions are not possible. To see why, it is easiest to consider a counter-example.

In Fig. 29.7 I have plotted a candidate solution $X(t)$ to differential equation Eq (29.12). However, such a solution is not possible. The reason is that the $X(t)$ curve visits the same X value multiple times. For example, $X(t)$ in the figure is equal to 40 at three different times, $t \approx 0.2, 0.35$, and 0.6. Each time $X(t)$ reaches 40 it has a different growth rate; its slope is different. But according to Eq. (29.12), this is impossible; the growth rate of $X(t)$ is determined solely by the the current value of X. An X of 40 corresponds to one and only one growth rate. So a curve such as that shown in Fig. 29.2 cannot be a solution to differential equations of the form Eq. (29.12).

The consequence of this is that the solution to a differential equation of the form of Eq. (29.12) can only increase or only decrease. It cannot do some of each.[1] So, there are a limited number of options for a solution to such a differential equation. It could tend toward positive or negative infinity. The only other possibility is that the solution approaches a

[1]It is possible, however, for a solution to be constant. This occurs if the initial condition is a fixed point.

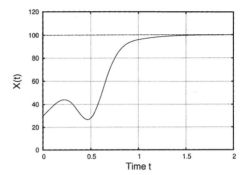

Fig. 29.7 A plot of a potential solution $X(t)$ to the differential equation Eq. (29.2). This function cannot be a solution, however, since it both increases and decreases.

fixed point, as is the case in Fig. 29.2. The orbit cannot turn around, for the reason given in the previous paragraph. And the orbit cannot cross the fixed point, since that point is fixed. If the solution hits the fixed point, it must remain there. So solutions to one-dimensional differential equations of the form of Eq. (29.12) cannot be periodic or chaotic.

Before concluding, I should mention that not all one-dimensional equations are of the form given in Eq. (29.12). For example, in the differential equation:

$$\frac{dX}{dt} = 10X(t) + \sin(2t) \, , \tag{29.13}$$

the growth rate $\frac{dX}{dt}$ is no longer a function only of X, since there is explicit t dependence as well. Thus, this equation is not of the form of Eq. (29.12). This means that solutions that both increase and decrease, such as that depicted in Fig. 29.7, are no longer forbidden. Hence, cycles and chaos are now a possibility. Equations of the form Eq. (29.12) are called **autonomous**, since the growth rate is independent (i.e., autonomous) of time .

To summarize, differential equations of the form Eq. (29.12) are rather limited in their dynamical behavior. A solution to such a differential equation can either increase, decrease, or remain constant. No other behavior is possible. In the next chapter we will examine two-dimensional differential equations, and we will see that they have a larger repertoire.

Exercises

(29.1) Consider a differential equation of the form

$$\frac{dX}{dt} = G(X) \, . \tag{29.14}$$

   (a) Sketch a possible $G(X)$ such that the differential equation has two attracting fixed points and one repelling fixed point.

(b) Sketch a possible $G(X)$ such that all solutions to the differential equation tend toward negative infinity.

(c) Sketch a possible $G(X)$ such that the differential equation has three attracting fixed points and no repelling fixed points.

(29.2) ⋆ Consider the following differential equation:

$$\frac{dX}{dt} = 2X . \qquad (29.15)$$

Without using Euler's method, find and classify all fixed points.

(29.3) Consider the following differential equation:

$$\frac{dX}{dt} = 3 + 2X . \qquad (29.16)$$

Without using Euler's method, find and classify all fixed points.

(29.4) Consider the differential equation:

$$\frac{dY}{dt} = G(Y) , \qquad (29.17)$$

where the function $G(Y)$ is shown in Fig. 29.8. Assume $Y > 0$ and that enough of $G(Y)$ is shown to determine the global behavior of the differential equation.

(a) How many fixed points does this differential equation have? Are they attracting or repelling?

(b) Sketch approximate solutions $Y(t)$ to the differential equation for the following initial conditions: $Y(0) = 0$, 2, 4, and 6.

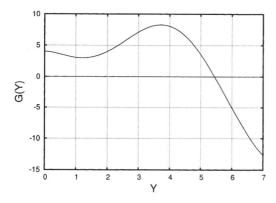

Fig. 29.8 The function $G(Y)$ for Exercise 29.4.

(29.5) Consider the differential equation:

$$\frac{dX}{dt} = F(X) , \qquad (29.18)$$

where the function $F(X)$ is shown in Fig. 29.9. Assume that enough of $F(X)$ is shown to determine the global behavior of the differential equation. How would you describe the global behavior

of the solutions? Are there any fixed points? Are they attracting or repelling?

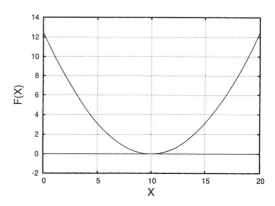

Fig. 29.9 The function $F(X)$ for Exercise 29.5.

(29.6) Consider the differential equation:

$$\frac{dX}{dt} = F(X) , \qquad (29.19)$$

where the function $F(X)$ is shown in Fig. 29.10. Assume that enough of $F(X)$ is shown to determine the global behavior of the differential equation. How would you describe the global behavior of the solutions? Are there any fixed points? Are they attracting or repelling?

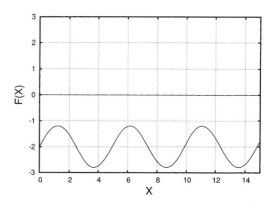

Fig. 29.10 The function $F(X)$ for Exercise 29.6.

Two-Dimensional Differential Equations

30

In the previous chapter we examined differential equations of the form $\frac{dX}{dt} = f(X)$ and saw that solutions $X(t)$ were limited in behavior. A particular solution can increase, decrease, or remain constant. But combinations of these behaviors are not possible, and hence cycles and chaos do not occur. In this chapter we will examine two-dimensional differential equations and will find that they are capable of a richer set of behaviors.

The initial example I will introduce is a model of two interacting populations, known as the **Lotka–Volterra** (LV) model or the Lotka–Volterra equations. Like the logistic equation, the LV model is very well known and thoroughly studied. It is a standard example of interacting populations in mathematical ecology and is also discussed in almost every textbook on differential equations.[1]

30.1 Introducing the Lotka–Volterra Model

Let us imagine that we have two populations of different creatures, perhaps rabbits and foxes. We will use $R(t)$ and $F(t)$ to denote the number of rabbits and foxes as a function of time t. First, let us think about the growth rate of the rabbits. The LV model assumes that the rabbit population grows in proportion to the rabbit population. This seems reasonable; the more rabbits there are, the more baby rabbits will get made, and thus the faster the population will grow. This statement is written as:

$$\frac{dR}{dt} = aR .\tag{30.1}$$

This says that the growth rate (i.e. the derivative of $R(t)$) is larger the larger the value of $R(t)$.[2] The symbol a in this equation is a parameter that sets the growth rate. The larger the value of a, the faster the rabbits will grow.[3] This differential equation yields unbounded growth. There are more rabbits that make more rabbits that make more rabbits, without end. We need a term in the model to limit the growth of rabbits.

In the LV model, the factor that keeps the rabbits from taking over the world is the presence of the foxes. The picture here is that foxes eat rabbits. How *many* rabbits do the foxes eat? This depends both on

[1] The model was introduced independently by Alfred Lotka and Vito Volterra in the mid-1920s.

[2] This differential equation was analyzed in Exercise 29.2.

[3] There will be quite a few variables and parameters in this section. I will use lowercase letters to refer to parameters and capital letters to refer to variables that change in time, such as $R(t)$, the number of rabbits.

the number of rabbits and the number of foxes. The more foxes there are, the more rabbits will get eaten. This seems straightforward enough. However, it is also the case that more rabbits will get eaten if there are more rabbits. The reason for this is that if there are more rabbits it will be easier for the foxes to find them and eat them.

Taking into account the foxes eating the rabbits, we modify Eq. (30.1) to obtain:

$$\frac{dR}{dt} = aR - bRF \ . \tag{30.2}$$

The last term on the right-hand side is the one that incorporates foxes eating rabbits.[4] This term is large when R and F are large. The negative sign in front of this terms indicates that the effect of the rabbit-fox interaction is to decrease the rabbit population. Here F is the fox population and b is a parameter that measures how effective the foxes are at hunting rabbits. The larger b is, the more deadly are the presence of foxes, from the rabbits' point of view.

Equation (30.2) describes the behavior of the rabbit population in the LV model. What about the foxes? There are two factors that influence the fox population. One is that they benefit from the presence of rabbits. This can be written:

$$\frac{dF}{dt} = cRF \ . \tag{30.3}$$

The quantity c is a parameter that captures how nutritious or beneficial rabbits are to foxes. The larger c is, the greater the growth rate of the foxes. The effect of this term is large if there are a large number of rabbits and foxes. If there are many foxes, there will be more baby foxes, and the growth rate of the population will be large. And if there are many rabbits, it will be easier for the foxes to find them, eat them, and gain the nutritional benefits of eating rabbits. This term is positive, because the rabbit-fox interaction is beneficial from the foxes' point of view.

How might the foxes die? If there are no rabbits, the foxes would starve to death since they would have nothing to eat. Is this the case according to Eq. (30.3)? Not quite. If there are no rabbits, $R = 0$ and the equation becomes:

$$\frac{dF}{dt} = 0 \ . \tag{30.4}$$

This says that the fox population does not grow, but it also does not shrink; the growth is constant. So in the model as it now stands, if all the rabbits suddenly died, the fox population would remain constant forever. This clearly is not realistic; the model needs to provide for some way for the foxes to die.

The simplest way to do this is to add a death term. Doing so, the equation for the fox growth rate is now:

$$\frac{dF}{dt} = cRF - dF \ . \tag{30.5}$$

[4]You may wonder why R and F are multiplied together instead of added. One possible response is that the growth rate should be proportional to both R and F. That is, doubling the rabbit population should double the effect of the foxes. Similarly, doubling the number of foxes should double the effect of the foxes. A term of the form bFR behaves in this way, while a term like $b(R + F)$ does not.

The second term on the right-hand side is the death term. It says that the rate at which foxes die is proportional to the number of foxes. This makes sense; the more foxes there are, the more foxes there are who will die. The parameter d is a measure of the death rate.[5] The larger d is, the larger the death rate of the foxes.

Taken together, Eqs. (30.2) and (30.5) constitute the Lotka–Volterra model:

$$\frac{dR}{dt} = aR - bRF\,, \qquad \frac{dF}{dt} = cRF - dF\,. \qquad (30.6)$$

This system is two-dimensional in the sense that there are two unknown functions: the rabbit population R and the fox population F. The LV model is an example of a *coupled* system of differential equations. The term coupled indicates that the growth rate of the rabbit population depends on both the rabbit population and the fox population, and the growth rate of the foxes depends on both the number of foxes and the number of rabbits. Thus, the two populations are coupled; they are not independent.

In the next section I will show how to adapt Euler's method to solve coupled equations such as Eq. (30.6). Before doing so, a few quick comments about the LV model. It goes without saying that the LV model is an extremely simple model of two interacting populations. There are many ways to make the model more realistic. However, that is not really the point. The LV model is designed to be a very simple model of interacting populations. It is not intended to be precise, but rather to give some general intuition about what might happen when two populations interact. There are many modifications one can make to the basic LV model which might make it more realistic and interesting.[6]

However, for our purposes, the main thing we are interested in is examining the global behavior of coupled two-dimensional systems such as the LV equations. So we are using the LV equations as a generic example of a coupled system of two differential equations; our main concern is not its use as a model in ecology. Since we are interested in the math of the situation and not actual rabbit and fox populations, I will not worry about the units on R and F. We will end up with fractional populations and populations less than 1, but we can assume that, as was the case with our study of the logistic equation, the populations are expressed as fractions of some maximum value.

[5] The parameter d should not be confused with the d in dF/dt. The quantity dF/dt is interpreted not letter-by-letter, but as a single quantity that represents the instantaneous growth rate of the foxes.

[6] Chapter 5 of Barnes and Fulford (2002) is a clear and accessible introduction to the LV model and variants thereof.

30.2   Euler's Method in Two Dimensions

How can we solve a coupled system such as Eq. (30.6)? Given the differential equation and starting values for the rabbit and fox populations, can we figure out the rabbit and fox populations at later times? It is not difficult to adapt Euler's method to get an approximate solution. For some sets of equations it is possible to use calculus to get exact solutions, but this is unusual. Most differential equations in two dimensions are solved using Euler's method or a related technique.

To illustrate Euler's method we need to select numerical values for the four parameters. Somewhat arbitrarily, I will choose $a = 1.0$, $b = 0.5$, $c = 0.2$, and $d = 0.6$. For these parameter values, the LV equations now read:

$$\frac{dR}{dt} = R - \frac{1}{2}RF, \qquad \frac{dF}{dt} = 0.2RF - 0.6F. \qquad (30.7)$$

I also need to choose starting values for the rabbit and fox populations. Again arbitrarily, I will choose $R(0) = 8$ and $F(0) = 4$. I will assume that the time t is measured in years. Finally, I will need to choose a step size Δt. We will then pretend that the rate of change is constant over time intervals of this length. I will start by choosing $\Delta t = 0.1$. Euler's method now proceeds almost exactly as it does for one-dimensional differential equations.

We begin with $R(0) = 8$ and $F(0) = 4$. What is the value of the populations at the next time step, 0.1 years later? The differential equation Eq. (30.7) tells us the rate of change of populations provided that we know the current population values. We can plug the initial values $R = 8$ and $F = 4$ into the right-hand side of Eq. (30.7) to obtain:

$$\frac{dR}{dt} = 8 - \frac{1}{2}(8)(4) = -8, \qquad (30.8)$$

and

$$\frac{dF}{dt} = 0.2(8)(4) - 0.6(4) = 4. \qquad (30.9)$$

The growth rates are not constant—the rates continually change as the population changes. The rates above are the exact rates only at $t = 0$. However, let us pretend that they are constant over the entire time interval of 0.1. We can then determine an approximate value for the rabbit and fox population at $t = 0.1$. Doing so, we obtain:

$$\begin{aligned}
R(0.1) \quad &= \quad R(0) + \text{growth rate} \times \Delta t & (30.10) \\
&= \quad 8 + (-8)(0.1) & (30.11) \\
&= \quad 7.2. & (30.12)
\end{aligned}$$

Note that this step is the same as Eq. (28.15) for the one-dimensional Euler's method. One does the same thing for the fox population:

$$\begin{aligned}
F(0.1) \quad &= \quad F(0) + \text{growth rate} \times \Delta t & (30.13) \\
&= \quad 4 + (4)(0.1) & (30.14) \\
&= \quad 4.4. & (30.15)
\end{aligned}$$

So after 0.1 years the rabbit population is 7.2 and the fox population is 4.4.

We repeat this procedure for the next time step. The rates of change are:

$$\frac{dR}{dt} = 7.2 - \frac{1}{2}(7.2)(4.4) = -8.64, \qquad (30.16)$$

and

$$\frac{dF}{dt} = 0.2(7.2)(4.4) - 0.6(4.4) = 3.696 \,. \tag{30.17}$$

Pretending these rates are constant from time $t = 0.1$ to $t = 0.2$, we obtain:

$$
\begin{align}
R(0.2) &= R(0.1) + \text{growth rate} \times \Delta t \tag{30.18} \\
&= 7.2 + (-8.64)(0.1) \tag{30.19} \\
&= 6.336 \,. \tag{30.20}
\end{align}
$$

And, for the foxes:

$$
\begin{align}
F(0.2) &= F(0.1) + \text{growth rate} \times \Delta t \tag{30.21} \\
&= 4.4 + (3.696)(0.1) \tag{30.22} \\
&= 4.7696 \,. \tag{30.23}
\end{align}
$$

Table 30.1 The first several rabbit and fox population values obtained by using Euler's method to solve Eq. (30.7) with $\Delta t = 0.1$.

t	$R(t)$	$F(t)$
0.0	8.0	4.0
0.1	7.2	4.4
0.2	6.336	4.7696
0.3	5.45859	5.08783
0.4	4.61583	5.33801

We continue in this manner and obtain a series of values for the rabbit and fox populations. The results for the first several values are shown in Table 30.1.

In summary, Euler's method works for two-dimensional equations almost exactly as it does for one-dimensional equations. The basic idea is identical. The differential equation tells us how the rates of change depend on the current populations, we pretend these rates of change are constant, and we then use these fictionally constant rates of change to determine the populations one time step later. Although this method yields approximate solutions, we can get as close as we like to the exact solution by choosing a sufficiently small time step Δt. The only limit is the computing power of the computer we use to perform Euler's method.[7]

[7] In practice, other techniques that are more efficient than Euler's method are often used. See Section 28.5.

30.3 Analyzing the Lotka–Volterra Model

In Fig. 30.1 I have plotted the rabbit and fox populations as a function of time for the initial populations $R(0) = 8$ and $F(0) = 4$. The solutions were generated via Euler's method with $\Delta t = 0.001$. In the figure we see that the populations cycle regularly. Note that the peak in the rabbit population occurs first and then is followed by a peak in the fox population. Then the rabbit population decreases and subsequently the fox population decreases as well. Then, after a little while where there are relatively few rabbits and foxes, the rabbit population peaks and the cycle begins anew.

In order to see the relationship between the rabbits and the foxes more clearly, one can plot the fox population and the rabbit population on the same axes, ignoring time. Such a plot is shown in Fig. 30.2. Here, the rabbit population is on the x axis and the fox population is on the y axis. This plot shows how the two populations cycle together. However, since this plot does not contain time, we cannot tell from looking at it how long it takes the population to complete one cycle.

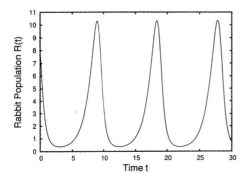 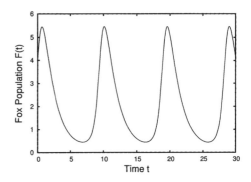

Fig. 30.1 A plot of the rabbit population $R(t)$ and the fox population $F(t)$ obtained by using Euler's method to solve Eq. (30.7). Both populations move in regular cycles.

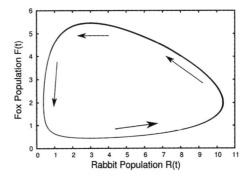

Fig. 30.2 A plot of the rabbit and fox populations $R(t), F(t)$ obtained by using Euler's method to solve Eq. (30.7), plotted on the phase plane.

One can tell a story to explain the cyclic behavior of the populations. We start in the lower left-hand corner of Fig. 30.2. Here there are few rabbits and few foxes. The rabbit population grows quickly, because there are not a lot of foxes around to eat them. The fox population does not grow much, because they do not have many rabbits to eat. Increasing the rabbit population corresponds to moving right on the plot. This brings us to the lower right part of the cycle. There are now many rabbits and the fox population begins to grow. An increasing fox population corresponds to upward motion on the plot. At around $R = 10$ the rabbit population reaches a maximum. The rabbit population starts to decrease, but the fox population continues to increase until around $F = 5.5$. This is the peak on the top of the cycle. At this point the fox population begins to decrease, because there are less and less rabbits for them to eat. On the left-hand side of the cycle the foxes decrease in number until we reach the lower left-hand corner of the cycle and the pattern begins again.

30.4   Phase Space and Phase Portraits

We summarized the solutions to one-dimensional differential equations with a phase line, such as Fig. 29.3. The phase line shows how solutions move and whether or not there are any fixed points. We cannot tell how fast a solution moves, since there is no time information on the phase line. For a two-dimensional differential equation such as the LV system there are two functions, and thus the solutions are visualized on a plane instead of a line. Thus, for two-dimensional systems one has a **phase plane** instead of a phase line. The solutions shown in Fig. 30.1 are plotted in the phase plane in Fig. 30.2. This is another way to see that the the two populations cycle together. I have added arrows to Fig. 30.2 to indicate the direction the populations move on the phase plane. I suggest taking a moment to convince yourself that the two curves plotted in Fig. 30.1 yield the shape shown in Fig. 30.2.

A more general notion than the phase plane is **phase space**, which is a geometric representation of the state variables of a system. In the LV example, the two state variables are the number of rabbits R and the number of foxes F. Knowing these two quantities at one point in time uniquely determines the values of R and F for all times in the future. This is what is meant by "state variable". The state (here R and F) determines the future. Depending on the system, there can be any number of state variables. Thus, the phase space could consist of a line, a plane, three-dimensional space, or a higher-dimensional space.

Moreover, the space does not need to be flat in the way that a line or a plane is. For example, suppose the system is the minute hand on a clock. The state variable here could be an angle between 0 and 360 degrees. Or a state variable could be the hours of the day instead, in which case the state variable is a number that starts at 0, increases, but resets to 0 the instant it reaches 24. In either case, the state space is a circle and not a line.

One can draw a particular solution to a differential equation as a curve through phase space, as was done in Fig. 30.2. However, often one is interested in a global view of the solutions to the differential equation. That is, we would like to understand the behavior of all the solutions—i.e., for all possible initial conditions—not just one. A diagram giving this information is called a **phase portrait**. To make a phase portrait for the LV model, we must determine solutions to Eq. (30.7) for several different sets of initial conditions. This will let us see if the cyclic behavior shown in Fig. 30.2 is stable, and whether or not there is any other stable behavior. In Fig. 30.3 I have plotted solutions to the LV equations for two different sets of initial conditions: $R(0) = 3, F(0) = 3$, and $R(0) = 5, F(0) = 4$. As was the case in Fig. 30.1 we see that the behavior of the populations are cyclic. Note, however, that different initial conditions give rise to different cycles. The period of the cycles are the same—i.e., it takes the same amount of time for the rabbit population to complete one cycle. However, the values of the populations are different. For different initial conditions, the maximum and minimum populations are different. In contrast, for the discrete logistic equation, periodic behavior was typically attracting; different initial conditions led eventually to the same periodic cycle.

We can get a better sense of the overall, global behavior of the LV system, Eq. (30.7), by plotting the solutions for all three initial con-

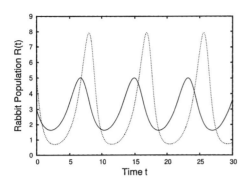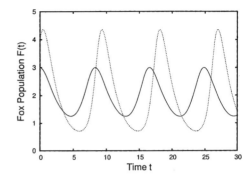

Fig. 30.3 A plot of the rabbit population $R(t)$ and the fox population $F(t)$ obtained using Euler's method to solve Eq. (30.7). The populations for two different initial conditions are shown. Both trajectories move in cycles; neither cycle is attracting.

ditions on the phase plane. The result is shown in Fig. 30.4. This figure is the same as Fig. 30.2 except that I have plotted the trajectories for the two additional initial conditions, $R(0) = 3, F(0) = 3$, and $R(0) = 5, F(0) = 4$. Figure 30.4 is the phase portrait for the LV equations. Looking at the phase portrait, the global structure of the LV model becomes clearer. We can infer that there is a cycle associated for every set of initial conditions. These cycles are neither attracting nor repelling. If the population is on one cycle and, say, the number of rabbits decreases slightly, the result is a new cycle, one that is slightly wider or narrower than the original cycle. The rabbit population is not pulled back to the original cycle, nor is it pushed away.[8] The phase por-

[8]Behavior such as this which is neither attracting nor repelling is called neutral.

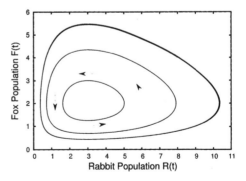

Fig. 30.4 The phase portrait for the Lotka–Volterra system of differential equations, Eq. (30.7). The trajectories for three different initial conditions is shown. Arrows indicate the direction of the trajectory.

trait thus consists of a nested set of ovals, like the layers in an onion. At the center of all the ovals is a fixed point, where the rabbit and fox populations do not change. It is not too difficult to determine the value of the fixed point. Exercise 30.7 leads you through this calculation.

30.5 Another Example: An Attracting Fixed Point

The LV system's global behavior consists of oscillations of neutral stability. Let us look at some other coupled two-dimensional differential
equations and see what other dynamical behaviors we find. Consider
the following set of differential equations:

$$\frac{dX}{dt} = Y , \qquad \frac{dY}{dt} = -0.5X - 0.4Y . \qquad (30.24)$$

Such an equation might arise in physics or engineering if one was studying the motion of an object that oscillates but is also subject to friction.
As was the case for the Lotka–Volterra equations, it is a straightforward,
if tedious, exercise to use Euler's method to find solutions $X(t)$ and $Y(t)$
to the differential equation. We just choose initial conditions and then
the differential equation determines the rate of change of those initial
conditions. Iterating forward, as per Euler's method, gives us a solution.
I chose the initial conditions $X(0) = 5$ and $Y(0) = 6$ and a step size of
$\Delta t = 0.0001$, and used a computer to determine approximate solutions
to the differential equation.

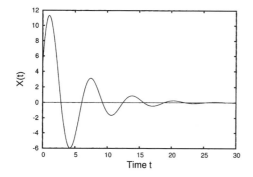 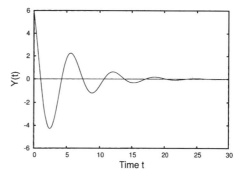

Fig. 30.5 Solutions $X(t)$ and $Y(t)$ to Eq. (30.24) obtained via Euler's method. The initial conditions are $X(0) = 5$ and
$Y(0) = 6$, and $\Delta t = 0.0001$.

The results are shown in Fig. 30.5, where we see that both the $X(t)$
and $Y(t)$ solutions oscillate. However, unlike the LV system, the am-
plitudes of the oscillations decay over time, and both $X(t)$ and $Y(t)$
approach zero. The point $X = 0, Y = 0$ is an attracting fixed point.
This can be seen more clearly in Fig. 30.6, which is the phase portrait for Eq. (30.24). Here I have plotted three different solutions to
Eq. (30.24) corresponding to three different sets of initial conditions.
The solid trajectory on the phase portrait, Fig. 30.6, is $X(t)$ and $Y(t)$
plotted in Fig. 30.5. The other two trajectories are for the initial conditions $X(0) = -10$, $Y(0) = 4$, and $X(0) = -7$, $Y(0) = -3$. We see
that different solutions all spiral toward the origin, confirming that the
origin is an attracting fixed point.

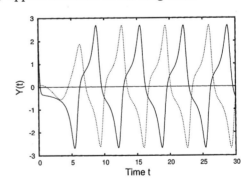

Fig. 30.6 The phase portrait for Eq. (30.24). Three solutions, obtained via Euler's method, are shown. All solutions spiral toward the attracting fixed point at the origin: $X = 0, Y = 0$.

30.6   One More Example: Limit Cycles

As a final example for this chapter we consider the van der Pol equation. This differential equation was introduced by Balthasar van der Pol in the mid-1920s while studying oscillations in electrical circuits.[9] It has subsequently found use in the physical and biological sciences. Van der Pol's equation is, like the other equations in this chapter, a two-dimensional differential equation:

[9]This equation played an important role in the early development of chaos and dynamical systems. See Section 1.5 of Aubin and Dahan Dalmedico (2002) and references therein.

$$\frac{dX}{dt} = Y , \qquad \frac{dY}{dt} = -X + (1 - X^2)Y . \tag{30.25}$$

I chose two sets of initial conditions $X(0) = 0.1, Y(0) = 0.1$, and $X(0) = 3, Y(0) = 3$ and used Euler's method with $\Delta t = 0.0001$ to generate solutions. The result is shown in Fig. 30.7. As was the case with the Lotka–Volterra equations, the trajectories are cyclic. However, unlike the solutions to Lotka–Volterra equations, it appears that the van der Pol solutions end up behaving the same. The solutions are out of phase, but otherwise they appear identical in the long term.

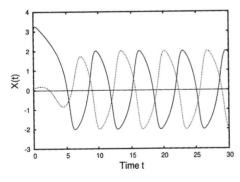

Fig. 30.7 Two solutions to Eq. (30.25) obtained via Euler's method. The initial conditions are $X(0) = 0.1, Y(0) = 0.1$, (dashed line) and $X(0) = 3, Y(0) = 3$ (solid line). The time step Δt is 0.0001.

This can be seen much more clearly by looking at the trajectories in the phase plane, shown in Fig. 30.8. The dashed trajectory from Fig. 30.7 starts at the point $0.1, 0.1$, indicated with a square. The trajectory spirals clockwise towards a cycle that appears sort of like a rounded

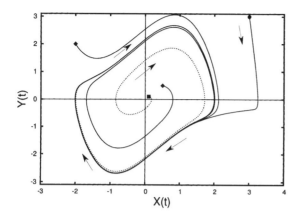

parallelogram. The solid trajectory from Fig. 30.7 starts at 3, 3, shown as a circle. This orbit is pulled almost straight down, and then left into the cycle. Two additional orbits are shown; these begin at the two points indicated with diamonds. The reason the cycle in the figure appears dark is that it is traced over multiple times as the trajectories cycle around it. As Fig. 30.8 suggests, the cycle is an attractor. The trajectories for all initial conditions eventually follow this periodic attractor. I have plotted this attractor without the transient behavior in Fig. 30.9. A closed attractor in a two-dimensional phase plane is often called a **limit cycle**.

30.7 Overview of Two-Dimensional Differential Equations

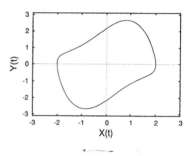

Fig. 30.9 The Van der Pol attractor.

We have seen two types of stable, attracting behavior: a fixed point, as in Fig. 30.6; and a limit cycle, as in Fig. 30.8. It turns out that these two phenomena are the *only* possible types of stable, global behavior exhibited by the two-dimensional differential equations of the sort we have been considering.

To see why this is so, let us think about what determinism implies for trajectories on the phase plane. The differential equations we are considering are of the form:

$$\begin{aligned}
\frac{dX}{dt} &= f(X, Y) \\
\frac{dY}{dt} &= g(X, Y) \,.
\end{aligned} \tag{30.26}$$

This means that dX/dt, the instantaneous growth rate of X, is a deterministic function of the current values of X and Y. The same is true for dY/dt, the instantaneous growth rate of Y. In other words, X and Y determine the growth rate. As a result, the current point in the phase plane determines the trajectory. At one level, this is nothing new. When considering discrete dynamical systems such as the iterated

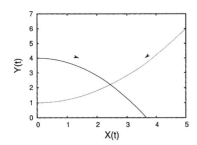

Fig. 30.10 An impossible situation for phase plane trajectories for solutions to a deterministic differential equation such as Eq. (30.26). Determinism means that two solutions to the same differential equation can never cross.

[10]You might want to scan back through this chapter at phase space trajectories and confirm that while trajectories may get close together, they never cross.

[11]This result is known as the Poincaré–Bendixson theorem. It was originally proved by Henri Poincaré in 1892. In 1901 a slightly stronger version of the theorem was proved by Ivar Bendixson.

logistic equation, the initial condition completely determines the orbit. The same is the case here—the initial condition uniquely determines the trajectory. However, because here the trajectory must trace out a continuous line through phase space, there is an additional geometric consequence of determinism: trajectories in phase space cannot cross.

Consider what would happen if two trajectories in phase space did somehow cross each other, as illustrated in Fig. 30.10. The trouble occurs where the two trajectories cross. According to Eq. (30.26), the rates of changes of X and Y are uniquely determined for each point X, Y. However, we can see that two trajectories emerge from the crossing, indicating that the rate of change is not unique. Hence, it cannot be the case that two trajectories ever cross.[10] By a similar argument, a given trajectory can never cross itself.

The fact that phase space trajectories cannot cross limits the possible long-term behaviors for two-dimensional differential equations.[11] Trajectories can get pulled into a fixed point or a cycle, but more complicated behavior is not possible. In particular, there is no possibility for bounded and aperiodic behavior. If the trajectory is bounded—i.e., it does not fly off to infinity—it must either cycle or get pulled into a point. The restriction that the trajectory cannot cross itself means that during its journey it quickly "runs out of room". Imagine drawing a continuous, curved line on a piece paper with the restriction that you can never have the line cross itself. Eventually you will block off enough regions of the paper that you will have nowhere else to go and you will stop. Thus, you will have reached a fixed point. The only way to avoid this fate is to move in a cycle, as in Fig. 30.8.

Since aperiodic, bounded behavior cannot occur for two-dimensional, continuous, deterministic dynamical systems, it immediately follows that such systems cannot exhibit chaos. We shall see in the next chapter that in order for a continuous dynamical system to be chaotic its phase space must have a least three dimensions. Further, we shall see that many such chaotic systems have a beautiful, fractal structure.

Exercises

(30.1) Verify that the values in Table 30.1 are correct.

(30.2) Consider the functions $X(t)$ and $Y(t)$ shown in Figs. 30.11 and 30.12. Sketch this trajectory in the X, Y phase plane.

(30.3) Consider a phase plane for the Lotka–Volterra system such as that shown in Fig. 30.4. Imagine instead of rabbits, we have aphids, which are considered an agricultural pest. And instead of foxes, we have wasps, which eat aphids.

(a) Suppose that in an effort to get rid of aphids, a farmer applies a pesticide that kills both aphids and wasps. How might this be represented on the phase plane? That is, in what direction on the phase plane does one move if the aphids and wasps both decrease?

(b) Suppose that the effect of the pesticide is to kill almost all the aphids and almost all the wasps. Argue that, based on the LV model, the result will be a larger aphid population than there was before the pesticide was applied. (This is sometimes used as an argument against broad-spectrum pesticides.)

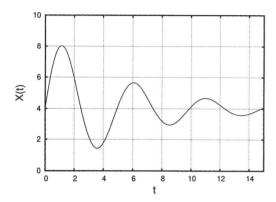

Fig. 30.11 The function $X(t)$ for problem 30.2.

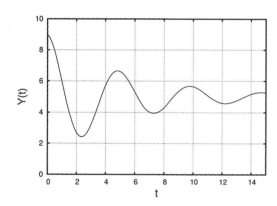

Fig. 30.12 The function $Y(t)$ for problem 30.2.

(30.4) Consider the trajectory in the X, Y phase plane shown in Fig. 30.4.

  (a) Sketch possible functions $X(t)$ and $Y(t)$ that could give rise to this trajectory.

  (b) Sketch possible functions $X(t)$ and $Y(t)$ if the arrows in Fig. 30.4 were reversed.

(30.5) Consider a deterministic, two-dimensional differential equation whose phase space is not a plane, but a torus. A torus is a donut shape. What sorts

of long-term solutions are possible for such a system? Is it possible to have an aperiodic orbit?

(30.6) Consider a deterministic, two-dimensional differential equation whose phase space is a sphere. What sorts of long-term solutions are possible for such a system?

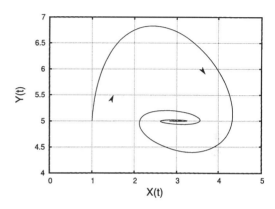

Fig. 30.13 The trajectory in the phase plane for problem 30.4. The arrows indicate the direction of motion.

(30.7) In this exercise you will investigate further the solutions to the LV system:

$$\frac{dR}{dt} = R - \frac{1}{2}RF ,$$
$$\frac{dF}{dt} = 0.2RF - 0.6F . \qquad (30.27)$$

These equations can be rewritten by factoring an R out of the first equation and an F out of the second:

$$\frac{dR}{dt} = R(1 - \frac{1}{2}F) ,$$
$$\frac{dF}{dt} = F(0.2R - 0.6) . \qquad (30.28)$$

  (a) Find the fixed points of the LV system. These are the R, F values for which both $\frac{dR}{dt}$ and $\frac{dF}{dt}$ equal zero. Set the right-hand side of both equations equal to zero, and solve for R and F. You should find two solutions. Discuss briefly the interpretation of each.

  (b) For what values of R and F is the rabbit population increasing? (I.e., for what R and F values is $\frac{dR}{dt} > 0$?) For what values is it decreasing? Indicate these regions on the R-F phase plane.

(c) For what values of R and F is the fox population increasing? For what values is it decreasing? Indicate these regions on the R-F phase plane.

(d) You should now have four regions on the phase plane, each of which has a different behavior. I.e., in one region both the rabbits and foxes increase, in another region the rabbits decrease while the foxes increase, and so on. Label these regions and compare to the phase portrait in Fig. 30.4.

Chaotic Differential Equations and Strange Attractors

In the previous chapter we explored two-dimensional differential equations and saw that aperiodic behavior was not possible. Trajectories in phase space cannot cross, and this restriction limits the behaviors that can occur on a two-dimensional surface. In three dimensions the situation is very different, as we will see in this chapter. We begin with an example which played a central role in the development of nonlinear dynamics as an area of study.

31.1 The Lorenz Equations

In the early 1960s, Edward Lorenz, a meteorologist at the Massachusetts Institute of Technology, was developing mathematical models of the weather and climate. Lorenz began with the Navier–Stokes equations, the fundamental equations describing how fluids behave. Initially, his models had many different variables: temperature, wind speed, humidity, wind direction, cloudiness, etc. These models proved difficult to understand, and so he sought to simplify the model so it would be more tractable, while still preserving some of the key features of weather dynamics. Eventually he was led to a system of three differential equations that now bear his name.

The **Lorenz equations** are:

$$
\begin{aligned}
\frac{dx}{dt} &= \sigma(y - x) \\
\frac{dy}{dt} &= x(\rho - z) - y \\
\frac{dz}{dt} &= xy - \beta z \, .
\end{aligned}
\tag{31.1}
$$

There are three variables, x, y, and z which describe the way hot air might move upward via convection, then cool, and then fall downward. In what follows, we will not worry about the physics these equations were originally designed to capture, and instead will focus on the mathematical properties of the equations. As usual, we will investigate the long-term behavior of trajectories and the stability of any fixed points or cycles. In Eq. (31.1) σ, ρ and β are parameters.[1] These are, re-

[1]This notation for the parameters is not entirely standard. Sometimes other letters are used, sometimes Greek, sometimes not. I will use Greek letters because it helps to distinguish parameters from the variables x, y, and z.

spectively, "sigma", "rho", and "beta", the eighteenth, seventeenth, and second lowercase letters of the Greek alphabet. We will see that by varying these parameters, solutions to the logistic equation display a range of different behaviors, including chaos.

As in the previous chapter where we studied two-dimensional differential equations, we will find solutions to the Lorenz equations via an approximate numerical method similar to Euler's method. I will not go into the details here, as it is conceptually the same as for two-dimensional differential equations but is even more tedious. So, let our exploration of the Lorenz equations begin.

31.2   A Fixed Point

Let us first try the following parameter values: $\sigma = 10.0$, $\rho = 8.0$, and $\beta = 2.667$. To find the solution to this differential equation—i.e., x, y, and z as a function of t—I used Euler's method with a time step of $\Delta t = 0.02$ and an initial condition of $x(0) = 1.0$, $y(0) = 0.0$, and $z(0) = 0.0$. The $x(t)$, $y(t)$, and $z(t)$ trajectories are shown in Fig. 31.1. We see that the solutions are pulled toward a fixed point at around $x = 4.25$, $y = 4.25$ and $z = 7.0$.

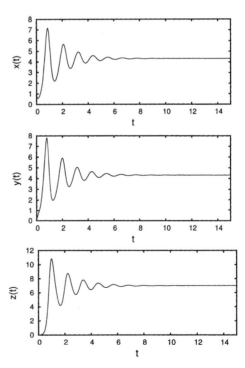

Fig. 31.1 The $x(t)$, $y(t)$, and $z(t)$ trajectories for the Lorenz equations with $\sigma = 10.0$, $\rho = 8.0$, and $\beta = 2.667$. The initial condition is $x(0) = 1.0$, $y(0) = 0.0$, and $z(0) = 0.0$.

As with the two-dimensional differential equations of the previous chapter, it is helpful to plot this trajectory in phase space. Here, however, phase space is three-dimensional, since there are three functions: $x(t)$, $y(t)$, and $z(t)$. This poses a challenge, since visualizing a three-dimensional object on a two-dimensional sheet of paper inevitably leads to some information loss. Nevertheless, there are a number of different strategies one can employ to get a useful view of a three-dimensional object.

One way of viewing the trajectory in three-dimensional phase space is to show all three variables but to draw them in perspective so that it appears three-dimensional despite being on a two-dimensional page. Most plotting and graphics programs do this automatically. Such a plot is shown in Fig. 31.2. One can clearly see the trajectory spiraling in toward the fixed point.

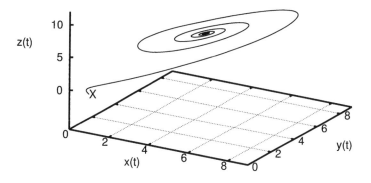

Fig. 31.2 The trajectory in phase space for the $x(t)$, $y(t)$ and $z(t)$ trajectories shown in Fig. 31.1. The initial condition is indicated by the "X" on the figure. The trajectory spirals in toward the attracting fixed point near $x = 4.25$, $y = 4.25$ and $z = 7.0$.

It is also possible to visualize a three-dimensional image by using a type of plot called a dual-image stereogram. We visually perceive the three-dimensionalness of the world because our two eyes see two slightly different images, owing to our two eyeballs' different positions in our head. So each eye sees something slightly different. You can easily verify this. Look at an object with just your left eye. Then, without moving your head, look at the object with just your right eye. Switch back and forth between the left- and right-eye views and you can see that the image you see shifts each time you change eyes. So the brain

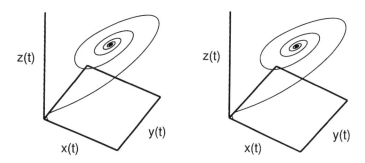

Fig. 31.3 A stereogram of the phase space trajectory shown in Fig. 31.2. To view the image, look straight at the figure from a distance of around 1.5 feet. Relax and defocus your eyes until the images merge.

receives two slightly different images, one from each eye. The neural circuitry of our brains then processes these images so that we perceive a single image that appears three-dimensional and allows us to experience depth.[2]

Figure 31.3 shows a dual-image stereogram of the phase space trajectory shown in Fig. 31.2. The stereogram works as follows. The two images in the stereogram look identical, but they are actually slightly different. The images show the object (in this case the spiraling trajectory) from two slightly different angles. These different angles correspond to the different views you would get from your left and right eyes. If you stare at the stereogram and relax and defocus your eyes, you should soon see a single image come into focus in between the two other figures. This single image will appear three-dimensional. Learning to view stereograms can take a little practice. The trick is to relax your eyes by imagining that you are focusing on an object far away. The two separate images will start to move closer together. Relax your eyes until the axes on the two plots line up. I find that it is helpful to be looking straight at the stereogram, not at an angle. It is also important to hold the image flat. I suggest holding the book flat on the table and looking straight down on the image from a distance of around 1.5 feet.

Returning to the Lorenz equations, we have seen that for $\sigma = 10.0$, $\rho = 8.0$, and $\beta = 2.667$, the solutions tend toward a fixed point at around $x = 4.25$, $y = 4.25$ and $z = 7.0$. One can show that other initial conditions are also pulled toward the fixed point, so it is indeed an attractor. This type of behavior is not at all new—we have seen many examples of attracting fixed points before. The only new twist is that now the trajectories are in three dimensions instead of one or two.

31.3   Periodic Behavior

Let us now consider the Lorenz equations for a different set of parameter values. As before, we will let $\sigma = 10.0$ and $\beta = 2.6667$. Now, however, we set $\rho = 160$. Choosing $x(0) = 50$, $y(0) = 50$, and $z(0) = 50$, I used Euler's method to find the solution for this initial condition. The results are shown in Fig. 31.4.

One sees that the x, y, and z trajectories all become periodic. By the time $t = 6$, the trajectories repeat. In some sense, this is similar to the trajectories for the van der Pol system, where we also saw periodic behavior. However, here the oscillations are more complicated—there are several peaks and valleys before the pattern repeats.

In order to see the relationship between the three variables we plot the trajectory in phase space. A stereogram of the trajectory is shown in Fig. 31.5. This figure only shows the long-term behavior of the orbit; it is a plot of the trajectory from $t = 50$ to $t = 55$, at which point the transient behavior has died away and one gets a clear view of the periodic attractor. Note that the trajectory does not cross itself. It appears to, but this is only because the three-dimensional trajectory is plotted on

[2]Seeing out of both eyes is essential to good depth perception. Try closing one eye and then have a friend gently toss and object at you. You will find that it is surprisingly difficult to catch.

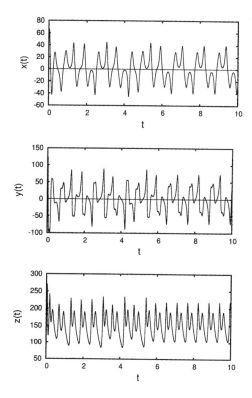

Fig. 31.4 The $x(t)$, $y(t)$, and $z(t)$ trajectories for the Lorenz equations with $\sigma = 10.0$, $\rho = 160$, and $\beta = 2.667$. The initial condition is $x(0) = 50$, $y(0) = 50$, and $z(0) = 50$.

a two-dimensional piece of paper. If you look at the stereogram and see the three-dimensional image, you should be able to observe that the trajectory loops through space without crossing itself.

Behavior like the closed, looped trajectory can not occur in a two-dimensional system. The phase space for a two-dimensional system is a plane, and so the only way to have a closed trajectory is if it is some sort of an oval, as we saw for the van der Pol equation—see Fig. 30.8. Recall that the determinism of the rule that generates the trajectory means that it is not possible for the curve to intersect itself. In three dimensions, however, there is room for a closed trajectory to curve above or below itself, and so complicated, twisted, or knotted structures are possible.

31.4   Chaos and the Lorenz Equations

We now consider one last set of parameters for the Lorenz equations: $\sigma = 10$, $\beta = 2.667$, and $\rho = 28$. These are the parameter values that Lorenz considered in his original work in the early 1960s. Trajectories for the initial condition of $x(0) = 50$, $y(0) = 50$, and $z(0) = 50$ are shown in Fig. 31.6. We are interested in the long-term behavior of the trajectories, so I have begun these plots at $t = 10$, not $t = 0$. One can immediately see that the trajectories are not periodic. This is especially clear for the x and y trajectories which wiggle up and down irregularly.

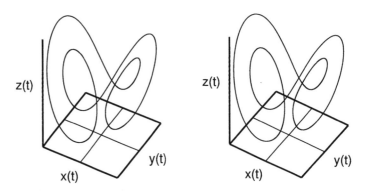

Fig. 31.5 A stereogram view of the trajectories of Fig. 31.4 plotted in phase space. The transient behavior is not shown. Only the trajectory from $t = 50$ to $t = 55$ is plotted. To view the image, look straight at the figure from a distance of around 1.5 feet. Relax and defocus your eyes until the images merge.

Note that the x and y trajectories are similar, but not identical. It appears, then, that the trajectory does not repeat, but also that the x, y, and z parts of the trajectory are closely related.

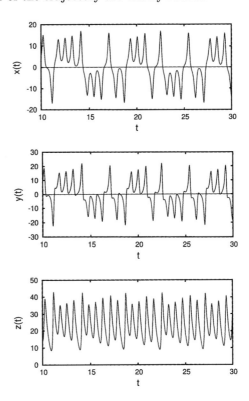

Fig. 31.6 The $x(t)$, $y(t)$, and $z(t)$ trajectories for the Lorenz equations with $\sigma = 10.0$, $\rho = 28$, and $\beta = 2.667$. The initial condition is $x(0) = 50$, $y(0) = 50$, and $z(0) = 50$. Note that the horizontal scale begins at $t = 10$.

To see this relationship, a stereogram of the trajectory in phase space is shown in Fig. 31.7. Looking at this picture, note that the trajectory never intersects itself. Rather, the two lobes are woven together in an intricate and complex way. It turns out that the trajectory does not repeat—it continues winding around the two lobes and weaving from side to side without ever crossing its path. Thus, the trajectory is aperiodic, one of the defining features of a chaotic system. A larger view of non-repeating path of the trajectory is shown in Fig. 31.9. Here once can perhaps see more clearly the way in which the trajectory moves from the left to the right lobe. The left node in this figure corresponds to negative values of x and y while the right node corresponds to positive values. The trajectory moves irregularly from the left to the right node.

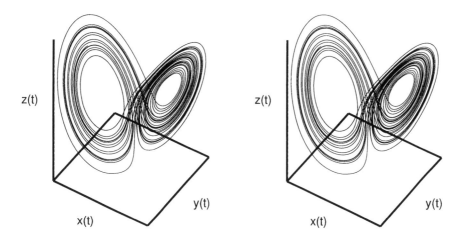

Fig. 31.7 A stereogram view of the trajectories of Fig. 31.6 plotted in phase space. The transient behavior is not shown. Only the trajectory from $t = 10$ to $t = 45$ is plotted. To view the image, look straight at the figure from a distance of around 1.5 feet. Relax and defocus your eyes until the images merge.

This can be seen in the $x(t)$ and $y(t)$ plots in Fig. 31.6, where one sees the x and y trajectories shifting together from positive to negative. Each peak or valley corresponds to one orbit around the right or left node, respectively. For example, starting at $t = 10$ in these figures, the phase-space trajectory visits the left (L) and right (R) lobes in the following sequence: RLRRRRLLLRLLRRRLRLLLLRLLRRR.

The trajectory for the Lorenz equation is aperiodic; it does not repeat. This is one of the four properties that a dynamical system must have in order to be considered chaotic. The other three are bounded orbits, determinism, and sensitive dependence on initial conditions. The orbit is indeed bounded, since the trajectory in Fig. 31.9 does not tend toward infinity. And the Lorenz equations, Eq. (31.1), are clearly deterministic. What about sensitive dependence on initial conditions? Does the Lorenz equation show the butterfly effect?

We test for evidence of the butterfly effect by plotting the trajectories for two different initial conditions that are very similar. The results of doing this are shown in Fig. 31.8. In these plots the trajectory plotted with a the solid line has an initial condition of $x(0) = 30$, $y(0) = 30$, and $z(0) = 30$. The dashed curve's initial condition is $x(0) = 30.01$, $y(0) = 30$, and $z(0) = 30$. Note that the two initial conditions differ by only a tiny bit in the x part, while their y and z parts are identical.

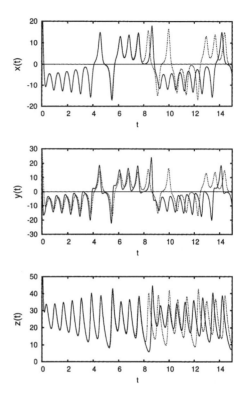

Fig. 31.8 Pairs of $x(t)$, $y(t)$, and $z(t)$ trajectories for the Lorenz equations with $\sigma = 10.0$, $\rho = 28$, and $\beta = 2.667$. The initial condition for the solid curve is $x(0) = 30$, $y(0) = 30$, and $z(0) = 30$. The dashed curve's initial condition is $x(0) = 30.01$, $y(0) = 30$, and $z(0) = 30$.

We can see that the two trajectories diverge significantly despite having very similar initial conditions. The two trajectories are very different by $t = 9$.

The trajectories of the Lorenz equation have sensitive dependence on initial conditions and are aperiodic. Thus, the Lorenz equation for the parameter values $\sigma = 10.0$, $\rho = 28$, and $\beta = 2.667$ is chaotic: it is a deterministic dynamical system whose orbits are bounded, aperiodic, and which has sensitive dependence on initial conditions. This is our first example of a chaotic differential equation. As we have seen in the previous two chapters, continuous systems cannot be chaotic in one or two dimensions, since determinism prevents trajectories from crossing, and hence bounded, aperiodic behavior is not possible. However, chaos is indeed possible for three-dimensional continuous systems.

31.5   The Lorenz Attractor

[3]In the rest of this section I will focus exclusively on the Lorenz equations with these parameters. So I will simply refer to it as *the* Lorenz equations from now on. It should be understood, though, that this refers to the Lorenz equation with these particular parameter values.

The dynamics of the Lorenz equation for the parameter values $\sigma = 10.0$, $\rho = 28$, and $\beta = 2.667$ is chaotic.[3] Moreover, the shape traced out by the trajectories in phase space shown in Fig. 31.7 and Fig. 31.9 is an attractor. This means that orbits will get pulled toward the shape, just as an attracting fixed point might pull in nearby orbits. However, in this case the attractor is much more complex than a fixed point—the motion on the attractor is chaotic.

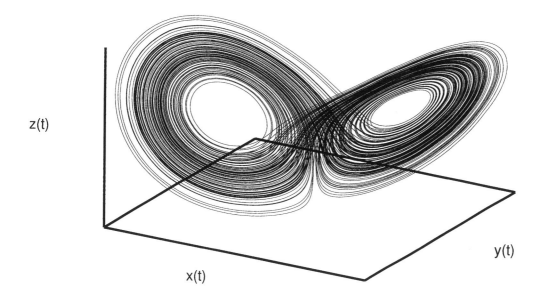

z(t)

y(t)

x(t)

Fig. 31.9 A large view of the Lorenz attractor, one of the icons of chaos.

Figure 31.10 demonstrates the attracting nature of the shape illustrated in Fig. 31.9. In this figure I have plotted snapshots of the evolution of 8000 different trajectories. The first frame of the figure in the upper left shows 8000 initial conditions, chosen to be uniformly spaced within a cube. I then have plotted the position of these 8000 points at $t = 0.1$, 0.2, 1.0, 2.0, and 10.0. One sees that the cloud of points contracts and gets pulled fairly quickly toward the lobed structure seen in the previous figures.

The shape of Fig. 31.9 is now commonly known as **the Lorenz attractor**. The Lorenz attractor is another example of a **strange attractor**. We encountered a strange attractor previously in Chapter 26 when examining the Hénon equation. As was the case for the Hénon attractor, the Lorenz attractor has three noteworthy features. First, it is an attractor; multiple initial conditions end up pulled into it. Second, the dynamics on the attractor itself are chaotic; orbits show sensitive dependence on initial conditions and are aperiodic. Third, the attractor is a fractal. The Lorenz attractor is almost two-dimensional, but not quite: the Lorenz attractor's dimension is estimated to be 2.05 (Grassberger and Procaccia, 1983). As noted in Section 26.6 there is not a standard technical definition of the term "strange attractor". Nevertheless, the above three features are generally taken to be the key characteristics of a strange attractor.

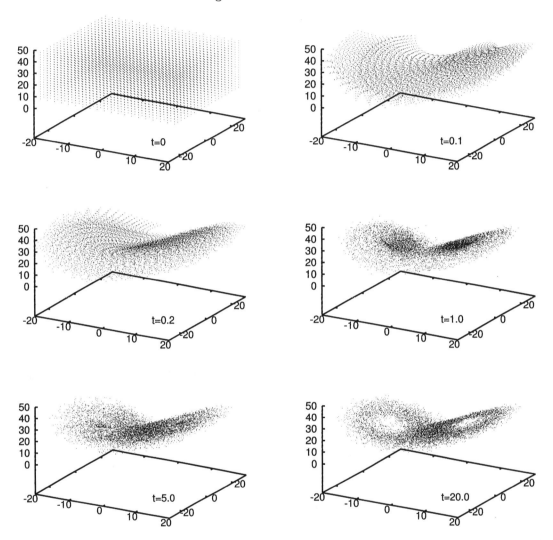

Fig. 31.10 At t=0, there are 8000 points distributed uniformly in a cube. These are taken as initial conditions and then the trajectories are calculated using the Lorenz equations. These 8000 points are shown at a series of subsequent times. One can see that the points get pulled toward the Lorenz attractor.

Strange attractors are mixtures of order and disorder. They are ordered in that almost all initial conditions get pulled into the attractor, and thus almost all initial conditions will trace out the same lobed shape in phase space. In this sense, the attractor is stable. A small change in the initial condition or a small perturbation while the trajectory is unfolding will not alter the overall shape of the attractor. Yet the behavior of individual trajectories on the attractor are unstable; they are sensitively dependent on their initial conditions. Thus, two orbits that begin close to each other on the attractor will soon diverge, as illustrated in Fig. 31.8. As a result, it is impossible to make accurate, long-term predictions for systems like the Lorenz equations. However, there is a certain structure or order to the unpredictability; one can be certain that the orbit will remain on the attractor, even though one is very uncertain what exact path the orbit will follow as it weaves its way across the attractor.

One cannot help but notice a resemblance between the Lorenz attractor and a butterfly's wings. It is occasionally thought that this resemblance is the reason why the phenomenon of sensitive dependence is known as the butterfly effect. This is not the case. In Lorenz's original work he did not produce a three-dimensional plot of the attractor that appeared butterfly-like. In fact, Lorenz originally used a seagull and not a butterfly as an example of the sort of small perturbation that could lead to large changes in the weather system. In 1972, Lorenz was scheduled to give a talk at a meeting of the American Association for the Advancement of Science (AAAS). A meeting organizer changed the title of Lorenz's talk while Lorenz was overseas and unreachable. The title of Lorenz's talk, unbeknown to him, was "Does the flap of a butterfly's wings in Brazil set off a tornado in Texas?" For a fascinating discussion of the origin of the butterfly metaphor, see the short essay by Robert Hilborn, "Sea gulls, butterflies, and grasshoppers: A brief history of the butterfly effect in nonlinear dynamics". (Hilborn, 2004)

Lorenz's work did not initially gain a wide audience, as he published mainly in atmospheric science journals. However, in the 1970s more physicists and mathematicians learned of his work, and the modern study of dynamical systems was well underway.[4] Since then, there have been many other strange attractors discovered in a wide range of systems. In many ways they can be viewed as a generic property of differential equations in three or more dimensions. There is nothing unique or unusual about strange attractors; they are a common phenomenon.

[4]Lorenz's results were by no means the only work that led to the work in dynamical systems in the 1970s and '80s and beyond, but it surely was a crucial piece of research and is generally credited as the source of one of the many streams of research that coalesced into the modern study of chaos and dynamical systems (Aubin and Dahan Dalmedico, 2002).

31.6   The Rössler Attractor

Before concluding, let us consider one more example: the Rössler equations, introduced in 1976 by Otto Rössler (1976) as a simplified version of the Lorenz equations. This dynamical system exhibits a strange attractor that is somewhat easier to visualize and analyze than the Lorenz attractor. Rössler's equations are:

$$\frac{dx}{dt} = -y - z$$
$$\frac{dy}{dt} = x + ay$$
$$\frac{dz}{dt} = b + z(x - c) . \tag{31.2}$$

We will study the behavior of these equations for $a = 0.1$, $b = 0.1$, and $c = 14$. Solving these equations using Euler's method with $\Delta t = 0.0001$

and initial conditions of $x(0) = 10$, $y(0) = 10$, and $z(0) = 10$, I obtain the trajectories shown in Fig. 31.11. The x and y trajectories appear somewhat similar to those of the Lorenz equation—a not-quite-regular oscillation. But the z trajectory is rather different. What do these trajectories look like plotted in phase space?

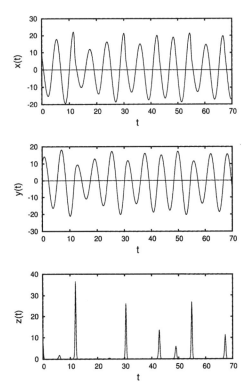

Fig. 31.11 Plots of $x(t)$, $y(t)$, and $z(t)$ trajectories for the Rössler equations. The initial condition is $x(0) = 10$, $y(0) = 10$, and $z(0) = 10$. These trajectories are plotted in phase space in Fig. 31.12.

The three trajectories plotted in Fig. 31.11 are shown in phase space in Fig. 31.12. This shape, known as the Rössler attractor, is a strange attractor. Trajectories on the attractor are chaotic, while orbits off the attractor are quickly pulled to it, as was the case for the Lorenz attractor. In Fig. 31.13 I have plotted a stereogram of the Rössler attractor to better illustrate its three-dimensional structure.

The motion on the Rössler attractor is as follows. Orbits move in a counter-clockwise circle in the x-y plane. Orbits toward the outside of the circle are stretched upward at the back of the figure. These orbits then rejoin the main circle, merging into the inner portion of the circle. The result is that the orbits fold over on themselves. We can thus think of the Rössler attractor as a system that stretches and folds repeatedly. This behavior is sketched in Fig. 31.14.

We can use Fig. 31.14 to help us think about what happens to a set of initial conditions under the Rössler equations. Picture these initial conditions as a blob of dough on the attractor. The dough rotates around the attractor, and in each cycle the dough is stretched and folded back on itself. After several iterations, the dough will be stretched into many

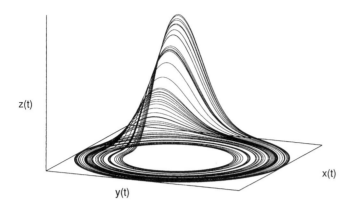

Fig. 31.12 The Rössler attractor.

thin sheets; it will resemble a flaky pastry.

These sheets of dough do not exactly merge. It is impossible for them to do so, since in a deterministic dynamical system trajectories cannot cross or merge. Instead, there remains a small space between the sheets as they are continually folded over on themselves. Remarkably, this spacing has the structure of a Cantor set. In other words, if one were to look at the outer edge of the lower band on the Rössler attractor, the spacing of the sheets would follow a Cantor set. This is illustrated in Fig. 31.15.

31.7 Chaotic Flows and One-Dimensional Functions

Looking at Fig. 31.13 or 31.14, it is clear that the Rössler attractor stretches and folds trajectories in phase space. In Section 10.5 I stated that stretching and folding were the essential geometric ingredients for chaos. Stretching is responsible for the butterfly effect; when a stretch occurs, nearby trajectories are pushed farther apart. Folding of some sort is necessary to keep orbits bounded. If there was not any folding, orbits would tend toward infinity. It is thus not surprising that we observe stretching and folding in a three-dimensional chaotic system such as the Rössler equations in addition to the one-dimensional logistic equation.

In Chapter 12 we saw that certain features of the period doubling route to chaos in the logistic equation were universal—the same for all systems that undergo period doubling. In Section 12.4 I sketched an argument designed to explain why almost all one-dimensional systems have the same basic behavior when undergoing period doubling. The question remained, however, how it is that multi-dimensional systems, such as dripping faucets and convection rolls in fluids, have the same universal properties as one-dimensional equations. The answer I gave in Section 12.5 is that low-dimensional systems like the logistic equation

Fig. 31.15 A view of the edge of the Rössler attractor. The sheets are arranged in a Cantor set.

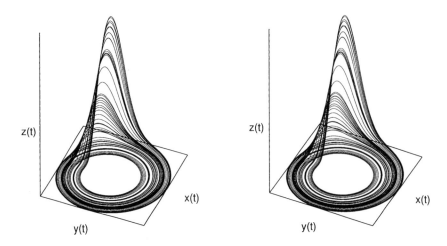

Fig. 31.13 A stereogram of the Rössler attractor. To view the image, look straight at the figure from a distance of around 1.5feet. Relax and defocus your eyes until the images merge.

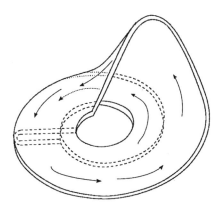

Fig. 31.14 A sketch of the flow of trajectories along the Rössler attractor. Adapted from a figure made by Christophe Letellier (2006). Used by permission.

capture the stretching-and-folding of higher dimensional systems. The Rössler attractor provides a vivid illustration of this. Figure 31.14 clearly exhibits a stretch-and-fold process.

It is possible to approximate higher-dimensional systems with lower-dimensional ones. The two-dimensional Hénon equations of Chapter 26 were introduced to approximate a cross section of the three-dimensional Lorenz equations. It is also possible to derive one-dimensional equations to capture one part of the three-dimensional motion of a system such as the Lorenz or Rössler equations. There are several ways to derive lower-dimensional systems from higher-dimensional ones. When one does so by making a two-dimensional slice through a high dimensional space, the resultant lower-dimensional equations are known as a **first recurrence map** or a **Poincaré map**. The slice that is made through the higher-dimensional object is known as a **Poincaré section**.

Further Reading

This chapter offers only a glimpse into the universe of strange attractors and multi-dimensional dynamical systems. The chapter titled "Strange Attractors" in Gleick (1987) and Chapter 6 of Stewart (2002) are good, non-technical discussions of strange attractors. For a more technical overview of strange attractors, I recommend Chapter 12 of Peitgen, Jürgens, and Saupe (1992). Chapter 9 of Strogatz (2001) contains a clear and thorough explication of the mathematics and physics of the Lorenz equations.

An important topic that I have not covered in this book is that of attractor reconstruction. The basic idea is that given a one-dimensional series of data—perhaps a measurement of temperature or the population of a certain species—one can reconstruct a higher-dimensional attractor. For example, even if one has access only to the $x(t)$ trajectory for the Lorenz equations, one can nevertheless reconstruct an attractor. This attractor will be similar to the full Lorenz attractor, and in many cases will have the same dimension. Attractor reconstruction is also referred to as the theory of embedding. Most references on this topic are fairly advanced. A clear, fairly accessible overview is Chapter 5 of Ott, Sauer, and Yorke (1994); see also Sauer (2006) and Section 12.4 of Strogatz (2001).

Exercises

(31.1) For the Lorenz equations with $\sigma = 10.0$, $\rho = 8.0$, and $\beta = 2.667$, verify that there is a fixed point at $x \approx 4.321$, $y \approx 4.3208$, and $x = 7$. I.e., plug all these values into the right-hand side of Eq. (31.1) and show that all the derivatives are zero.

(31.2) Consider the the Lorenz equations with $\sigma = 10.0$, $\rho = 28.0$, and $\beta = 2.667$. These parameter values yield the famous Lorenz attractor, Fig. 31.9. At the center of each lobe there is an unstable fixed point. Verify that the coordinates of these fixed points are $x = 6\sqrt{2}, y = 6\sqrt{2}, z = 27$, and $x = -6\sqrt{2}, y = -6\sqrt{2}, z = 27$.

Part VI

Conclusion

Conclusion

32

Having completed our overview of chaos and fractals, I conclude by summarizing, synthesizing, and taking stock. I begin by quickly reviewing the main topics covered in the text. In the subsequent few sections I highlight some of the key themes and lessons of chaos and fractals and offer some thoughts on how to characterize their impact.

32.1 Summary

The central items of study in this book have been dynamical systems—mathematical systems that change over time. The first two parts of the book focused on iterated one-dimensional functions, primarily the logistic equation. We encountered chaotic behavior: bounded orbits generated by a deterministic equation that are aperiodic and have sensitive dependence on initial conditions. For chaotic systems long-term prediction is not possible. Such systems behave as if they are random, despite the fact that they are deterministic. Iterated functions are capable of a wide range of behavior. One particularly useful way to visualize this is via a bifurcation diagram. We also saw that quantitative features of the period-doubling route to chaos were the same for almost all systems, including higher-dimensional physical systems. Lastly, we observed that chaotic behavior can be statistically stable. Sensitive dependence on initial conditions makes detailed prediction impossible, but a histogram formed from a chaotic orbit takes on a predictable shape, enabling one to make accurate predictions about the long-run average behavior of a chaotic orbit.

In Part III we turned our attention to fractals, self-similar geometric objects. We saw that fractals can be generated by regular, deterministic procedures, as well as via random processes. Fractals can be characterized by their self-similarity or box-counting dimensions. Such dimensions are usually more meaningful descriptors of an object or process than specifying an average or typical size. In fact, we saw an example of a simple process for which an average does not exist. We also looked at power-law distributions—the type of distribution that describes a scale-free or fractal phenomenon.

Julia sets and the Mandelbrot set were the topics of Part IV. Here we saw again that simple iterated systems produce images and forms of remarkable intricacy and beauty. In Part V we examined a number of other dynamical systems, including two-dimensional iterated functions, cellular automata, and one-, two-, and three-dimensional systems of differential equations. We encountered strange attractors: complex, stable attractors in phase space on which the dynamics are chaotic.

32.2   Order and Disorder

One of the central lessons of the study of chaos and fractals is that a phenomenon can be generated by a process that seems opposed to or contrary to the behavior it produces. For example, throughout the book we have seen that deterministic dynamical systems can generate unpredictable and seemingly random outcomes. The logistic equation, the Hénon map, the Lorenz equations, and many others, all exhibit chaotic behavior. Ironically, it is the determinism of the equations that is responsible for sensitive dependence on initial conditions. The system is *so* deterministic that small imprecisions in our knowledge of the initial condition are quickly amplified, foiling attempts at prediction.

We have also seen many examples of simple rules that give rise to complex behavior. The bifurcation diagram for the logistic equation, Julia sets and the Mandelbrot set, cellular automata rule 110, and the Lorenz and Rössler attractors are all rich and complex images that are produced with very simple iterated rules or equations. It can be difficult to believe that the Mandelbrot set is based solely on iterating the equation $f(z) = z^2 + c$. There is an intrinsic creativity associated with iteration.

We have also encountered many ways to make intricate and elaborate fractals. Fractals such as the Sierpiński triangle, the Koch curve, and the Cantor set can be generated by the repeated application of simple deterministic procedures. However, it is also the case that a random process can produce a fractal that is as intricate and precise as one made by a deterministic rule. An example of this was given in Section 17.4, where we saw that the chaos game—where moves on a triangle are chosen via a random process—generates a crisp and predictable Sierpiński triangle.

Taken together, the examples described above illustrate that order and disorder are not complete opposites but are interrelated. Deterministic systems can produce behavior that is indistinguishable from randomness, and random systems can produce precise, symmetric structures that appear carefully and deliberately constructed. In each case one could say that a phenomenon is caused or explained by its opposite. However, I prefer to view these examples as telling us that randomness and order are best not thought of as opposites—they are not mutually exclusive qualities and they need not be viewed as being in opposition to one another. At the risk of oversimplifying, randomness can be caused by order, and vice-versa.

In addition, randomness and order mix and mingle together in all sorts of interesting and complex ways. A trajectory of the logistic equation for $r = 4.0$ is chaotic. Two orbits that start off close to each other will soon be very different, but histograms built from each of their itineraries will, in the long run, be essentially identical. The motion of orbits on a strange attractor for the Lorenz or Hénon equations is chaotic. But the attractor is stable; almost all orbits are quickly pulled into the attractor. Both of these examples—the strange attractor and the logistic equation—combine local unpredictability or instability with global predictability and regularity. And then there are structures like the space-time diagram of cellular automata rule 110 in Fig. 27.12. It is hard to know how to characterize this pattern, but it surely combines elements of randomness and order. The picture that emerges, then, is that randomness and order are subtle and interrelated.

32.3   Prediction and Understanding

One of the conclusions that results from the study of chaos is that there are limits on our knowledge. Long-term prediction is impossible for a chaotic dynamical system. It is important to underscore that not *everything* is chaotic; there are many predictable, stable systems both in mathematics and, more importantly, in the physical world. Nevertheless, chaos is not an unusual state of affairs. And so the study of chaos suggests that there are bounds and limits to what we will be able to predict. We can make very accurate tide tables years into the future, but we will never be able to make accurate, detailed weather predictions beyond a few weeks. So chaos spells the end of the Laplacian dream of prediction.

Yet there is much more to the study of chaos than this negative result. For while chaos closes the door on the idea that everything can be predicted, it opens several paths for other types of understanding. Even though a dynamical system may have sensitive dependence on initial conditions, it is certainly not order-less. Although unpredictable in detail, the orbit may unfold along a strange attractor, a stable and predictable structure. Also, even systems with sensitive dependence on initial conditions can be accurately predicted in the short term.

The study of dynamical systems shows us that complicated behavior can have simple origins, and thus it is possible in some cases to explain and understand complex phenomena with simple equations. Perhaps the most striking example of this is the property of universality. We have seen that some features of the period-doubling route to chaos are universal—quantitatively the same for almost all functions and physical phenomena. There are thus commonalities across seemingly different systems, and these common features can be understood.

32.4   A Theory of Forms

Let us turn our attention for the moment to fractals. By almost any measure, the idea of fractals—self-similar objects—are a standard part of science and popular culture. Fractal structures can be quantified by their dimension. There are several different definitions of dimension that are applied to fractals, but all capture a constant scaling relationship: a relation that stays the same as the scale of analysis changes. Scientists routinely calculate or measure dimensions, much as they would measure height or weight or color. While fractals are a new idea and fractional dimensions may initially stretch the imagination, in some regards I think fractals have become just another standard tool in the scientists' toolbox. A researcher might measure the weights of cats and then try to determine how diet is related to weight. Or a researcher might measure the dimension of a mountain ridge and then try to determine how this dimension is related to the age of the mountain.

More broadly, fractals give one a language and conceptual framework for understanding and describing a class of shapes that are not well described by the regular forms—circles, lines, cubes—of Euclidean geometry. An appreciation of fractals leads us to look for similarities across scales. Looking out my office window I see very few straight lines or regular circles. But in the trees and sky and landscape I can see many fractal forms. It would seem that fractals are not exceptional objects, but are quite ordinary.

One can then ask where fractals come from. How can such seemingly complicated shapes be made? We have seen many dynamical systems, both deterministic and random, that produce fractals. Viewing fractals dynamically—thinking of, say, a tree not as a static object but as something that grows over time—one sees that fractals are actually quite easy to make. Complex fractals can be made via very simple rules. This suggests to me that fractals are a sort of generic shape. It is almost as if they occur by default.

It is striking that one sees very similar forms arising in very different systems. For example, the branching structures of trees, blood vessels, and river basins, look very similar. What to make of this? One approach would be to seek to understand the shape of a river by analyzing the forces on the material on the riverbank and the speed and shape of the water flow. One might seek to understand the shape of a tree by considering the chemical and biological details of tree growth, or by considering the biological function of tree branches and analyzing this in light of evolution via natural selection. These approaches surely have merit, but there is also much to be gained by considering the branching structures themselves.

That is, rather than studying trees or rivers or blood vessels, we can study *branches*. What are the different processes that lead to fractal branching structures? To what extent are branching structures typical? The study of fractals and dynamical systems suggests that we can gain insight by studying features of the fractal forms themselves, independent of their material origins in trees, blood vessels, or rivers. This approach complements and enriches a more traditional view in which one understands structures and forms via an understanding the structure's constituents and their interactions. This geometric and less reductive approach suggested by the study of fractals is very much in keeping with the geometric view underpinning many approaches to dynamical systems.

32.5 Revolution or Reconfiguration?

Is chaos a scientific revolution on par with the development of quantum mechanics, relativity, and calculus? What impact has chaos had, and what impact will it have in the future? I think it is difficult to argue that chaos as an area of study is not here to stay. There are books and courses on the subject. Many researchers consider chaos to be their primary field of research; many more use ideas and analytical tools from chaos and apply them to their areas of study. There are scholarly journals and conferences devoted to chaos and its applications. So I think the question is not whether or not chaos is a big deal, but what kind of big deal it is.

I think the name "chaos" sets unreasonable expectations. Stephen Kellert lays this out in the prologue to his book *In the Wake of Chaos*:

> Chaos theory is not as interesting as it sounds. How could it be? After all, the name "chaos theory" makes it seem as if science has discovered some new and definitive knowledge about utterly random and incomprehensible phenomena.
>
> Actually, what seems to be going on is a kind of magic trick like the one Ludwig Wittgenstein described as putting something in a drawer and closing it, then turning around and opening the drawer, and removing the object with an expression of surprise. By calling certain physical [and mathematical] systems "chaotic", scientists lead us to think that they are totally unintelligible—just a muddle of things happening with no connections or structures. So when they find interesting mathematical patterns in these unpredictable systems, they can exclaim that they have discovered the secrets of "order within chaos", even though only by christening these systems chaotic in the first place can they make such an impressive result possible. (Kellert, 1993, p. ix).

It is certainly not the case that interest in or excitement surrounding chaos is purely a linguistic trick. Nevertheless, the phrase "chaos theory" can lead to impossibly high expectations for the field of study.

That said, is chaos a scientific revolution? Chaos does not oblige us to revise or discard Newton's laws. In this sense it is a scientific advance very different from special relativity or quantum mechanics, both of which tell us that Newton's laws and other equations of classical physics are an inaccurate description of the behavior of objects at speeds approaching that of light or of objects that are smaller than typical molecules. Quantum mechanics and relativity are thus often referred to as scientific revolutions: advances that occur relatively suddenly and which require the rejection of past knowledge or theories. In my view, chaos is not a revolution of this sort.

Rather, chaos is a conceptual realignment and a cultural shift. It is not a sudden event; the history of chaos follows a long arc, starting with the work of Henri Poincaré at the turn of the nineteenth century. Historians David Aubin and Amy Dahan Dalmedico (2002) argue compellingly for a *longue durée* approach to the history of chaos, tracing several strands of its development over the better part of a century. Aubin and Dahan Dalmedico "take the emergence of 'chaos' as a science of nonlinear phenomena not as the mere development and wide application of a certain mathematical theory but as *a vast process of sociodisciplinary convergence and conceptual reconfiguration*" (emphasis in original) (Aubin and Dahan Dalmedico, 2002, p. 3).

Chaos challenges some of the assumptions of classical physics: namely that a simple system should exhibit simple behavior, and that complicated behavior must have a complex origin. This notion is not a part of classical physics in the sense that there is an equation that embodies this assumption. But I think it was a central, if unspoken, premise of much of science before the blossoming of chaos and dynamical systems research in the 1970s and '80s. The study of chaos and fractals shows us that these assumptions are not true, compelling one toward a more nuanced and complicated view of the relationship between simplicity and complexity. This is the sort of conceptual reconfiguration to which Aubin and Dahan-Dalmedico refer. "[C]haos has definitely blurred a number of old epistemological boundaries and conceptual oppositions hitherto seemingly irreducible such as order/disorder, random/nonrandom, simple/complex, local/global, stable/unstable ... " (Aubin and Dahan Dalmedico, 2002, p. 53).

There is another type of conceptual reconfiguration or shift associated with chaos. The study of chaos suggests a type of understanding that does not place "knowing the equations" as the central goal. In much of traditional physics, understanding a phenomenon is almost synonymous with being able to write down equations for it. Once one has the equations figured out, the problem is solved. The equation, both literally and metaphorically, encodes the solution. But we have seen many times that even if we know the equation that governs a dynamical system it is still not easy to deduce the dynamical behavior. For example, consider the two equations

$$f(x) = 3.62x(1 - x) \,, \tag{32.1}$$

and

$$f(x) = 3.7x(1 - x) \,. \tag{32.2}$$

The first equation is periodic; the second is chaotic. The only way I know how to determine this is to iterate both equations using a calculator or computer. Knowing the equation is, of course, necessary in this context, but it is not enough. One needs to do something more to determine the long-term dynamical behavior.

Similarly, in an experimental context where one has access to data—the orbit or trajectories of one or more dynamical variables—one may wish not to focus on deducing the equations that govern these variables. After all, even if one could figure out the equations, it may not shed light on the dynamical behavior. Instead, one makes a plot of the variables in phase space and looks for a strange attractor. This strange attractor, a geometric object, may be much more informative than the equations themselves. In this sense, the style of understanding associated with chaos is often geometric. One is interested in how an attractor folds and stretches in phase space, or the rate at which nearby orbits are pulled apart, instead of an exact, numerical prediction of an orbit.

Chaos also is a confluence of a complex of methods, techniques, and ideas—what Aubin and Dahan-Dalmedico refer to as a "sociodisciplinary convergence" (2002, p. 3). The study of chaos and dynamical systems arose out of similar concerns in different disciplines, including meteorology, population biology, physics, and mathematics. The formation of chaos as a field or area of study required researchers from these different disciplines to adopt a common language and perspective. There is now a shared understanding across traditional disciplines of terms like strange attractor, fractal, and chaos. There is also a more or less common set of techniques and agreed-upon quantities that are interesting to calculate, including fractal dimensions and Lyapunov exponents. More fundamentally, though, I think there is some shared sense of what is scientifically interesting and what constitutes research that is worth pursuing. It is almost an aesthetic choice. Researchers in many fields have opted to grapple with the strange and fun mixtures of predictability and unpredictability, simplicity and complexity, that are at the heart of chaos.

Further Reading

Much has been written that aims to assess, analyze, and characterize the meanings and impact of chaos and fractals. Not surprisingly, not everyone agrees. These are young and still developing fields; the history of chaos and fractals is still being written. There are some analyses and assessments of chaos and fractals that I particularly recommend. The article by David Aubin and Amy Dahan Dalmecido (2002) is an impressively thorough examination of several strands of research that congealed to form the field of chaos. I also recommend Dahan Dalmedico (2004). Stephen Kellert's *In the Wake of Chaos* (1993) is a lucid and balanced overview of what chaos is, what it is not, and why it matters. His later book, *Borrowed Knowledge* (2008), takes a critical and thoughtful look at how and why chaos theory has been put to use in other academic fields. The last chapter of *Does God Play Dice?* by Ian StweartStewart (2002) is an enthusiastic but balanced overview and assessment of the impacts of chaos and fractals.

Part VII

Appendices

Review of Selected Topics from Algebra

A.1 Exponents

In this appendix I briefly review some of the properties of exponents. Our starting point—and really the only thing you have to memorize—is the definition of an exponent: an exponent indicates successive multiplication. That is, x^n means x multiplied by itself n times:

$$x^n \equiv \overbrace{xx \cdots x}^{n \text{ times}} . \tag{A.1}$$

So,

$$x^3 = xxx , \tag{A.2}$$

and

$$x^6 = xxxxxx , \tag{A.3}$$

and so on.

Rules for Multiplication, Division, and Exponentiation

What does $x^a x^b$ equal? Let us apply the definition, Eq. (A.1), and see. We start with an example.

$$x^3 x^5 = (xxx)(xxxxx) = xxxxxxxx = x^8 . \tag{A.4}$$

In words, x multiplied by itself three times, times x multiplied by itself five times, is the same as x multiplied by itself eight times. So, in general,

$$x^a x^b = x^{a+b} . \tag{A.5}$$

What about x^{a^b}? We will follow the same approach: let us apply the definition Eq. (A.1) and see what we get. Again, we try an example.

$$x^{3^5} = (xxx)^5 = (xxx)(xxx)(xxx)(xxx)(xxx) = x^{15} . \tag{A.6}$$

In general, then,

$$x^{a^b} = x^{ab} . \tag{A.7}$$

Next, we consider $\frac{x^a}{x^b}$, again applying the definition of Eq. (A.1) and considering a particular case:

$$\frac{x^5}{x^3} = \frac{xxxxx}{xxx} = xx = x^2 . \tag{A.8}$$

Note that there are five x's on the top and three on the bottom. The three on the bottom cancel three on the top, leaving $5-3 = 2$ on the top. In general, then, it follows that:

$$\frac{x^a}{x^b} = x^{a-b} . \tag{A.9}$$

Zero as an Exponent

We can use Eq. (A.9) to figure out the meaning of a zero exponent. Consider the following expression:

$$\frac{x^3}{x^3} = ??? . \tag{A.10}$$

One way to evaluate this is as follows:

$$\frac{x^3}{x^3} = \frac{xxx}{xxx} = 1 . \tag{A.11}$$

But we could also evaluate this using the exponent rule, Eq. (A.9):

$$\frac{x^3}{x^3} = x^{3-3} = x^0 . \tag{A.12}$$

We are now in a bit of a quandary: what does x^0 mean? Well, in Eq. (A.11) we see that

$$\frac{x^3}{x^3} = 1 . \tag{A.13}$$

And in Eq. (A.12) we see that

$$\frac{x^3}{x^3} = x^0 . \tag{A.14}$$

For these two equations to be consistent, it must follow that

$$x^0 = 1 . \tag{A.15}$$

Equation (A.15) is somewhat counter-intuitive. Appealing to the definition of exponentiation, Eq. (A.1), one could argue that x^0 equals 0. The reasoning behind this is that any number multiplied by itself zero times yields zero. This does make sense. But, as one thinks about it, it is not at all clear what it means to multiply something by itself zero times. The fact that $x^0 = 1$ is really just a convention. But hopefully the preceding argument motivates why mathematicians have chosen to set $x^0 = 1$: doing so makes Eq. (A.9) consistent for all possible values of a and b.

There is, however, one important exception, namely,

$$0^a = 0 \text{ for all } a . \tag{A.16}$$

This seems reasonable; zero multiplied by itself any number of times is still zero.

We can also use the rule of Eq. (A.9) to figure out what x^1 means. To do so, we will consider $\frac{x^3}{x^2}$:

$$\frac{x^3}{x^2} = \frac{xxx}{xx} = x \; . \tag{A.17}$$

And we know from Eq. (A.9) that

$$\frac{x^3}{x^2} = x^{3-2} = x^1 \; . \tag{A.18}$$

So, for the above two equations to be consistent, it must be that

$$x^1 = 1 \; . \tag{A.19}$$

Square Roots and Such

We now turn our attention to square roots. The square root of a number x is defined as follows: the square root of x is another number which, when multiplied by itself, returns x.[1] This is perhaps either to think about symbolically with an equation:

$$x = \sqrt{x} \times \sqrt{x} \; . \tag{A.20}$$

I tend to think of this equation as defining what the symbol $\sqrt{x}$, the square root of x, means.

We now seek a way to write $\sqrt{x}$ as x to some power. That is, we are looking for n in the following equation:

$$\sqrt{x} = x^n \; . \tag{A.21}$$

Let us start this quest by rewriting Eq. (A.20):

$$x^1 = \sqrt{x} \times \sqrt{x} \; . \tag{A.22}$$

We now plug in Eq. (A.21), to obtain

$$x^1 = x^n \times x^n = x^{2n} \; , \tag{A.23}$$

where in the last step I have used Eq. (A.5). We now solve for n in Eq. (A.23). Since $x^1 = x^{2n}$, it must be the case that

$$1 = 2n \; , \tag{A.24}$$

so $n = \frac{1}{2}$. We have thus solved for n in Eq. (A.21), succeeding in our goal of expressing the square root as an exponent:

$$\sqrt{n} = x^{\frac{1}{2}} \; . \tag{A.25}$$

The same thing holds for other roots. The cube root of a number x is defined as a number which, when multiplied by itself three times, returns x:

$$x = \sqrt[3]{x} \times \sqrt[3]{x} \times \sqrt[3]{x} \; . \tag{A.26}$$

[1] For example, 6 is the square root of 36, since $6^2 = 36$.

By a similar argument, one can show that

$$\sqrt[3]{x} = x^{\frac{1}{3}} .$$
(A.27)

And, in general,

$$\sqrt[a]{x} = x^{\frac{1}{a}} ,$$
(A.28)

where $\sqrt[a]{x}$ is the ath root of x; $\sqrt[a]{x}$ multiplied by itself a times returns x:

$$\left(\sqrt[a]{x} \right)^a = x .$$
(A.29)

Summary

For convenience, I collect the main results of this section:

$$x^a x^b = x^{a+b} .$$
(A.30)

$$x^{a^b} = x^{ab} .$$
(A.31)

$$\frac{x^a}{x^b} = x^{a-b} .$$
(A.32)

$$x^{-a} = \frac{1}{x^a} .$$
(A.33)

$$x^{\frac{1}{a}} = \sqrt[a]{x} .$$
(A.34)

$$x^0 = 1 \text{ for } x \neq 0 .$$
(A.35)

$$0^a = 0 \text{ for all } a .$$
(A.36)

These are formulas that you will probably want to remember. However, this does not mean that you should memorize them. The only thing to memorize is the definition of an exponent, Eq. (A.1). All of these other properties follow directly from this basic definition. You will not want to necessarily have to rederive these results every time you need them. But if you practice using the formulas, and you understand where they come from and why they are the way they are, I think they will settle into your consciousness without much explicit effort spent memorizing.

A.2   The Quadratic Formula

Suppose that we need to solve an equation of the form

$$Ax^2 + Bx + C = 0 ,$$
(A.37)

for x, where A, B, and C are constants. The quadratic formula gives us the solution to such an equation. The solution(s) are:

$$x = \frac{-B \pm \sqrt{B^2 - 4AC}}{2A} .$$
(A.38)

Usually there are two solutions to a quadratic equation. In the formula these are given by the two possibilities for the $\pm$. However, if the term inside the square root, $B^2 - 4AC$, is zero, then there will be only one solution.

Also, it could be that the roots are not real numbers, but instead are complex, or "imaginary". This occurs if the term inside the square root is negative.

Some Examples

Suppose we need to solve the following equation for x:

$$3x^2 - 10x + 8 = 0 . \tag{A.39}$$

In this case, $A = 3$, $B = -10$, and $C = 8$. Plugging into the quadratic formula, Eq. (A.38), we obtain:

$$x = \frac{10 \pm \sqrt{10^2 - 4(3)(8)}}{2(3)} . \tag{A.40}$$

Simplifying some, we get:

$$x = \frac{10 \pm \sqrt{100 - 96}}{6} , \tag{A.41}$$

$$x = \frac{10 \pm \sqrt{4}}{6} , \tag{A.42}$$

$$x = \frac{10 \pm 2}{6} . \tag{A.43}$$

There are two solutions. For the "$-$" in the quadratic formula, we have:

$$x = \frac{10 - 2}{6} = \frac{8}{6} = \frac{4}{3} \approx 1.33 . \tag{A.44}$$

And for the "$+$":

$$x = \frac{10 + 2}{6} = \frac{12}{6} = 2 . \tag{A.45}$$

We can easily verify that 2 and $\frac{4}{3}$ are solutions to Eq. (A.39) by plugging these x values in and seeing if they make the equation true. For example, plugging $x = 2$ into Eq. (A.39), we obtain

$$3(2)^2 - 10(2) + 8 = 0 ? \tag{A.46}$$

Simplifying,

$$3(4) - 20 + 8 = 12 - 20 + 8 = 0 , \tag{A.47}$$

thus confirming that $x = 2$ is indeed a solution of the original equation.

Here is another example. Suppose that we wish to solve the following equation for x:

$$x^2 + 5 = -4x . \tag{A.48}$$

Our first step is to manipulate this equation so that it is of the same form as Eq. (A.37). Doing so, we obtain:

$$x^2 + 4x + 5 = 0 .$$ (A.49)

So $A = 1$, $B = 4$, and $C = 5$. Plugging into Eq. (A.38), we obtain

$$x = \frac{-4 \pm \sqrt{4^2 - 4(1)(5)}}{2(1)} .$$ (A.50)

Simplifying, we get

$$x = \frac{-4 \pm \sqrt{-4}}{2} .$$ (A.51)

At this point it appears that we are in a quandary, as we have to take the square root of a negative number. There is no real square root of a negative number. Thus there are no real solutions to Eq. (A.49).

There are, however, complex solutions. The square root of negative four is:

$$\sqrt{-4} = 2i ,$$ (A.52)

where i is defined as the square root of negative one:[2]

$$i \equiv \sqrt{-1} .$$ (A.53)

[2] Complex numbers are discussed more fully in Chapter 23.

Using Eq. (A.52) in Eq. (A.51), we obtain

$$x = \frac{-4 \pm 2i}{2} = -2 \pm i .$$ (A.54)

There are thus two solutions to Eq. (A.49),

$$x = -2 - i ,$$ (A.55)

and

$$x = -2 + i .$$ (A.56)

A.3   Linear Functions

A linear function is a function of the form

$$f(x) = mx + b .$$ (A.57)

The number b in the above equation is the y-intercept—the value of the function when x is 0. The number b is thus the point at which the function intersects the y-axis. This is illustrated in Fig. A.1.

The quantity m is the slope; it measures how steep the line is. The slope is given by the "rise over run". To determine the slope m one chooses any two points on the line. The "rise" is the difference in "altitude" (i.e., the y value) between the two points, and the "run" is the difference in the x values. That is,

$$m = \frac{y_2 - y_1}{x_2 - x_1} ,$$ (A.58)

where x_1, y_1 are the coordinates on one point on the line, and x_2, y_2 are the coordinates of any other point. It does not matter which points you chose; you will get the same slope no matter what.

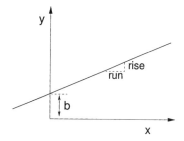

Fig. A.1 A linear function. The quantity b is the y-intercept. The slope, denoted m, is the "rise" divided by the "run".

Some Examples

Let us illustrate some properties of linear functions with some examples. To begin, consider Fig. A.2. The y-intercept can be read directly off the graph; we see that the function has a value of 3 when $x = 0$. Thus, $b = 3$. The slope can also be determined from the graph. To do so, choose any two points on the graph. For example, we could use $(2, 7)$ and $(3, 9)$. The rise between these two points is $9 - 7 = 2$, while the run is $3 - 2 = 1$. Thus the slope, which is the rise divided by the run, is 2/1 or simply 2. A slope of 2 means that the line rises two units for every unit that we move to the right. Thus, the equation for the line of Fig. A.2 is

$$f(x) = 2x + 3 \ . \tag{A.59}$$

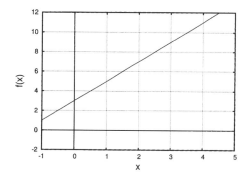

Fig. A.2 A linear function. The y-intercept is 3. The slope m is 2; moving to the right by one unit corresponds to moving two units up.

For our next example, consider the linear function shown in Fig. A.3. Again reading off the graph, we see that the y-intercept b is 6. The slope is $-\frac{1}{2}$. To see this, note that the line decreases by 1 unit for every 2 units one moves to the right. Thus, the rise is -1 and the run is 2. So the equation for this line is

$$f(x) = -\frac{1}{2}x + 6 \ . \tag{A.60}$$

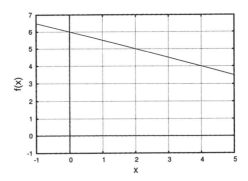

Fig. A.3 A linear function. The y-intercept is 6. The slope m is $-\frac{1}{2}$.

A.4   Logarithms

Logarithms Defined

Suppose we have an equation of the form

$$10^x = 25 \, , \qquad\qquad\qquad (A.61)$$

and we need to solve for x. At first blush, there may not appear to be any way to isolate x. So let us guess different x values instead of trying to deduce the answer using algebra. If $x = 1$, we have $10^1 = 10$ on the left-hand side of Eq. (A.61), and if $x = 2$, then we have $10^2 = 100$ on the left-hand side. So the x value must be between 1 and 2.

We can also see this graphically. In Fig. A.4 I have plotted a graph of the function 10^x. We are looking for the x such that $10^x = 25$. From looking at the figure it appears that this occurs somewhere between $x = 1.3$ and 1.5. As expected based on our argument in the above paragraph, the x value we seek is indeed between 1 and 2.

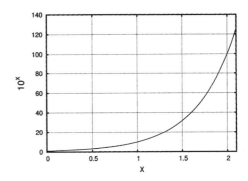

Fig. A.4 The function 10^x. We are looking for the x value that makes $10^x = 25$. It appears that this occurs somewhere between 1.3 and 1.5.

In Fig. A.5 I have again plotted the function 10^x, but this time the x axis ranges from 1.3 to 1.5. Reading off the graph, it appears that when $x = 1.4$, 10^x is very close to 25. Let us try this out. Using my calculator, I find that $10^{1.4} \approx 25.12$. So $x = 1.4$ is a little bit too large. So let us try $x = 1.39$. Doing so, I get $10^{1.39} \approx 24.55$. So this x value is too small. I could keep experimenting with different x values. Eventually I would find that $10^{1.398} \approx 25.003$.

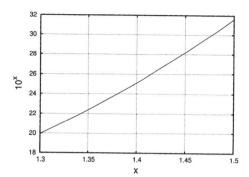

Fig. A.5 The function 10^x. We are looking for the x value that makes $10^x = 25$. It appears that this occurs close to $x = 1.4$.

So by using a few graphs and some repeated guessing and checking, we have found that 1.398 is that number, which, when 10 is raised to it, gives us 25. In other words,

$$10^{1.398} \approx 25 . \tag{A.62}$$

We have thus solved Eq. (A.61) for x. Doing so was fairly straightforward, but certainly a little cumbersome. It would be nice if there was a simpler method. In particular, it might be nice if there was a name for the number 1.398 that appears in Eq. (A.62), as it appears that it could be moderately handy.

More generally, we might be interested in a number x that makes the following equation true:

$$10^x = y . \tag{A.63}$$

(In the above example, $y = 25$ and $x \approx 1.398$.) Clearly, the number x must be related to y somehow. This relationship is known as the **logarithm**. Specifically, if Eq. (A.63) is true, then we say that x is the logarithm of y, and we denote this as $x = \log(y)$. Using this, we can write Eq. (A.63) as

$$10^{\log(y)} = y . \tag{A.64}$$

I tend to think of this equation as being the relationship that defines the logarithm. When I am trying to remember or prove a relationship about logarithms, this is where I usually start.

Equation (A.64) shows us that the logarithm is the inverse of the exponential function. That is, the log "undoes" exponentiation.[3] I.e.,

$$\log(10^y) = y . \tag{A.65}$$

[3] By exponentiation, I mean the act of taking a number and using it as an exponent. For example, exponentiating the number 7 gives 10^7.

Properties of Logarithms

The defining equation for logarithms, Eq. (A.64), lets us quickly derive some properties of logarithms. For example, suppose that we are interested in the log of AB, the product of two numbers. Then Eq. (A.64) tells us that

$$10^{\log(AB)} = AB . \tag{A.66}$$

However, we also know that

$$A = 10^{\log(A)} , \tag{A.67}$$

and

$$B = 10^{\log(B)} . \tag{A.68}$$

Using Eqs. (A.67) and (A.68) on the right-hand side of Eq. (A.66), we get

$$10^{\log(AB)} = 10^{\log(A)}10^{\log(B)} . \tag{A.69}$$

Using the fact that $x^a x^b = x^{a+b}$, we rewrite the right-hand side of the above equation to obtain

$$10^{\log(AB)} = 10^{\log(A)+\log(B)} . \tag{A.70}$$

Hence, we have that

$$\log(AB) = \log(A) + \log(B) . \tag{A.71}$$

This equation tells us how to take the logarithm of a product of two numbers.

What if we want to take the log of a number that is itself raised to a power. For example,

$$\log(A^3) = ??? \tag{A.72}$$

To approach this we only need remember that an exponent means successive multiplication and then make use of the property we just obtained in Eq. (A.71). Doing so, we find

$$\begin{aligned}
\log(A^3) &= \log(AAA) & \text{(A.73)}\\
&= \log(A) + \log(A) + \log(A) & \text{(A.74)}\\
&= 3\log(A) . & \text{(A.75)}
\end{aligned}$$

Note that I used Eq. (A.71) to go from Eq. (A.73) to (A.74). The general result is:

$$\log(A^n) = n\log(A) . \tag{A.76}$$

This property is particularly useful because it lets us easily solve equations for variables that are "upstairs" in the exponent. An example of this is given below.

We can combine the properties for logarithms of products and exponents in Eqs. (A.71) and (A.76) to obtain an expression for the logarithm of a quotient:

$$\begin{aligned}
\log\left(\frac{A}{B}\right) &= \log(AB^{-1}) & \text{(A.77)}\\
&= \log(A) + \log(B^{-1}) & \text{(A.78)}\\
&= \log(A) - \log(B) . & \text{(A.79)}
\end{aligned}$$

Finally, note that there is not an expression for $\log(A+B)$. In particular,

$$\log(A + B) \neq \log(A) + \log(B) . \tag{A.80}$$

Summary of Properties

For reference, here are a few of the key properties of logarithms:

$$\log(AB) = \log(A) + \log(B) . \tag{A.81}$$

$$\log(A^n) = n \log(A) . \tag{A.82}$$

$$\log\left(\frac{A}{B}\right) = \log(A) - \log(B) . \tag{A.83}$$

Some Examples

Suppose we wish to solve the following equation for x:

$$5 = 3^x . \tag{A.84}$$

To do so, first take the logarithm of each side:

$$\log(5) = \log(3^x) . \tag{A.85}$$

Using Eq. (A.82) on the right-hand side, we get

$$\log(5) = x \log(3) . \tag{A.86}$$

We now solve for x by dividing both sides by $\log(3)$:

$$x = \frac{\log(5)}{\log(3)} . \tag{A.87}$$

The above expression is an exact answer. We can approximate it by using a calculator. Doing so, we obtain

$$x \approx 1.46497 . \tag{A.88}$$

You might wish to take a moment and verify that you get this number using your calculator; it is easy to accidentally enter things incorrectly. In particular, please note that

$$\frac{\log(5)}{\log(3)} \neq \log\left(\frac{5}{3}\right) . \tag{A.89}$$

As a second example, suppose we wish to solve the following equation for x:

$$5^{3x} = 100 . \tag{A.90}$$

First, we take the log of each side:

$$\log(5^{3x}) = \log(100) . \tag{A.91}$$

Using the fact that $\log(A^n) = n \log(A)$, we get

$$3x \log(5) = \log(100) . \tag{A.92}$$

Solving for x, we obtain

$$3x = \frac{\log(100)}{\log(5)} , \tag{A.93}$$

$$x = \frac{\log(100)}{3 \log(5)} . \tag{A.94}$$

Evaluating the right-hand side using a calculator, we obtain

$$x \approx 0.9538 . \tag{A.95}$$

Exercises

(A.1) Simplify:

   (a) $\sqrt{2^8}$

   (b) $x^{a^b} x^{-b}$

   (c) $\frac{x^a y^a}{x^{2a} y^{-b}}$

   (d) $z^b z^{-b}$

   (e) $\sqrt[3]{y^{15}}$

   (f) $x^a x^{\frac{1}{a}}$

(A.2) Evaluate the following using a calculator:

   (a) $7^{\frac{2}{3}}$

   (b) $\sqrt[4]{20}$

   (c) $(-14)^4$

   (d) 4^{-14}

   (e) $\frac{3^4}{5^6}$

   (f) $\frac{2^{\frac{1}{2}}}{4^{\frac{1}{3}}}$

   (g) 3^{4^5}

(A.3) Evaluate the following without using a calculator:

   (a) 2^3

   (b) $9^{\frac{1}{2}}$

   (c) 0^4

   (d) $16^{\frac{1}{4}}$

   (e) 7^1

   (f) $\frac{17^9}{17^8}$

   (g) 4^{-2}

   (h) 0^0

(A.4) Solve for x:

   (a) $x^2 - 36 = 0$

   (b) $2x^2 + 1 - 10x = 0$

   (c) $3x^2 + 2x = 10$

(A.5) Let $f(x) = 3x(1-x)$. Find the fixed point(s) of f.

(A.6) Let $f(x) = 3x^2 - 2$. Find the fixed point(s) of f.

(A.7) Let $g(x) = x^2 + c$.

   (a) For what value(s) of c does g have two real fixed points?

   (b) For what value(s) of c does g have one real fixed point?

   (c) For what value(s) of c does g have no real fixed points?

(A.8) Draw the graph of the following functions:

   (a) $f(x) = 3x + 1$

   (b) $f(x) = -3x + 1$

   (c) $f(x) = \frac{1}{2}x - 1$

   (d) $f(x) = 2$

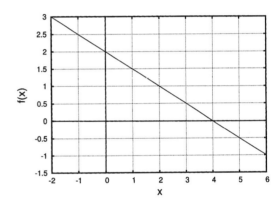

Fig. A.6 The function for Exercise A.9.

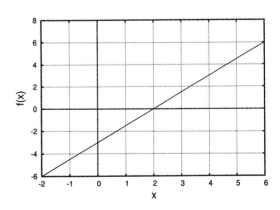

Fig. A.7 The function for Exercise A.10.

(A.9) Determine the formula for the linear function plotted in Fig. A.6.

(A.10) Determine the formula for the linear function plotted in Fig. A.7.

(A.11) To get firewood delivered to your house costs $150 a cord, plus an additional $50 for a delivery fee. The delivery fee is $50 no matter how many cords you order. Write down a formula for $f(c)$, the cost of c cords of wood.

(A.12) For all of these exercises, solve for x. Express your answer both as an exact value, such as $\log(2)$, and as an approximate number, such as 0.301.

(a) $72 = 10^x$

(b) $3 = 6^x$

(c) $3^x = 6$

(d) $2^{5x} = 1000$

(e) $7^{3x} = 2^{4x}$

(f) $44 = 7 \times 10^x$

(A.13) Explain why $\log(1) = 0$.

(A.14) Explain why $\log(x)$ approaches $-\infty$ as x approaches 0.

(A.15) Explain why $\log(x)$ is not defined for negative values of x.

(A.16) You deposit $100 in an account that earns 5% interest yearly. The amount of money $M(t)$ in your account is thus given by the function:

$$B(t) = 10(1.05)^t . \tag{A.96}$$

(a) How much money do you have after two years?

(b) How long would you have to wait for your money to double?

(c) If you wanted your money to double in ten years, what interest rate would you need to earn?

Histograms and Distributions

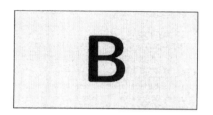

B.1 Representing Data with Histograms

This appendix is a brief introduction to histograms and distribution functions, which are used extensively in Chapters 13 and 20. Histograms are type of graph used to summarize the frequencies of outcomes in a data set. They are very widely used; they are by no means particular to the study of chaos or fractals. I suspect that most readers will have seen histograms before in a previous math or science class, or perhaps when reading the newspaper or a magazine.

As usual, we start with an example. Let us suppose we are studying the heights of a certain variety of tomato plants. Suppose my garden has ten such plants, and they have the following heights, measured in meters:

$$1.21, 1.13, 1.18, 0.92, 0.96, 1.14, 1.21, 1.38, 0.84, 1.04 . \qquad \text{(B.1)}$$

How can we summarize this data? One thing we could do would be to calculate the average height. The average turns out to be 1.065. But this single number is a rather coarse summary of my garden full of tomato plants.

If we want to give more information about the heights of the tomato plants, one option is to simply list the heights of all of the plants, as was done in Eq. (B.1). For a small set of data such as this, listing them all is a viable option. However this is not feasible for large data sets, where such a listing might take many pages. In such an instance we seek a way of summarizing the data that is more compact than a long, exhaustive listing, but which is more informative about the different data values than stating the average value.

A histogram is a type of graph that meets these criteria. The idea is to report not every single data point, but rather the number of data points that fall inside a particular range. A histogram for the tomato heights listed in Eq. (B.1) is shown in Fig. B.1. Each box in the histogram corresponds to one of the data points. For example, Fig. B.1 tells us that there is one tomato plant whose height is between 0.8 and 0.9 meters, there are two tomato plants between 0.9 and 1.0 meters, and so on. If you have not seen histograms before, take a moment to make sure you see how to go from Eq. (B.1) to Fig. B.1.

Usually the individual boxes on a histogram are not shown, as in

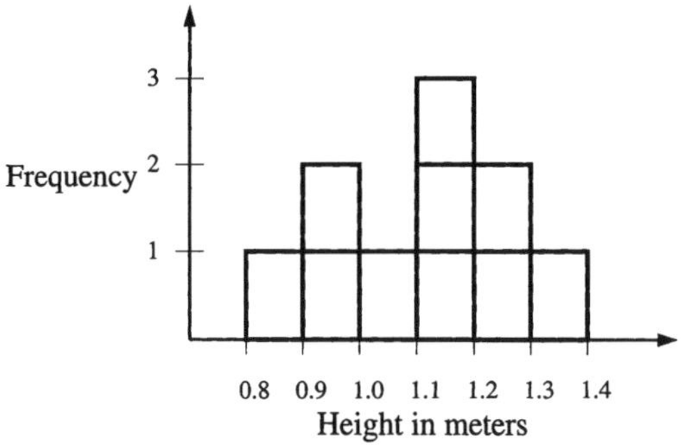

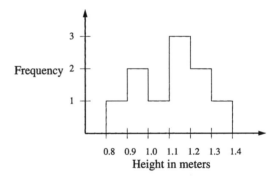

Fig. B.1 A histogram for the heights of the tomato plants. Each box corresponds to one of the data points listed in Eq. (B.1).

Fig. B.2. When reading this histogram, one determines the number of data points in a given range not by counting boxes but instead by reading the corresponding value off the vertical axis. For example, we see that there are three plants between 1.1 and 1.2 meters, while there are two plants between 1.2 and 1.3 meters.

Fig. B.2 A histogram for the heights of the tomato plants. This is the same as the previous figure, except that the individual boxes for each data point are not shown. Instead, the number of boxes can be read off the vertical axis.

B.2   Choosing Bin Sizes

I hope that the above example has convinced you that making a histogram is not a difficult a task. However, there are two subtleties associated with making and interpreting histograms. The first, discussed this section, is the selection of a bin size for the histogram. The second concerns how to normalize and interpret histograms in terms of probabilities; this is the topic of the next two sections.

When forming a histogram we must choose a range over which we group the data values. For example, in the previous section I used a range of 0.1 meters. That is, in forming the histogram of Fig. B.1, I collected the data values into groups with a range of 0.1. We interpret the histogram as telling us, for example, that there are three plants whose height is between 1.1 and 1.2, while there is just one plant whose height is between 1.0 and 1.1. The histogram does not tell us the exact values of the data, just how many fall in each range.

The intervals that are used to group the data are usually referred to as **bins**. The width of each bin is called the **bin size**. In Fig. B.1, the bin size is 0.1. But there is nothing special about the bin size of 0.1. We could just as well choose a bin size of 0.05. Doing so yields the histogram of Fig. B.3. Note that Figs. B.1 and B.3 look quite different. Which is

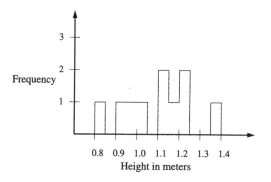

Fig. B.3 A histogram for the heights of the tomato plants. This is identical to Fig. B.2, except that a bin size of 0.05 has been used.

correct? They both are. Which is more useful? It depends on what features of the data we are trying to illustrate with the histogram. In general, there is no single correct bin size. However, choosing a "wrong" bin size can make the histogram misleading.

This is illustrated in rather dramatic fashion in Fig. B.4. The four histograms in this figure look different, but I made them using the same data. For all figures the same 200 data points were used. In the first figure the bin size is fairly large, 0.05. In the second plot, I made it smaller, 0.01. In the first plot it looks like the data are distributed fairly evenly from about 0.42 to 0.63. However, in the second plot, with a smaller bin size, we see that this is not the case. The histogram has two quite distinct peaks, one near 0.5 and the other near 0.6.

This illustrates again that histograms can look quite different depending on the bin size one chooses. In this instance, the second plot in Fig. B.4 is clearly better, as it shows the concentration of data around 0.5 and 0.6, while this cannot be seen in the top plot. However, this begs the question: what is the optimal size for histogram bins? It might seem that we should choose bins as small as possible so as to see all of the structure that is present in the data. However, this is not a good idea. The problem is that if the bin size is too small, then there will be, on average, very few data that fall in each bin. In the extreme case, there may be only one or two data points for each bin.

An example of this is shown in the third histogram from the top in Fig. B.4 where I have used a bin size of 0.001. This plot is, at best, somewhat confusing. One can see that there are two peaks in the data, but it is much less clear than in the previous figure. Finally, the bottom plot in Fig. B.4 was made with a bin size of 0.0001. Now the bin size is so small that no more than two data points are in any one bin. The result is that the two peaks, evident in the previous two histograms, are essentially gone.

This illustrates that there is a tradeoff. One wants to use as small a bin size as possible to learn as much as one can about the details of the way the data are distributed. But the data impose limits to how small

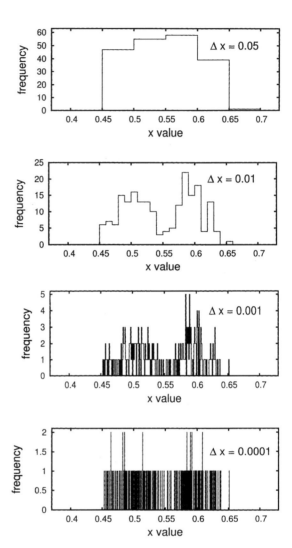

Fig. B.4 Four different histograms formed with identical data but using different bin sizes. In the top histogram, the bin width is 0.05, in the next histogram the bin width is 0.01, in the next histogram the bin width is 0.001, and in the bottom histogram the bin width was 0.0001.

the bins can be. If the bins are too small, there may not be enough data points to fill most of the bins more than once or twice. The result, then, will be a histogram that looks like the bottom plot in Fig. B.4.

This fundamental tradeoff is common in data analysis and statistics. Any data set is finite, and this thus imposes limits on what one can infer about the data. The goal is to infer as much as one can, but not so much that one is essentially drawing inferences from one or two observations. Much of advanced statistics is concerned with the optimal way to make this tradeoff. As a practical matter, the best thing to do when making a histogram is to is experiment with different bin sizes and look at the resultant histograms. The fact that histograms can look quite different for different bin sizes is something to bear in mind when making histograms of your own[1] or when looking at histograms that are reported in research papers or in newspapers or magazines.

[1]Most spreadsheet programs such will quickly make histograms for you.

B.3 Normalizing Histograms

A histogram lets us see not just the average value of the data, but how the data are distributed. That is, with a histogram we can observe how the data values are spread out or divided among different ranges of values. It is often the case that this distribution is the feature of the data that we are interested in. The total number of data points that fall in any one bin will depend on the number of data points in our sample. In contrast, the distribution, since it is expressed in terms of fractions, does not depend on the total number of data points.

Consider again the histogram of Fig. B.3. Here we can see that two tomato plants are between 1.20 and 1.25. Since there are ten total plants, would say that the fraction of the plants that are between 1.20 and 1.25 is 0.2. The reason for this is that there are ten total data points, and so the fraction between 1.20 and 1.25 is $\frac{2}{10} = 0.2$.

It would be nice to have a histogram from which we can determine frequencies directly, without having to do the division required in the above paragraph. For this simple example the division was straightforward, but in general, for a large data set, it may be a little messy. More importantly, if we can read frequencies directly off of histograms, we can more readily compare two different data sets that might be of different sizes.

One can modify the histogram to express frequencies in terms of fractions. This process is a little bit subtle. Let us look again at Fig. B.3. On this histogram each box represents one data point. Note that there are ten boxes, and there are ten data points listed in Eq. (B.1). But now we want each box to represent a fractional part of the entire data set. In this case, since there are ten data points, each box should represent $\frac{1}{10}$, or 0.1. It is customary to have the *area* of the box equal this fraction. We achieve this by an appropriate rescaling the vertical axis.

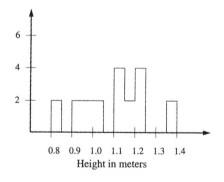

Fig. B.5 The normalized histogram for the heights of the tomato plants. This is identical to Fig. B.3, except that the scale has been normalized. The height of each box is now 2. The total area under the histogram is 1.

In this case, we want each rectangular box to have an area of 0.1. The base of the rectangle is 0.05. What must its height be? Well, the formula for the area of a rectangle is,

$$\text{Area} = \text{Base} \times \text{Height} . \tag{B.2}$$

Solving for height, we get:

$$\text{Height} = \frac{\text{Area}}{\text{Base}} \, . \tag{B.3}$$

Plugging in, we find:

$$\text{Height} = \frac{0.1}{0.05} = 2 \, . \tag{B.4}$$

Thus, each box should now have a height of 2 and not 1.

Such a histogram is shown in Fig. B.5. We can use this plot to determine the fraction of our data in a given range. For example, suppose we want to know the fraction of our data between 1.10 and 1.15. To determine this, we look at the area under the graph between 1.10 and 1.15. This area is a rectangle with a base of 0.05 and a height of 4. Since $0.05 \times 2 = 0.4$, we conclude that the fraction of the data between these two values is 0.2.

The process used to make the histogram of Fig. B.5 is known as **normalization**. What we have have done is ensured that each box has an area equal to the fraction of the data it represents; if there are N data points, each box must have an area of $\frac{1}{N}$. As a result, since there are N total data points, the total area of all the boxes is 1. Geometrically, this means that the total area under the histogram equals 1.

It is crucial to remember that in normalized histograms is the *area* that we interpret as fractions, not the value on the vertical axis. For example, in the previous example had we used the value on the vertical axis, we would have concluded that the fraction of tomato plants between 1.10 and 1.15 is 4, which clearly is nonsense, since the fraction must between 0 and 1.

The process used above to normalize the histogram is generalized as follows. Let us assume that we have N data points and that we have chosen a bin size of Δx. Each box should have an area of $\frac{1}{N}$. We determine the height for the boxes by plugging in to Eq. (B.3). Doing so, we obtain:

$$\text{Height} = \frac{\text{Area}}{\text{Base}} = \frac{\frac{1}{N}}{\Delta x} \, . \tag{B.5}$$

Simplifying, this may be written as

$$\text{Height} = \frac{1}{N \Delta x} \, . \tag{B.6}$$

Note that this equation says that the height of each box will get smaller as the total number of data points N gets larger. In addition, as we make our bin size Δx smaller, the height of the box needs to get larger to ensure that each box has the proper area.

A final technical issue concerns the units on the vertical axis of a normalized histogram such as Fig. B.5. The key thing to remember with a normalized histogram is that areas of rectangles are interpreted as frequencies or probabilities. In other words

$$\text{Frequency} = \text{Height} \times \text{Base} \, . \tag{B.7}$$

The units on the base of the rectangle are meters for the example of the heights of tomato plants. Thus, the units for the height of the rectangle must be frequency/meters. This way, multiplying the base by the height yields a frequency, as it should. So the units on the vertical axis of Fig. B.5 are frequency/meter. However, I have not included units on the figure. While I tend to believe that axes should always have their units labeled, it is not unusual to leave off vertical units for a normalized histogram, since what is meaningful on such a histogram is the area— interpreted as a frequency or probability—and not the vertical scale. By indicating that the histogram is normalized, one knows that the total area under the curve must equal 1. This is all that is needed to interpret the histogram quantitatively.

B.4 Approximating Histograms with Functions

There is one more aspect of histograms that bears discussion. Frequently, one approximates the staircase-like curve of a histogram with a continuous function. For the examples we have considered thus far, this might not appear to make much sense, as the histograms seem inescapably bumpy. Hence, it appears unwise to approximate them with something that is smooth. However, in many cases such an approximation does make sense. Here is one example.

Let us imagine that you are interested in the size of cats, and thus determine the masses of a great many cats. Their average mass turns out to be close to five kilograms. But you are interested in more than just the average; you want to know how little the little ones are, how big the big ones are, and so on. So you plot your data in a histogram. The (imaginary) results of doing this are shown in Fig. B.6. The top histogram shows 100 measurements, the middle plot 1,000 measurements, and the bottom plot 10,000 measurements. All histograms are normalized, and for all I used a bin size of 0.05.

In Fig. B.6 one can see that the histogram is getting less jagged and bumpy as the number of data points increases. It is natural to ask, then, whether or not we can approximate this histogram by a smooth function. In this instance, it turns out that we can. The tops of the histogram are very well approximated by the following function:

$$p(x) = \frac{1}{\sigma\sqrt{2\pi}} e^{\frac{-(x-a)^2}{2\sigma^2}} , \qquad (B.8)$$

where $a = 5.0$ and $\sigma = 0.5$. As I include more and more data points, and as the bin size gets smaller and smaller, the histogram looks more and more like a smooth curve and less like a staircase. And the curve that the histogram resembles is $p(x)$.

Where does Eq. (B.8) come from? It turns out that this is an equation that describes a vast number of distributions. You have probably seen it before: it is known as a **normal** or **Gaussian** distribution. Is is

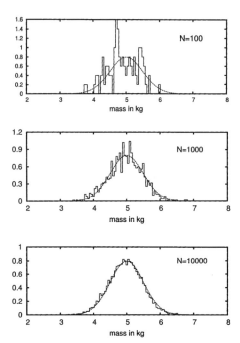

Fig. B.6 Three different normalized histograms for increasing amount of data. The number of data points is 100, 1,000, and 10,000, top to bottom. For each histogram a bin size of 0.05 was used. The dashed line is the distribution function $p(x)$, Eq. (B.8).

[2]Determining an algebraic expression for such a curve as in Eq. (B.8) is, however, another matter. In some cases, as with the cats in this example, there may be a good theory that tell us what sort of function to use. In other cases there may be multiple possible functions that fit that data fairly well, in which case one would need to use some techniques from statistics in order to decide which function to chose. For now, we will assume we have a correct (or good enough) $p(x)$. The focus of this section is how to interpret a probability distribution once one has been found.

also often referred to as a **bell curve**. In Eq. (B.8) the quantity a is the average value of x, and σ is a measure of how spread out the data points are. A larger σ means there is more variation around the mean; geometrically this has the effect of widening the bell curve.

The normal distribution is discussed in more detail in Section 20.1. For now, let it suffice to mention that this distribution function can be shown to apply to essentially any situation in which the variable in question—in this case x, the mass of the cats—depends in an additive way on many other variables. In this example, these other variables might be things like how much food the cats ate when they were kittens, the quality of their food, how much other cats beat them up, how often they got sick, the size of their parents, and so on. By additive, I mean that these effects can be added together (as opposed to multiplied) to yield a prediction of the cat's mass. But the main goal of this section is to learn how to interpret $p(x)$.

So, what does Eq. (B.8) mean? The quantity $p(x)$ is known as the **probability density**. It is also referred to as a **probability distribution** or a **distribution function**. Equation (B.8) is plotted in Fig. B.7. The procedure for forming $p(x)$ is, I hope, clear: one builds histograms with smaller and smaller bin sizes and with more and more data points. Usually the histogram gets smoother and smoother and resembles a continuous curve. The resultant curve is $p(x)$.[2] However, interpreting $p(x)$ requires some care.

When interpreting a probability density the key thing to keep in mind is its genesis as a histogram. We use a normalized histogram to inquire about the fraction of the data which fall in a *range* of values. And

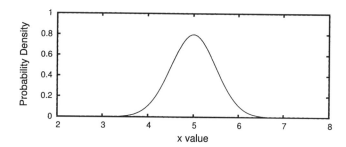

Fig. B.7 The probability density for a variable distributed according to $p(x)$, Eq. (B.8), a normal distribution with an a of 5.0 and a σ of 0.5.

we determine this fraction by calculating the *area* under the histogram. Probability density functions are interpreted in exactly the same way.

For example, suppose we are interested in the fraction of cats between 5.0 and 5.2 kilograms. This quantity is given by the area underneath the $p(x)$ curve between 5.0 and 5.2. We can figure this out by approximating this area by the dashed rectangle shown in Fig. B.8. The base of this rectangle is 0.2. The height of the rectangle is 0.8. Multiplying these two numbers together yields 0.16. Thus, we can infer that approximately 16% of the data fall in this range. This is an approximate statement because the area under the $p(x)$ curve is only approximately equal to the rectangle. The actual area will be slightly less.

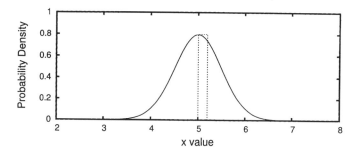

Fig. B.8 The probability density for a variable distributed according to $p(x)$, Eq. (B.8), a normal distribution with an a of 5.0 and a σ of 0.5. The fraction of data between 5.0 and 5.2 is approximately given by the area of the dashed rectangle.

It is tempting to interpret Fig. B.7 as telling us that, for example, the probability that a cat has a mass of 5.0 kilograms is 0.8, or that the fraction of cats with mass 4.5 kilograms is around 0.4. But this is not correct. One cannot read the vertical axis in this way; it is the area under the curve that is the probability. In fact, strictly speaking it makes no sense to say that the probability that a cat is exactly 5.0 kilograms is 0.8. There are an infinite number of different masses that a cat can be. So the probability that it is any one, exact, particular value is zero.

The units on the probability density function for this example are probability per kg. To see this, recall that the area of the rectangle in Fig. B.8 is interpreted as a probability. The base of this rectangle has units of kilograms, since x is a mass. The height, then, must have units of probability/kg, so that the height times the base yields an area that has units of probability. The units on $p(x)$ help to justify us calling it a probability density. A mass density has units of mass per volume, perhaps kg/m^3. In order to get a mass from a mass density, one has to multiply the mass by a volume. Similarly, to get a probability from a probability density $p(x)$, one needs to multiply $p(x)$ by whatever units are on x.

Exercises

(B.1) Form a histogram and a normalized histogram for the following data set:

$$1.81, 2.55, 2.32, 2.48, 2.41,$$
$$1.80, 1.92, 2.03, 1.99, 2.26,$$
$$2.38, 1.92, 2.29, 2.18, 1.88,$$
$$2.14, 2.18, 1.78, 2.19, 2.37 \ .$$

(B.2) Form a histogram and a normalized histogram for the following data set:

$$4, 3, 3, 3, 2, 2, 4, 17, 4,$$
$$5, 2, 2, 3, 6, 3, 5, 7, 11,$$
$$5, 8, 6, 4, 5, 7, 3, 2, 2 \ .$$

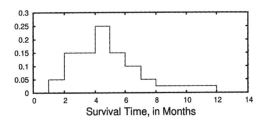

Fig. B.9 A normalized histogram for survival times, in months, after diagnosis of a disease.

(B.3) Use Fig. B.8 to estimate the probability that a cat is between 4 and 5 kilograms.

(B.4) A study is done with 100 patients to see how long they live after being diagnosed with a fatal disease. A normalized histogram for survival times is shown in Fig. B.9.

(a) What fraction of the patients live between 5 and 6 months after diagnosis?

(b) What fraction of the patients live between 6 and 8 months?

(c) What fraction of the patients live more than 9 months?

(d) What fraction of the patients live between 0 and 14 months?

(e) What fraction of the patients live between 5 and 5.5 months?

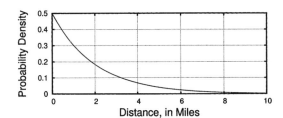

Fig. B.10 The distribution of distances between roadkill.

(B.5) It has been determined that on a road in a national park, the distance between roadkill is approximately distributed by the function shown in Fig. B.10.

(a) What is the approximate probability that two roadkills are found between 1 and 1.5 miles apart?

(b) What is the approximate probability that two roadkills are found between 1 and 2.0 miles apart?

(c) What is the approximate probability that two roadkills are found at least 3 miles apart?

(d) What is the approximate probability that two roadkills are found exactly 2.1 miles apart?

(B.6) What are the units for the probability density function shown in Fig. B.10?

(B.7) What are the units for the probability density function shown in Fig. B.11?

(B.8) Suppose a certain type of snake ranges in length from 1 meter to 3 meters. All lengths in between these extremes are equally likely. Sketch a probability distribution function that describes the distribution of snake lengths.

(B.9) The distribution of trees in a certain forest is given by the distribution function shown in Fig. B.11. Why is it it not true to say that most trees are around 1 meter tall?

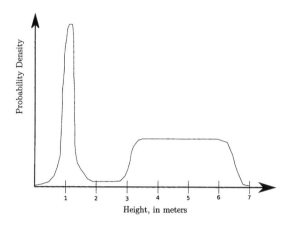

Fig. B.11 The distribution of the heights of trees.

Suggestions for Further Reading

The topics covered in this book are just the beginning. The fields of chaos and fractals have many more delights and surprises than I have been able to cover here. There are also many additional applications and extensions of chaos and fractals. In this appendix I aim to provide some suggestions for those who wish to explore further.

A vast amount has been written about chaos and fractals and related areas. And there there is much interesting research being done on chaos and fractals and their application to fields ranging from physics and physiology to literature and political science. I think most would agree that the bulk of this work ranges from rigorous and solid, to usefully speculative and provocative. However, not all would agree as to where in this spectrum a particular piece of work belongs. Also, I think there are some writings about chaos and fractals that are too sensational and/or are based on fundamental misunderstandings of what chaos and fractals actually are. So you should be aware that, in my view, a small but not negligible fraction of what has been written about chaos and fractals stretches metaphors, stretches the truth, or is just plain wrong. I hope this book has prepared you to approach chaos and fractals with an informed, critical, but open-minded perspective.

Many of the references and resources that I have listed below are at a more advanced level than this book. Do not let that prevent you from investigating them. Yes, some portions may be difficult to follow. But you can definitely get a lot out of them regardless. Learning how to read technical material and skim or skip parts that do not make sense is an important and useful skill.

C.1   (Mostly) Books

Below are a number of additional references, mostly books. I have not tried to make an exhaustive list. Rather, I have chosen books which I think are particularly strong or that have been particularly influential. Please take this list as a starting point, not the final word.

Gleick's *Chaos: Making a New Science*

One book is so important that it gets a section of its own. James Gleick's *Chaos: Making a New Science* (Gleick, 1987) is an accessible, non-technical survey of chaos and fractals. Gleick, a science writer, does an extraordinary job of capturing the fun and excitement of the early years of research and discovery in chaos and fractals. The book not only was a popular success, but it arguably also helped to synthesize the emerging area of chaos and dynamical systems (Lewenstein, 2007) and draw young researchers to the field.

When I teach chaos and fractals, I have always had students read Gleick's book, and almost all find it readable, interesting, and thought-provoking. In many ways, this textbook was designed to be an introductory technical companion to Gleick's book.

Gleick's book received decidedly mixed reviews from mathematicians. A series of reviews, responses, and counter-responses in the *Mathematical Intelligencer* in 1989 make fascinating reading and provide insight into divisions within the mathematics community as well as distinctions between mathematics and other areas of science. John Franks published a largely negative review (1989c) of *Chaos: Making a New Science* that was accompanied by a response from James Gleick (1989a) and a response by Franks to Gleick's response (1989b). An essay by Morris Hirsch (1989) critical of some aspects of Gleick's book appeared subsequent issue of the *Mathematical Intelligencer*, again accompanied by a response from Gleick (1989b). This issue also contained an essay by Benoit Mandelbrot (1989a) in which he took issue with some of the comments about fractals that Franks made in his review of *Chaos: Making a New Science*. The opinion section of this issue concluded with Franks (1989a) responding to Hirsch, Gleick, and Mandelbrot. In the same issue, letters to the editor by Keith Devlin (1989) and Ronald Douglas (1989) were supportive of Gleick and took exception to Franks' review. The reception of Gleick's book in the physics community was much more positive (Shlesinger, 1988; Glazier and Gunaratne, 1988).

In a related vein, the pointed commentary on fractals by Steven Krantz (1989), while only tangentially about Gleick's book, is a similar illustration of the hostility and backlash to some of the hype around fractals. Mandelbrot responded to Krantz's essay in a predictably prickly fashion (Mandelbrot, 1989b).

Popular Books

- Philip Ball, *Nature's Patterns: A Tapestry in Three Parts* consists of three short books: *Shapes* (2011c), *Flows* (2011b), and *Branches* (2011a). These are fascinating and well written explorations of some of the common forms and structures in the natural and built worlds.

- Nigel Lesmoir-Gordon, *Introducing Fractal Geometry* (2006) and Ziauddin Sardar *Introducing Chaos* (2005). These are illustrated cartoon-style books. Both are accessible, fun, and technically sound.

- Nigel Lesmoir-Gordon, *The Colours of Infinity: The Beauty, The Power and the Sense of Fractals* (2004). A nicely illustrated volume with contributions about different aspects of fractals. There is an accompanying documentary with the same title, narrated by Arthur C. Clarke. Both the video and the book are informative, engaging, and entertaining.

- Melanie Mitchell, *Complexity: A Guided Tour* (2009). Focuses on complexity and complex systems. Contains excellent discussions of chaos and power laws.

- Ian Stewart, *Does God Play Dice: The New Mathematics of Chaos* (2002). Excellent, highly readable non-technical explanations of the key ideas of chaos. Does a very good job of putting ideas from chaos into a broader scientific context.

- Mitchell Waldrop, *Complexity: the Emerging Science at the Edge of Order and Chaos* (1992). More about complexity and not so much about chaos and fractals. Relevant in so far as complexity is, arguably, an intellectual descendant of chaos and fractals. Similar in style to Gleick's chaos book.

Books at Roughly the Same Level as This Text

- *A Tool Kit of Dynamics Activities* consists of four workbooks: (Devaney, 2000), (Devaney and Choate, 2000), (Choate, Devaney, and Foster, 2000*a*), (Choate, Devaney, and Foster, 2000*b*). Each workbook contains lesson plans and worksheets on chaos and fractals designed for use in high school math and science classes. A very useful set of resources.

- Nina Hall, *Exploring Chaos: A Guide to the New Science of Disorder* (1994). A collection of short non-technical essays giving overviews of different areas of application of chaos and fractals.

- Richard Kautz, *Chaos: The Science of Predictable Random Motion* (2010). A clear and well written introduction to chaos. Somewhat more advanced than this text.

- Benoit Mandelbrot and Michael Frame, *Fractals, Graphics, and Mathematics Education* (2002). A collection of short articles containing many interesting ideas and useful resources for teaching fractals to high-school and college students.

- David Peak and Michael Frame, *Chaos Under Control: The Art and Science of Complexity* (1994). Somewhere between a popular science book and an elementary textbook.

More Advanced Texts

- Ralph Abraham and Christopher Shaw, *Dynamics: The Geometry of Behavior (Studies in Nonlinearity)* (1992). Excellent illustrations are used to explain the geometry of chaos and strange attractors.

- Michael Barnsley, *Fractals Everywhere* (2000). The standard reference on iterated function systems.

- J. M. Cushing, et al, *Chaos in Ecology: Experimental Nonlinear Dynamics* (2002). An overview of applications of nonlinear dynamics in ecology.

- Gary William Flake, *Computational Beauty of Nature.* (1999). Flake does a great job of giving an overview of chaos, fractals, and many topics and themes from complex systems. Highly recommended. Not too much more advanced than this text. Very clear.

- Daniel Kaplan and Leon Glass, *Understanding Nonlinear Dynamics.* (1995). A very clear, interdisciplinary textbook on chaos and fractals. More of an emphasis on biology than other texts. For science majors. Assumes a knowledge of calculus.

- Heinz-Otto Peitgen, Hartmut Jürgens, and Deitmar Saupe, *Chaos and Fractals: New Frontiers of Science* (1992). Do not be intimidated by the size of this book. It is immense. But it is very clear, and there is no need to read it from cover to cover. Strikes an excellent balance between intuition and rigor. Highly recommended. One of my favorite books on chaos and fractals.

- Robert Devaney, *An Introduction to Chaotic Dynamical Systems* (1989). A standard text on dynamical systems for junior/senior level math majors.

- Kenneth Falconer, *Fractal Geometry: Mathematical Foundations and Applications* (2003). The standard advanced undergraduate text on fractals. Written for math majors.

- Larry Liebovitch, *Fractals and Chaos Simplified for the Life Sciences* (1998). This is more of an outline and a collection of overhead slides than a textbook. Nevertheless, a useful reference.

- Steven Strogatz, *Nonlinear Dynamics And Chaos: With Applications To Physics, Biology, Chemistry, And Engineering* (2001). This is a standard text for applied math and physics classes on chaos and nonlinear dynamics.

- Edward Ott, *Chaos in Dynamical Systems* (2002). A well-written textbook with a physics emphasis. Uses more advanced and formal mathematics than Strogatz.

History and Philosophy of Science

- David Aubin and Amy Delmedico, "Writing the History of Dynamical Systems and Chaos: *Longue Durée* and Revolution, Disciplines and Cultures" (2002). A long, thoroughly researched article discussing the history of the study of chaotic dynamics. One of the best and most nuanced histories of chaos I have read.

- Stephen Kellert, *In the Wake of Chaos* (1993). A very clear and well written exploration of the epistemological and philosophical implications of chaos.

- Stephen Kellert, *Borrowed Knowledge: Chaos Theory and the Challenge of Learning Across Disciplines* (2008). A critical look at different attempts to use ideas from chaos in other fields.
- Peter Smith, *Explaining Chaos* (1998). A book about the implications of chaos for both the philosophy and practice of science. At times assumes a knowledge of calculus and differential equations.

Films

- *The Colours of Infinity* (1995). A film about fractals, focusing on the Mandelbrot set. Narrated by Arthur C. Clarke, directed by Nigel Lesmoir-Gordon, and co-written by Clarke and Lesmoir-Gordon. Engaging and informative.
- *The Secret Life of Chaos* (2010). Narrated by Jim Al-Khalili and directed by Nic Stacey. A fascinating and well-produced overview of chaos. Includes interviews with many scientists.

Online Resources

- *Stanford Encyclopedia of Philosophy.* `http://plato.stanford.edu`. A great place for essays and reviews on, among other things, the philosophy of science.
- Wolfram Mathworld. `http://mathworld.wolfram.com`. "A free resource from Wolfram Research...created, developed, and nurtured by Eric Weisstein with contributions from the world's mathematical community." An excellent, highly technical resource.
- WolframAlpha. `www.wolframalpha.com`. A "computational engine." A remarkably useful website. "WolframAlpha introduces a fundamentally new way to get knowledge and answers–not by searching the web, but by doing dynamic computations based on a vast collection of built-in data, algorithms, and methods."

C.2 Peer-Reviewed Papers

Another place to learn more about chaos and fractals is the scientific literature—peer-reviewed articles that are published in scientific journals. This section contains some general remarks on journal articles and then some suggestions on how to search for and obtain articles.

Articles in scientific journals are peer-reviewed. The process usually works as follows. An author or a group of authors submits a paper for publication, and an editor of the journal sends the article out for peer review. This means that the article is evaluated by peers—other scientists working in the same area as the topic of the paper. Often referees are chosen based on who is cited in the paper. In some cases the authors' names are removed from the paper before it goes out for review. This seems to be more common in the biological and social sciences. In physics, this is rarely done. Usually there are two to four referees for a paper. Each writes a short report on the paper recommending publication, rejection, or most commonly the referees make suggestions and then ask to see the paper again. Ultimately, it is the editor of the journal who takes the referee recommendations into account and makes the final publication decision. Referees are anonymous; authors do not know who reviews their papers. Acceptance rates for journals vary widely, but most are in the range from 10% to 50%.

Peer review does not guarantee that a paper is correct. Poor papers get published, and good papers get rejected. But the peer review process does provide some sort of a filter, albeit an imperfect one. It helps avoid egregiously wrong papers, those which are so poorly written as to be unreadable, and those that put forth results that are not novel.

Peer-reviewed publications are a big deal to scientists. A scientist's publication record is a key part of most hiring and promotion decisions. A strong publication record is essential for being competitive for grants, which is how much of science—including scientists' salaries—is funded. Perhaps even more important than the number of publications is a scientist's citation count: the number of times each of her or his papers has been cited by other peer-reviewed papers. The logic is that those papers that are cited are seen as more valuable or important. They certainly are having a larger impact on the field than those papers which are never cited at all.[1]

I mention all this because I think knowing about peer review is helpful when reading the peer-reviewed literature. Peer-reviewed scientific articles are a very different genre of writing than textbooks or even newspaper articles about science. They have a different audience—first the peer reviewers, and then other scientists—with the ultimate goal of getting cited.

[1] There is no guarantee that an often-cited paper is making a positive contribution. It could be a paper with a bad mistake which subsequent authors corrected, and in so doing, provided a citation to the original, wrong paper.

Early Papers in Chaos and Fractals

Robert Hilborn and Nick Tufillano have assembled a "resource letter" in the *American Journal of Physics* (1997) in which they have listed most of the influential early papers on chaos and fractals. They have also compiled many other resources for teaching and learning about these topics. (Their paper was published in 1997, so it does not include any reference more recent than this.)

Google Scholar and Other Databases

You will need a way to search the scientific and scholarly literature. The best free way to do this is via Google Scholar: `http://scholar.google.com/`. This is a version of Google's search engine that is limited to scholarly and academic works. For searching the scientific literature, use Google Scholar and not regular Google. There are also other databases that you can search, but these are not free. Consult your college or university library for details.

A particularly good way to find papers is to search citations. Suppose you find a paper from 2002 about fractal patterns on turtles that you think is really interesting. You might wonder if anyone has done any follow-up work. You can find this out by seeing if anyone has cited the 2002 turtle article. On Google Scholar, the list of references citing a paper can be found on the lower left of its listing.

How to Access Journal Articles

Unfortunately, many paper published in journals are not freely available. Here are some steps you can follow that should help you find many of the journal articles you are looking for.

(1) Google Scholar may point you to a free version of the pdf. Or, your library may have access to it. Google Scholar can be set up to include links directly to your library's online resources. To do so, select "scholar preferences" from the top menu, and then edit the section titled Library Links.

(2) Go back to regular Google (or some other search engine) and do a search for the exact title of the article. I.e., if the title of the article is "Chaotic Dynamics of Cats", enter that exactly into the search box, including the quotations.

(3) Find the personal web pages of the paper's author(s). Many scientists have links to copies of their papers on their website.

(4) Send an email to the author(s) of the paper asking for a copy, explaining very briefly why you are interested in their work. It is very flattering to get such an email, and you will quite often get a reply within a day or two.

(5) If none of these techniques work, it may be possible to get a copy of the paper from your library via Inter Library Loan. For students at a university library this service is often free, but if you are not an enrolled student this service may not be available or it may not be free.

(6) Travel to a library that has a subscription to the journal you are seeking. Even if you do not have a library card, you should be able to enter the library and make photocopies. You may also be able to gain library privileges to a university or other large library, even if you are not enrolled.

C.3   Suggestions for Further Reading

In the spirit of fractals, I cannot resist having a small section on further reading in a chapter about further reading. Hull, Pettifer, and Kell's review article, "Defrosting the Digital Library: Bibliographic Tools for the Next Generation Web," (2008) contains an excellent overview of the current state of digital libraries and also a number of applications for managing and sharing bibliographic materials online. Miller, Chabot, and Messina's, "A student's guide to searching the literature using online databases", (2009), is, as the title suggests, an overview of how to efficiently and effectively carry out searches of the scientific literature online.

References

Abraham, Ralph and Shaw, Christopher D. (1992). *Dynamics: The Geometry of Behavior* (2nd edn). Addison Wesley Publishing Company.

Aubin, David and Dahan Dalmedico, Amy (2002). Writing the history of dynamical systems and chaos: *Longue durée* and revolution, disciplines and cultures. *Historia Mathematica*, **29**(3), 273–339.

Ball, Philip (2011*a*). *Branches: Nature's Patterns: A Tapestry in Three Parts*. Oxford University Press.

Ball, Philip (2011*b*). *Flow: Nature's Patterns: A Tapestry in Three Parts*. Oxford University Press.

Ball, Philip (2011*c*). *Shapes: Nature's Patterns: A Tapestry in Three Parts*. Oxford University Press.

Barnes, Belinda and Fulford, Glenn R. (2002). *Mathematical Modelling with Case Studies: A Differential Equation Approach Using Maple*. Taylor & Francis.

Barnsley, Michael F. (2000). *Fractals Everywhere*. Morgan Kaufmann.

Barrow, John D. (1992). *PI in the Sky: Counting, Thinking, and Being*. Back Bay Books.

Blanchard, Paul, Devaney, Robert L., and Hall, Glen R. (2006). *Differential Equations* (3rd edn). Brooks/Cole.

Boccara, Nino (2004). *Modeling Complex Systems*. Springer.

Choate, Jonathan, Devaney, Robert L., and Foster, Alice (2000*a*). *Fractals: A Tool Kit of Dynamics Activities*. Key Curriculum Press.

Choate, Jonathan, Devaney, Robert L., and Foster, Alice (2000*b*). *Iteration: A Tool Kit of Dynamics Activities*. Key Curriculum Press.

Clauset, Aaron, Shalizi, Cosma R., and Newman, M. E. J. (2009). Power-Law Distributions in Empirical Data. *SIAM Review*, **51**(4), 661+.

Crutchfield, James P. (1994). Is anything ever new? Considering emergence. In *Complexity: Metaphors, Models, and Reality* (ed. G. Cowan, D. Pines, and D. Melzner), Volume XIX of *Santa Fe Institute Studies in the Sciences of Complexity*, pp. 479–497. Addison-Wesley.

Cushing, J. M., Costantino, Robert, Dennis, Brian, Desharnais, Robert, and Henson, Shandelle (2002). *Chaos in Ecology: Experimental Nonlinear Dynamics*. Academic Press.

Cvitanović, Predrag (1989). *Universality in Chaos* (2nd edn). Taylor & Francis.

Dahan Dalmedico, Amy (2004). Chaos, disorder, and mixing: A new fin-de-siècle image of science? In *Growing Explanations: Historical Perspectives on Recent Science* (ed. M. N. Wise), pp. 67–94. Duke University Press.

Devaney, Robert L. (1989). *An Introduction to Chaotic Dynamical Systems*. Perseus Publishing.

Devaney, Robert L. (1996). *Professor Devaney Explains The Fractal Geometry of the Mandelbrot Set (VHS Tape)*. Key Curriculum Press.

Devaney, Robert L. (2000). *The Mandelbrot and Julia Sets: A Tool Kit of Dynamics Activities*. Key Curriculum Press.

Devaney, Robert L. and Choate, Jonathan (2000). *Chaos: A Tool Kit of Dynamics Activities*. Key Curriculum Press.

Devlin, Keith (1989). Letter to the editor. *The Mathematical Intelligencer*, **11**(3), 3.

Douglas, Ronald (1989). Letter to the editor. *The Mathematical Intelligencer*, **11**(3), 3–4.

Epstein, Joshua M. (2008). Why Model? *Journal of Artificial Societies and Social Simulation*, **11**(4), 12.

Falconer, Kenneth (2003). *Fractal Geometry: Mathematical Foundations and Applications* (2nd edn). Wiley.

Flake, Gary W. (1999). *The Computational Beauty of Nature: Computer Explorations of Fractals, Chaos, Complex Systems, and Adaptation*. MIT Press.

Ford, Joseph (1983). How random is a coin toss? *Physics Today*, **36**(4), 40–47.

Frank, Steven A. (2009). The common patterns of nature. *Journal of Evolutionary Biology*, **22**(8), 1563–1585.

Franks, John (1989*a*). Comments on the responses to my review of Chaos. *Mathematical Intelligencer*, **11**(3), 12–13.

Franks, John (1989*b*). Response to James Gleick. *Mathematical Intelligencer*, **11**(1), 70–71.

Franks, John (1989*c*). Review of James Gleick, *Chaos: Making a New Science*. *Mathematical Intelligencer*, **11**(1), 65–69.

Glazier, James and Gunaratne, Gemunu (1988). The fascinating physics of everyday complexity, beautifully portrayed, Review of *James Gleick, Chaos: Making a New Science*. *Physics Today*, **41**, 79+.

Gleick, James (1987). *Chaos: Making a New Science*. Penguin Books.

Gleick, James (1989*a*). Response to John Franks. *Mathematical Intelligencer*, **11**(1), 69–70.

Gleick, James (1989*b*). Response to Morris Hirsch. *Mathematical Intelligencer*, **11**(3), 7–8.

Glenday, Craig (ed.) (2010). *Guinness World Records 2010*. Bantam.

Godfrey-Smith, Peter (2003). *Theory and Reality: An Introduction to the Philosophy of Science* (1st edn). University of Chicago Press.

Gowers, Timothy (2002). *Mathematics: A Very Short Introduction* (1st edn). Oxford University Press.

Grassberger, Peter and Procaccia, Itamar (1983). Measuring the strangeness of strange attractors. *Physica D: Nonlinear Phenomena*, **9**(1-2), 189–208.

Hall, Nina (ed.) (1994). *Exploring Chaos: A Guide to the New Science of Disorder*. W. W. Norton & Company Inc.

Hayes, Brian (2001). Computing science: Randomness as a resource. *American Scientist*, **89**(4), 300+.

Hilborn, Robert C. (2004). Sea gulls, butterflies, and grasshoppers: A brief history of the butterfly effect in nonlinear dynamics. *American Journal of Physics*, **72**(4), 425–427.

Hilborn, Robert C. and Tufillaro, Nicholas B. (1997). Resource Letter: ND-1: Nonlinear Dynamics. *American Journal of Physics*, **65**(9), 822–834.

Hirsch, Morris W. (1989). Chaos, rigor, and hype. *Mathematical Intelligencer*, **11**(3), 6–8.

Hirsch, Morris W., Smale, Stephen, and Devaney, Robert (2004). *Differential Equations, Dynamical Systems, and an Introduction to Chaos* (2nd edn). Academic Press.

Hoefer, Carl (2010). Causal Determinism. In *The Stanford Encyclopedia of Philosophy* (Spring 2010 edn) (ed. E. N. Zalta).

Hull, Duncan, Pettifer, Steve R., and Kell, Douglas B. (2008). Defrosting the digital library: Bibliographic tools for the next generation web. *PLoS Computational Biology*, **4**(10), e1000204+.

Kaplan, Daniel and Glass, Leon (1995). *Understanding Nonlinear Dynamics*. Springer-Verlag.

Kautz, Richard (2010). *Chaos: The Science of Predictable Random Motion*. Oxford University Press.

Keller, Evelyn F. (2005). Revisiting "scale-free" networks. *Bioessays*, **27**(10), 1060–1068.

Kellert, Stephen H. (1993). *In the Wake of Chaos: Unpredictable Order in Dynamical Systems*. University of Chicago Press.

Kellert, Stephen H. (2008). *Borrowed Knowledge: Chaos Theory and the Challenge of Learning across Disciplines*. University of Chicago Press.

Kim, Sang-Hoon (2005). Fractal structure of a white cauliflower. *Journal of the Korean Physical Society*, **46**(2), 474–477.

Kingsland, Sharon E. (1995). *Modeling Nature: Episodes in the History of Population Ecology* (2nd edn). University of Chicago Press.

Krantz, Steven (1989). Fractal geometry. *The Mathematical Intelligencer*, **11**(4), 12–16.

Laplace, Pierre-Simon (2009). *Essai philosophique sur les probabilités* (5th edn). Cambridge University Press.

Lesmoir-Gordon, Nigel (1995). *The Colours of Infinity*. Gordon Films.

Lesmoir-Gordon, Nigel (2004). *The Colours of Infinity: The Beauty, The Power and the Sense of Fractals*. Clear Books.

Lesmoir-Gordon, Nigel (2006). *Introducing Fractal Geometry* (3rd edn). Totem Books.

Letellier, Christophe and Rössler, Otto E. (2006). Rössler attractor. *Scholarpedia*, **1**, 1721+.

Levins, Richard (1966). The strategy of model building in population biology. *American Scientist*, **54**(4), 421–431.

Levins, Richard (2006). Strategies of abstraction. *Biology and Philosophy*, **21**(5), 741–755.

Lewenstein, Bruce V. (2007). Why should we care about science books? *Journal of Science Communication*, **6**(1), 1–7.

Lewis, Ronald (2002). Exploring fractal dimensions by experiment. In *Fractals, Graphics, and Mathematics Education* (ed. M. Frame and B. B. Mandelbrot), pp. 117–137. Cambridge University Press.

Li, Tien-Yien and Yorke, James A. (1975). Period three implies chaos. *The American Mathematical Monthly*, **82**(10), 985–992.

Liebovitch, Larry S. (1998). *Fractals and Chaos Simplified for the Life Sciences*. Oxford University Press.

Mandelbrot, Benoît B. (1975). *Les Objets Fractals: Forme, Hasard et Dimension*. Flammarion.

Mandelbrot, Benoît B. (1989*a*). Chaos, Bourbaki, and Poincaré. *Mathematical Intelligencer*, **11**(3), 10–12.

Mandelbrot, Benoît B. (1989*b*). Some "facts" that evaporate upon examination. *The Mathematical Intelligencer*, **11**(4), 17–19.

Mandelbrot, Benoît B. and Frame, Michael (2002). *Fractals, Graphics, and Mathematics Education*. The Mathematical Association of America.

Martin, Robert (2008). The St. Petersburg paradox. In *The Stanford Encyclopedia of Philosophy* (Fall 2008 edn) (ed. E. N. Zalta).

May, Robert M. (1976). Simple mathematical models with very complicated dynamics. *Nature*, **261**, 459–467.

May, Robert M. (2002). The best possible time to be alive. In *It Must be Beautiful: Great Equations of Modern Science* (ed. G. Farmelo), pp. 28–45. Granta Books.

May, Robert M. (2004). Uses and abuses of mathematics in biology. *Science*, **303**(5659), 790–793.

McDowell, Margaret A., Fryar, Cheryl D., Ogden, Cynthia L., and Flegal, Katherine M. (2008). Anthropometric Reference Data for Children and Adults: United States, 2003–2006. U.S. Department of Health and Human Services, Centers for Disease Control and Prevention.

Mermin, N. David (1992). The (non)world (non)view of quantum mechanics. *New Literary History*, **23**(4), 855–875.

Miller, Casey W., Chabot, Michelle D., and Messina, Troy C. (2009). A student's guide to searching the literature using online databases. *American Journal of Physics*, **77**(12), 1112–1117.

Mitchell, Melanie (2009). *Complexity: A Guided Tour*. Oxford University Press.

Mitzenmacher, Michael (2004). A brief history of generative models for power law and lognormal distributions. *Internet mathematics*, **1**(2), 226–251.

Newman, M. E. J. (2005). Power laws, Pareto distributions and Zipf's law. *Contemporary Physics*, **46**, 323–351.

Ott, Edward (2002). *Chaos in Dynamical Systems* (2nd edn). Cambridge University Press.

Ott, Edward, Sauer, Tim, and Yorke, James A. (1994). *Coping with Chaos: Analysis of Chaotic Data and The Exploitation of Chaotic Systems* (1st edn). Wiley-Interscience.

Peak, David and Frame, Michael (1994). *Chaos Under Control: The Art and Science of Complexity* (illustrated edn). W. H. Freeman & Company.

Peitgen, Heinz-Otto, Jürgens, Hartmut, and Saupe, Dietmar (1992). *Chaos and Fractals: New Frontiers of Science*. Springer-Verlag.

Poincaré, Henri (2001). *The Value of Science: Essential Writings of Henri Poincaré*. Modern Library.

Rood, Will (2004). Fractal limits: The Mandelbrot Set and the self-similar tilings of M. C. Escher. In *The Colours of Infinity* (ed. N. Lesmoir-Gordon), Chapter 5, pp. 82–95. Springer.

Ross, Sydney (1962). Scientist: The story of a word. *Annals of Science*, **18**(2), 65–85.

Rössler, Otto E. (1976). An equation for continuous chaos. *Physics Letters A*, **57**(5), 397–398.

Ruelle, David (1980). Strange attractors. *The Mathematical Intelligencer*, **2**(3), 126–137.

Ruelle, David (1993). *Chance and Chaos*. Princeton University Press.

Ruelle, David and Takens, Floris (1971). On the nature of turbulence. *Communications in Mathematical Physics*, **20**(3), 167–192.

Russell, David A., Hanson, James D., and Ott, Edward (1980). Dimension of strange attractors. *Physical Review Letters*, **45**(14), 1175–1178.

Sardar, Ziauddin (2005). *Introducing Chaos* (new edn). Totem Books.

Sauer, Timothy D. (2006). Attractor reconstruction. *Scholarpedia*, **1**(10), 1727+.

Sethna, James P. (2006). *Statistical Mechanics: Entropy, Order Parameters and Complexity*. Oxford University Press.

Shaw, Robert (1984). *The Dripping Faucet as a Model Chaotic System*. Aerial Press.

Shishikura, Mitsuhiro (1998). The Hausdorff dimension of the boundary of the Mandelbrot set and Julia sets. *The Annals of Mathematics*, **147**(2), 225–267.

Shlesinger, Michael F. (1988). Book review: *Chaos: Making a new Science*. *Journal of Statistical Physics*, **50**(5), 1285–1286.

Smith, Leonard (2007). *Chaos: A Very Short Introduction*. Oxford University Press.

Smith, Peter (1998). *Explaining Chaos*. Cambridge University Press.

Sprott, Julien C. (1993). Automatic generation of strange attractors. *Computers & Graphics*, **17**(3), 325–332.

Stacey, Nic (2010). *The Secret Life of Chaos*. British Broadcasting Corporation (BBC).

Stewart, Ian (2002). *Does God Play Dice? The New Mathematics of Chaos* (2nd edn). Wiley-Blackwell.

Strogatz, Steven (2001). *Nonlinear Dynamics and Chaos: With Applications to Physics, Biology, Chemistry and Engineering*. Perseus Books.

Stumpf, Michael P. H. and Porter, Mason A. (2012). Critical truths about power laws. *Science*, **335**(6069), 665–666.

Styer, Daniel (2000). *The Strange World of Quantum Mechanics*. Cambridge University Press.

Thompson, Silvanus P. and Gardner, Martin (1998). *Calculus Made Easy* (revised, updated, expanded edn). St. Martin's Press.

van Buuren, Armin (2010). This Light Between Us (featuring Christian Burns). In *Mirage*. Ultra Records.

Waldrop, M. M. (1992). *Complexity: The Emerging Science at the Edge of Order and Chaos*. Simon and Schuster.

Walters, Peter (2000). *An Introduction to Ergodic Theory (Graduate Texts in Mathematics)*. Springer.

Watts, Duncan J. (2004). *Six Degrees: The Science of a Connected Age*. W. W. Norton & Company.

Wolfram, Stephen (1984). Universality and complexity in cellular automata. *Physica D: Nonlinear Phenomena*, **10**, 1–35.

Wolfram, Stephen (2002). *A New Kind of Science* (1st edn). Wolfram Media.

Index

Made in the USA
San Bernardino, CA
04 June 2014